KINGS ACADEMY

By
Alan Stroup

I0476760

ISBN-13:978-1512224337

PRELUDE

This book is dedicated to males, past, present and future, who have, who are, or who will dedicate their lives to make society a more enriching place to live in by their talents and energies and, who just happen to be homosexual.

Historical Name	Born – Died	Kings' surname
Alexander the Great (King)	356 B.C. – 323 B.C.	Alex Kerho
James Barrie (Writer)	1860 - 1937	Tad Barrie
Lord Byron Poet	1788 – 1824	Brett Byron
Caravaggio (Artist)	1573 – 1610	Cole Marion
Charles XII (King)	1682 – 1718	Leif Kerho
Frederick the Great (King)	1712 – 1786	Eric Marion
Nathan Hawthorne (Poet)	1804 – 1864	Will Hawthorne
Lawrence of Arabia (Adventurer)	1888 – 1935	Bode Lawrence
Leonardo de Vinci Artist/inventor	1452 – 1519	Randi de Vinci

Abraham Lincoln President	1809 – 1865	Jon Lincoln
Alain Locke Writer/Scholar	1886 – 1954	Michael Locke
William Marshall Knight	1147 – 1219	Alan Marshall
Michelangelo Artist	1475 – 1564	Toni Simoni
Miyato Musashi Samurai	1589 – 1667	Aki Musashi
Isaac Newton Scientist	1642 – 1727	Bobby Newton
Peter the Great King	1672 – 1725	Trevor Alekseyevich
Herman Melville Writer	1819 – 1891	Tim Melville
Cole Porter Composer	1891 – 1964	Zach Porter
Richard II King	1157 – 1197	Rom Kerho
Friedrich Steuben General	1730 – 1794	Tomas Steuben
Bill Tilden Athlete	1893 – 1953	Jacob Tilden

Walt Whitman Poet	1819 – 1892	Dow Whitman
Oscar Wilde Writer	1854 – 1900	Tommy Wilde
Peter Tchaikovsky Composer	1840 – 1893	Alexie Tchaikovsky
Rudolf Nureyev Dancer	1938 – 1993	Dane Nureyev
Vaslav Nijinsky Dancer	1890 – 1950	Trent Nijinsky
Horatio Alger Writer	1832 - 1899	Marc Alger
Thomas Eakins Artist	1844 - 1916	Luke Eakins
Wildred Owens Poet	1893 - 1918	Spirit Owens
Alan Turing Scholar	1912 - 1954	Johnny Turing

When you are in a relationship with another, that relationship has only one purpose. It exists as a vehicle for you to decide and to declare, to create and to express, to experience and to fulfill your highest notion of Who You Really Are

Chapter One
(London, England)

Alfred Janson was quickly approaching retirement at the end of the year. At 65 years of age, his whole life had revolved around one building in London, England--Westminster Abbey. Without the most prominent education or illustrious grades, he'd followed in the footsteps of his father as a custodian in one of the most famous churches in the world. After twenty obedient years he was finally promoted as the night security supervisor in the protection of England's heroes in death and reverence to its long history of world leaders in the arts, sciences, religion and politics.

On this night Alfred had far more on his mind than his nine to six shift. He couldn't remember the last time he'd skipped an evening at the pub with his many cronies, rehashing the day's news or England's chances in the upcoming Wimbledon Tennis Championships. On this evening the events would require Alfred's complete sobriety and concentration. The fact that he'd never in his life done anything risk taking had his temperament giddy with excitement and anxiety. Actually, in his 65 years on earth, he'd never done anything that required risk or something as daring as the challenge before him.

Alfred sat before his vanity mirror and reflected on his appearance. All but bald headed, except for two streaks of frost, like white Nike stripes above his ear lobes, he saw a man who had never quite dwelt with life with a sense of reality. He had never married, never even dated a woman, as far as that went. His whole life had been spent in adoration of England's soccer and Rugby clubs, but he knew deep down that this adoration was for the players, not the sport. Being recognized as this fervent fan had given him the only satisfaction and pride he'd ever known.

The collection of soccer and Rugby memorabilia was a treasure chest of memories and a life's dedication to this hobby. Whatever money he had earned through his job usually went to buy season tickets or to gamble on his favorite team. Alfred figured he'd probably come out a little behind on all those transactions through the years. In all reality, he'd lost far more than he'd ever won.

Circling his bushy mustache his eyes went to a collection of various articles of clothing and pennants on his bedroom wall. Jerseys, shorts, emblems and flags were his favorites. Among these items hung six jock straps and all of them were autographed by special players who had found humor in Arthur's request for these undergarments of sport clothing. Alfred had stated they were to be gifts for his wife, though no one truly believed he was married. Alfred smiled with these memories, a fascination with men's athletic supporters. A smile creased his lips, before a laughter interrupted this validation of this bizarre hobby. It wasn't unusual for Alfred to take these items from his trophy case and hold them close to his face in order to smell the aroma of masculinity. He wondered if other men had such fetishes.

These athletic supporters were probably the closest thing Alfred ever had to a relationship, though in his younger years he'd hung out in the toilets where athletes congregated before and after competition. He knew he had serviced more than his share of his heroes through glory holes between stalls, and he had often fantasized to the face he wished to place on the party hidden from view. These recipients of Alfred's generosity were the same type of thugs who had tormented him in school by calling him names like poof, queer, or fairy. They were now at his mercy when he had their maleness to pleasure.

With these memories the last forty-five years felt more like a decade to this senior citizen. Such a pensive moment gave him an odd emotion that he was going off to war, rather than a simple mission that was quite rewarding. He had accomplished so very little in his life and time had disappeared without much to show for his efforts. His job felt more like home to him than labor, a way to hide from others his loneliness. What better way to harbor resentments and despair than with the dead. Fortunately it had few challenges, and the boredom was only interrupted once in a while by some humorous attempt, usually a teenager in their puerile acts to seek solitude after hours with their obsession of the dead.

Alfred could tell his stories about stupid kids who wanted to spend the night with Handel or Charles Dickens or Geoffrey Chaucer. This security guard never allowed such pleading and obsessions to interfere with his responsibility for the Abbey. Souvenir hunters were the worse and the most dangerous of these obsessed fans. It was frequently around the Poets' Corner of the Abbey where most of these freaks would hang out to give Alfred the most grief.

All this would end soon, but Alfred had little to show for the years he'd spent protecting the 3,000 people entombed or enshrined within Westminster Abbey. With an odd sort of bond Alfred felt a kindred spirit to the remains of kings, statesmen, scientists, poets, and musicians, whose tombs were packed into every last niche and alcove. The tombs were often times simple, others more grandeur from the most regal of mausoleums—Queen Elizabeth I, whose canopied sarcophagus inhabits its own private, apsidal chapel—down to the most modest etched floor tiles whose inscriptions have worn away with centuries of foot traffic, leaving it to one's imagination whose relics might lie below the tile in the undercroft.

His mere pension would barely afford what he had come to enjoy, traveling throughout Europe to watch England play in World Cup matches and international play. It was the opportunity of a lifetime that had presented itself to him a few weeks earlier which had given Alfred a new lease on life and a vision that his retiring years could be spent with pride and self-esteem.

Alfred often ruminated with this memory of that unusual day when he had met a stranger. It had been at the end of his shift as he was returning the six blocks to his home on a very crisp fall morning. He had been wearing his favorite brown gibus-hat with a tan trench coat. One thing Alfred did pride himself on—he dressed like a true English gentleman. An older man than Alfred was sitting on a park bench; his eye contact never wandered off of this elderly security guard's approach. This scrutiny made Alfred uneasy. When the stranger patted the bench to have Alfred join him, there was a hesitation, then a compliance.

"Grand morning, don't you think?" the stranger had asked.

"It is. I haven't noticed you around here before," Alfred replied curiously while keeping his distance a body length from this suspicious character. Nonetheless, he still took a seat on the other side of the park bench.

"I'm not from this area but I'm pleased to make your acquaintance, Mr. Janson. To get right to the point, I'm here to offer you 25,000 pounds for a task I believe you have the skills to perform."

"Who do you want me to knock off?" Alfred humored as his vision was lost to the mind's focus of instant wealth. "I do pray it's not an English soccer star? You've barked up the wrong tree for that one." Alfred finally twisted to confront this stranger. "How is it that you know my name and I don't know yours?"

"I must apologize for my rudeness, sir. I'm Dr. Jolson. You may call me Jacob, if you wish. I can assure you I mean you no harm, nor is murder part of this request."

Alfred placed his cane across his lap, like this was a truce between two old men. Though he had a most bemused expression on what this could possibly be all about, his patience was wearing thin. "Okay, Bloke, you have my attention."

"To get to the point, I'm in need of a human hair, a nail clipping or a piece of bone," the elderly, Jewish dignitary stated without the least bit of discomfort in being so blunt.

"Don't all the tourists!" Alfred snapped back, but then remembered that a very large sum of money was offered. "And whose sample are we talking about here?"

"Isaac Newton's would be my selection."

"It's not that easy, my dear fella. These aren't caskets that one simply raises the lid and pulls out what they want, like you're at some bloomin' supermarket. Aren't we of a relic yourself, sir, to be hankering for souvenirs?"

"I'm not into souvenirs, Mr. Janson. I'm a scientist who has visions exceeding your common laboratory inventor. I can assure you I'm very reputable."

The stranger pulled out a drawing from his pocket with an exact detail drawing on how an extraction could be done. Alfred examined it closely and, at the same time, he wondered how 25,000 pounds could fit into his retirement plans.

8

"I suppose you furnish all the equipment," Alfred said, figuring this dilemma would put a halt to the discussion.

"Every piece that you'll need, with instructions. I've examined the contents of the tomb through my means of a fluoroscope and have contrived a laser and a video camera that will suffice. A small repair job to the marble will mask any damage."

Alfred curled the tip of his mustache while he perused the drawing. "I have to question why I'm the lucky one you have sought out to do your dirty work?"

"A fair question. I suppose I could have asked your new custodian you recently hired; the young soccer player who was injured."

"Only temporary. I believe the young man has a great future in the game. He needed a part-time job to tie him over while his injury mends."

"Yes, I can imagine how you assist his motivation to recover. I'm amazed how often you two find time together to visit the men's room."

"Ah, so you're resorting to bloody blackmail. I hadn't figured you to be the type."

"It's not my nature. Who buggers who isn't my concern, Mr. Janson. I'd rather have someone who's mature and motivated. To offer this young man this financial opportunity would likely get you a few thousand pounds. That's only if he asked you to cover for him at all while he fumbles through the process and makes a mess of the sarcophagus."

"I can see your point. It wouldn't be fair if he'd kept all these pounds and offered me a pittance to cover his arse. I'm sorry I don't reach your level of normal."

"Au contraire. I can assure you, Mr. Janson, I don't stand in judgment of your physical desires. Normal human behavior is not natural at all but rather a habitual behavior that over a period of time has become typical in a particular society. I might even commend your choice of pleasure." A business card was handed to Alfred with an Israeli phone number. "You have twenty-four hours to make a decision. If you are successful we might be able to negotiate further business opportunities. I must warn you, though. Don't think that any

9

sample will suffice. I have the means to test the authenticity of the origin, just so we understand each other. It's none of my business but I've noticed you lack the niceties of life to live comfortably as you're retiring soon, and I'm sure your gambling debts bring discomfort."

"Don't you fret about me, old boy. I'm jolly fine and can handle my affairs quite nicely." Alfred knew he was lying as he said it, though he was quite relieved that his financial stress would be diminished with this much money.

The squinting Englishman scanned the silver card with the name, Dr. Jacob Jolson, M.D., Geneticist. He flipped the card over, as if there was going to be a contract printed that agreed to pay 25,000 pounds in return for a specified assignment payable to one Alfred Janson. There were no such words.

Seconds passed when Alfred glanced up. "So you're one of those guys who figure out those family trees."

"Something like that," Dr. Jolson said with the utmost refrain from rolling his eyes at such ignorance. "The World Cup is in Brazil next year, old boy. I'm sure you'd love to be there."

"Bloody right, there! You wouldn't be a soccer fanatic yourself, would you?"

"I'm too old to be a ruffian or a hooligan like you, Alfred. My role is to design the athletes than to enjoy their expression on the field of play. I'm much more comfortable watching the boys in tights at the ballet. One of my few delicacies I allow myself."

Alfred chuckled, the robust words of masculinity in the soccer world made him proud in being a loyal fan of English football. The implication made his mind puzzled to the man's intentions. What exactly did this old fart mean by design? The pondering passed quickly with the vision of 25,000 pounds dancing about in his head. He had made his decision before the two senior gentlemen shook hands upon their departure.

Three weeks later he was now on the verge of doing something his training had taught him to prevent. He couldn't have devised a more brilliant method of extracting a few fragments of a cadaver than this doctor had. It was simple, but brilliant. No one would miss a few fragments of a long-deceased Isaac Newton. The key was in masking this tampering to where someone would never even notice the extraction, let alone, question the integrity of the Abbey's chief security official.

As agreed, half of the payment was discovered in Arthur's locker at work the night before the excavation; the other half was to be paid upon delivery.

The day of the grand caper, Alfred's anxiety was beyond a child's holiday exuberance. His entrance into the Abbey was two hours earlier than his usual arrival. It was as if he hadn't seen Westminster Abbey in over a decade; his eyes absorbed the magnificence of this great cathedral, a structure that was considered neither a cathedral nor parish church. Instead it exhibited the classification of royal peculiar, subject only to the Sovereign.

Since its hosting of the coronation of William the Conqueror on Christmas Day in 1066, the dazzling sanctuary which has witnessed an endless procession of royal ceremonies and affairs of state—from the canonization of Edward the confessor, to the marriage of Prince Andrew and Wendy Ferguson, to the funerals of Henry V, Queen Elizabeth I, and Lady Diana. Even a recent burial of Richard the Second, centuries after the king had died in battle, had brought the Abbey high recognition. To discover the king's bones buried beneath a parking lot was a godsend.

Alfred whispered as he strolled by the gray stone columns, ascending like redwoods into the shadows, arching gracefully over dizzying expanses, then shooting back down to the stone floor. His whispers weren't so much for his own listening, as they were to justify his actions and convince this great Abbey that its future rested in great minds like himself and this scientist who had visions of what?...design? It wasn't for Alfred to define or question the man's motives. Before him, the wide alley of the north transept stretched out like a deep canyon, flanked by sheer cliffs of stained glass. Alfred knew, on sunny days, the Abbey floor was a prismatic patchwork of light.

A docent, unlike most churches, had its entrance on the side, rather than the standard rear of the church via the narthex at the bottom of the nave. The Abbey had a series of sprawling cloisters attached. One false step through the wrong archway and a visitor was lost in a labyrinth of outdoor passageways surrounded by high walls. Such was the case as Alfred meandered his way in contemplation. He stopped to assist a couple of lost French tourists, then redirected their progress by showing them the route back to the entrance.

Alfred arrived in front of Newton's tomb; at its feet stood two winged boys holding a scroll. Behind Newton's recumbent body rose an austere pyramid. The sarcophagus was recessed in a niche, obscured from this oblique angle. A black marble encased the target from which he was to be paid a king's wages, as far as Alfred was concerned. The excavating equipment was already inside his locker. It was now simply a matter of gaining the courage and putting his experience of knowing where his night workers were during any given time to allow him the freedom to accomplish the task. The trouble was, it all seemed too easy.

Before him he laid out the equipment prepared by the doctor. The Abbey was stone silent, its eerie ambiance would have spooked most people but those who had worked here for so many years. Even then the ghost stories from employees could feel a book of its own for those aficionados of the afterlife. Alfred stood beside the baroque monument, erected by Newton's heirs.

The monument dominated an area which has since become known as Scientists' Corner—a final resting place which Isaac Newton shares with the likes of Charles Darwin, James Clerk Maxwell, Michael Faraday and other illustrious British scientists. Designed by William Kent, the monument before him illustrated the many facets of Newton's life. Cherubs play with a prism, a telescope and newly minted coins; a celestial globe shows the path of the 1681 comet. Dominating the monument is a relief of Newton himself, with still more cherubs in attendance. Regally, Isaac leans on a pile of four books, labeled 'Divinity', 'Chronology', 'Optica' and "Phil Princ. Math'.

In a state of fascination Alfred picked up the laser and released the safety, only to discover his other finger was on the trigger. A cherub wing from above fell to the floor with a loud clang. The startling sound even made Alfred jump. His forehead was already dripping with a fever sweat; his whole body reacted with a flush of heat.

The panicky Englishman was close to feint for what he'd gotten himself into. He quickly corrected his thoughts and settled the laser back into its cradle, then examined the cable with a miniature camera at one end. Assured and with a newfound sense of confidence he began to calm his mind and began to locate the exact area that Dr. Jolson had diagramed.

By eight o'clock the following morning Alfred waited patiently at Chelsea's Café, a tea house on the corner of Whitewall and Downing, only a block from the Abbey. Under his gray jacket he held tightly to a plastic bag with the contents he had selected with the assistance of a miniature light attached to a micro camera. Alfred was amazed and overwhelmed that he, and he alone, had seen the remains of the great Isaac Newton. Thoughts of being cursed, or the stories of those who uncovered the remains of King Tut, flashed through his mind. These he dismissed quickly as nonsense.

The arrival of his benefactor brought a nervous smile to the Englishman's face. There was, after all, a sense of accomplishment, a pride of a deed well done. Alfred acknowledged the probabilities that he was dealing with a Jew, but such prejudice had never bothered him before, and he had often heard how brilliant Jewish scientists were.

The two men greeted each other with warm expressions. Alfred, somewhat too eager to consummate the deal, passed the plastic bag underneath the table just as a waitress approached with two menus. Dr. Jolson ignored the object on his knee and ordered without blinking an eye. Alfred's posture was more embarrassing than suspicious.

After a tense moment and the waitress's departure the doctor accepted the package, then slid an envelope with the remaining pounds across the table in full view. Alfred, with a jittery right hand, hustled the money to his jacket pocket, revealing a T-shirt that said, "I'm in No Shape to Exercise." It was the needed humor that allowed this Israeli scientist to laugh and break the aura of two spies in total paranoia. The tenseness evaporated with their first cup of tea, a glimmer of hope for a better future shined in both their countenances.

"I assume everything went as planned," Dr. Jolson asked.

"A professional could not have done better," Alfred lied and knew he'd been a total head case through most of the ordeal.

They ate a hearty breakfast as old friends, talking about England's chances in the upcoming World Cup of soccer. Alfred was not disappointed when the doctor picked up the tab for the meal. He offered his services for future acquisitions, just in case this scientist needed anymore souvenirs.

The successful business deal was, at least to this renowned geneticist, the first of many transactions which would change the way human beings examined their own mortality, yet these plots had all the criteria and volatility to set off a crisis of ownership and entitlement within the world community. Dr. Jolson knew he was directing the ultimate Rosetta's Stone, if not a Pandora's Box of religious dogma and exasperation.

The most interesting aspect of all this, his own desolate longings for ideal beauty and intellectual brilliance through design, not the least forgetting lasting notoriety, far exceeded the possible reputation of stealing the great icons of history and rewriting history as we know it.

Chapter 2

At the ripe age of 76, Dr. Jacob Jolson was the archetypal adventure scientist. He had earned his ultimate kudos, a Nobel Prize, for his research in the field of preventative genetics—more accurately stated as stem cell research. Ostracized by his own Jewish community in the 1960's as a blatant homosexual, Jacob had few equals in his brilliance. He often joked that his childhood was blessed by an accepting mother, if not one who smothered him with affection. His often related joke was that he'd come out to his mother at an early age with the self-diagnosis that he was attracted to other boys. His mother had thus replied, "We shouldn't rush into rash decisions, honey. Let's see what second grade offers."

Dr. Jolson's wealth wasn't inherited, but as a result of several inventions and noteworthy scientific discoveries, especially in the field of preimplantation genetic diagnosis, in vitro fertilization, and eugenics. His support of the gay community worldwide had brought him both fame and ridicule. Jacob's many books guided the human psyche into unmapped territory, bordering the murkiest corners of the human mind—with equally harrowing results. He reinforced with verifiable data that homosexuals were indeed born with this genetic preference, and such a preference presented a gifted state for athletes to some of the most brilliant minds of science, music, and the arts.

In frail health Dr. Jolson had one more vision in mind, a lifetime legacy of leaving something to the world of such an enormous magnitude his name would live in infamy. He had begun his research five years earlier by studying over two-hundred famous men: Their childhoods, education, parents, athletic potential, health problems, sexual habits, siblings, and their state of mind throughout life.

The doctor's trek had begun in London, but his journey was far more reaching. In his first year of recruiting men to assist him, Jacob had many disappointments and rejections; many rejected his offer immediately; others played games or attempted to give false samples. The deceased men, whom he was in search of to accomplish his goal, proved either inaccessible or the results were negligible for the doctor's aims.

Great masters like writers Henry James and Jonathan Swift; the most influential 20th-century playwright, George Bernard Shaw; the French symbolist, Charles Baudelaire; the French underground writer, Samual Beckett; the quintessential writer of the neoclassical period in Europe, Alexander Pope; a famous comical writer of world literature, François Rabelais; influential French writer, Albert Camus; French neoclassical drama writer, Jean Racine; another French writer buried in the Pantheon, Emile Zola; Chinese novelist, Cao Xueqin; the great literary artist, Voltaire; one of Jacob's favorite poets, Federico Garcia Lorca; 20th century's fictional master, E. M. Forster; and the ultimate French novelist, Marcel Proust, these men had all slipped through Dr. Jolson's fingers in his travels to procure his dream children.

Dr. Jolson adjusted his ambitions as he traveled. Deciding on the greatest castrati of all time, Carlo Broschi, known as Farinelli, was second-guessed at the last moment. Would this child grow as his predecessor and be able to span 3 octaves and hold a note for 60 seconds if he wasn't castrated? Jacob certainly wasn't going to castrate any boy for such a purpose as to procure a eunuch. He debated and kept his selection to thirty finalists; thus, it was like choosing members of a family that would live, others would perish. In this case, the others already had perished and would remain so. There were athletes versus musicians, dancers versus scientists, and writers versus military geniuses. Jacob had spent many a late night debating with himself his final selections.

Many of Jacob's idols were not attainable under designated burial plots, yet this did not stall his aims to find some remnant of the man's existence. Jacob was well aware that the Queen of England during Sir Francis Drake's period was given an article of clothing and strands of hair of this great sea adventurer. It was Jacob's goal to lay hands on these samples to snag a few strands of genetic material.

16

Only after discovering that William Shakespeare had put a curse on his own burial plot did the good doctor think twice about violating this foreboding curse. Late discoveries in Egypt and Italy allowed this famous geneticist a means to achieve even greater accessibility to the greats he wished to possess.

Though the doctor had his disappointments, his successes far outnumbered the losses. Six samples in three weeks, all in and around London. To say he was quite pleased would be minimizing his exuberance. Two of his grandest goals were in St. Petersburg, then Sweden. With his treasure chest in tow he flew out of Heathrow that evening with his best public relations smile and his credo for finding the right men for the job: "Find something to grab their balls with and their heart and mind will follow," was Jacob's creed. Money was always a steadfast motivator.

One of St. Petersburg's worse winters greeted Dr. Jolson as he stepped off the plane. The vast city of nearly 5 million was blanketed with snow, and the decaying remnants of this historical location were showing its age under such harsh conditions. Being the hometown of Russia's leaders, the city had taken on a fresh paint job with pastel colors, the museums and building facades gleamed with an old European charm.

His appointment with Yuri Borzov, curator at the Cathedral of Peter and Paul, was scheduled for the afternoon at the Heritage Museum, the pride of St. Petersburg and a place where the two of them could get lost in the crowd. The doctor had previously set the groundwork for such a conversation with several phone calls, as Jacob had come across as a historical biographer of Peter the Great. The two men had both laughed at the audacity of this ruler of Russia, a man who had killed his own son, fearing betrayal, then buried the boy beneath the steps of the church. Peter believed his son's spirit would never be able to rest, hearing the sounds of pounding footsteps for eternity.

Dr. Jolson had studied extensively the history of each of his historical figures so he could debate the finer nuances of history with anyone. The fact that Peter was a precocious child at the age of nine, the age he had been declared Tsar, was of extreme importance in the man's selection for the doctor's objectives.

Peter the Great had Westernized Russia with the help of influential friends and male admirers. Jacob had no doubts that men like Lefort had taken advantage of the boy's beauty and admiration of adults from the Western world. Likewise, when Peter became an adult he well knew the beauty of boys, often using naked ones for his pillows during afternoon naps, if napping was the appropriate intention. With boys by his side his Grand Embassy expedition within Europe had secured the advancement of a backward country like Russia and advanced the country's goals for the future.

By the end of three hours of sightseeing and constant bantering of the past, an agreement had been reached that even exceeded the doctor's expectations. Besides the guaranteed 1 million-- 250,000 rubles-- Yuri managed to secure a promise from the doctor that, when this "new" Peter turned twelve, or about, he would return to St. Petersburg for a visit. The promise was secured with a handshake.

The Cathedral of Peter and Paul was closed for renovation for three days. By the fourth day Dr. Jolson was on his way to Stockholm, Sweden, the burial place of Charles the Twelfth, one of the doctor's top three objectives. It was no coincidence that Charles and Peter shared many of the same attractions.

Chapter Three
Washington D.C.

One year previous the President of the United States contrived with his vice-president to give his constituents something in return for their invaluable support and his consequent election. As part of the ultra-right conservative movement, the president knew that geneticists had found the specific gene for homosexuality. He decided that the United States should be the first country to incorporate a massive movement to rid its children from this orientation.

Selecting a geneticist from Dallas, Texas, the president proposed to this scientist, Dr. Steven Kerho, a grand scheme to be carried out immediately in every school system in America—all totaled, forty-five million children from ages five to eighteen. The substance used would never be revealed and no one born with the genetic disposition of homosexuality would ever discover that their orientation had changed to one of heterosexuality.

What the president hadn't foreseen was the son of Dr. Kerho, a fifteen-year old boy, who just happened to be gay. Toby Kerho was fortunate enough to overhear his father discuss this diabolical plot with Mrs. Kerho, and Toby took it upon himself to make sure such a ploy would never be successful. Dr. Kerho had no intention of allowing the plot to succeed either when he contrived scientific codes and various computer schemes to guarantee that no child would have his genetics changed. His own plan backfired when his own son hacked the laboratory's computers and switched the genetic printouts. What Toby did was accidentally make 45 million children gay.

Toby Kerho had survived what most kids his age would describe as a Rambo-comes-to-life movie. On the run for three months from the United States government, Toby, 15 years of age, was shot at, nearly drowned, and had helicopters chasing him on land, water, and snow. The teenager had enough information to bring down the Presidency and make the United States' government the laughing stock of the world.

During Toby's run from the law he became friends with England's

Prince William, during his visit to America. Toby politely informed the prince that he was likely gay if he'd stopped and visited any school in America. As a teenager himself, Prince William, in true English spirit, found the humor in all this and was eventually returned to his original orientation.

With the assistance of Prince William, Toby managed to bring down a corrupt president and vice-president and a scheme that would've destroyed the future of a gay culture for a younger generation. The teenager became both a hero and a despised teenager by millions of Americans. The demise of what was known as the Hyacinthus Project was considered the most diabolical ruse ever set forth by an American president.

Now as a sixteen-year old Toby was no longer the blond-haired towhead who often accompanied his father, America's leader in genetic science, to his laboratory every chance the boy had. To say that the doctor's only son was following in his footsteps would be an understatement. Toby, or, as his parents and friends had known him for years, Toy, loved the science of life and playing with the components of the human genome.

At seven Toby would impress his teachers with writing an abbreviated thesis on Molecular Structure of Nucleic Acids, or, as we know it, DNA. He'd impressed his mentors with drawings of a double helix, combined with the building blocks called nucleotides. It only seemed logical that he relished DNA's four nitrogenous bases, represented by the letters A, T, C, and G.

The nation was slightly better balanced when Toby was found and reunited with his father. Authorities had arrested those responsible and the country was assured that the balance of sexuality would soon be restored to the vast population: those who were genetically born gay would remain so; others would soon have their heterosexuality back.

On this pleasant afternoon father and son had finally reunited in Washington D.C., and were relaxing in a restaurant outside the Smithsonian Museum—a welcome part of a tour of the Capitol, compliments of America's newest president and first black leader of the nation.

Across the aisle sat what appeared to be another father and son combination. Kami Marion and Mr. Kamito had also recently partaken in an adventure of their own. Kami had just turned sixteen, blond and blue-eyed, without the slightest hint of his platinum hair turning brown any time soon. He looked forward in the next few months to getting his driver's license now that he'd turned sixteen. Kami looked upon his sensei as both his adopted father and mentor in all aspects of life. It wasn't secret that the two males had a devoted love for each other that involved honor and the possible sacrifice of their own life to preserve the other.

It didn't take long for the two boys to catch the other staring their direction. The first couple of times there were hastily dropped heads or eyes that immediately were redirected back to the menu. God forbid another teenager would think of them as weird. The third glance became a charm, a perfect chemistry when their eyes locked and two smiles commenced. They both nodded, as if this was an acceptable greeting between strangers. A few seconds later there were repeated smiles, coyness this time with a hint of attraction and a true antenna of gaydar. Their brains registered that there was mutual attraction in the other; yet, gaydar was, at least at this age, considered prone to error.

Both Dr. Kerho and Mr. Kamito noticed this nonverbal communication between the two teenagers. A glance by this astute Japanese gentleman caught a similar turn of the head by the father of Toby; the doctor was a decade younger than Kamito.

"Good morning," both adults said at the same time. They each chuckled with this camaraderie.

"From out of town?" Dr. Kerho asked.

"Yes. A trip up from North Carolina, though I live in Tokyo, myself. You, sir?"

"From Dallas, Texas, sir. It appears our two teens have caught each other's eye." Dr. Kerho knew he'd embarrassed his son with this remark, but he rarely could get one over on his son at this age.

He'd long been aware that Toby was gay, a fact known since Toby had come out to the family when he was twelve. His older sister, Brandy, had accepted it without opinion or judgment, but that had never stopped the gay jokes.

"Dad!" Toby said with a childish whine and a blush when he glanced again at this handsome boy from North Carolina.

"I'd guess that the two of them would rather have dinner with someone their own age than with two adults," Mr. Kamito suggested, but this assertion received an annoyed stare from Kami.

Named after his sensei by his father, who was once a student of Mr. Kamito in Japan, Kami loved his sensei way too much to ever have him think he'd take second to anyone. Yes, there was a bit of truth to this Japanese wisdom, but he couldn't admit it right there.

"Would you care to join us, or we can join you," Dr. Kerho suggested.

"It would be our pleasure," Mr. Kamito said.

He and Kami slid out of their booth and into the one across the aisle. Kami made sure he sat across from this interesting teenager from Texas.

"This is my son, Toy, and I'm Steve."

"Toby, Dad." The correction was whispered to his father.

It was the first time in his life he'd thought that his boyhood name was too inappropriate for the occasion. At school and play all his teachers and friends called him Toy. It had been a name that had stuck since the age of two, thanks to his grandmother.

"I'm Kamito, and this is my adopted son, Ka-mi-chan."

Kami beamed with the introduction. He'd never heard his sensei introduce him with such pride and dignity. "Toy?" Kami asked with a smile, then received a vivid explanation from Dr. Kerho on how his son received his moniker.

"I like it," Kami admitted and decided to call his new friend by this nickname. When Toy grinned, Kami's heart did an extra flip and his mind felt dizzy with adoration.

Their conversation evaded the particulars of why the four of them were in Washington, but other aspects of their lives captured each other's curiosity with attentive interest. The boys were put on the spot when asked to describe their loves and interests and hobbies and, most importantly, if they had a girlfriend. Kami held his breath in pray that

22

Toy would say he didn't.

"I'm not really interested in girls," Toy responded to Kami's question.

"Ara utsukushiikoto!" Kami said louder than he should have.

"Huh?" Toy asked, concerned that the words were a form of judgment.

Mr. Kamito laughed and swept his hand through his son's long blond locks. "You must excuse my son. He is most satisfied that you do not have an interest in girls."

"You don't either?" Toy quizzed with great hope, but sensed he was likely responsible for this new orientation.

"I'm gay," Kami flat out admitted and expected a shocked look from these new acquaintances.

"May I ask when you discovered this?" Toy asked without hesitation.

"I've known most of my life, I guess," Kami answered with utmost honesty.

"That's the best news I've ever heard," Toy admitted and received a curious look from his friend. "I'm gay, too."

"Would you like to move to South Carolina or Tokyo?" Kami asked and had the four of them laughing.

Through lunch and dessert, then numerous cups of coffee, stories were passed and adventures told that made the four of them think they'd known each other for years. It was nearly four o'clock, three hours later, when they decided that this restaurant might charge them also for dinner if they didn't get a move on. Though they were staying in different hotels, the Kerho father and son team invited Mr. Kamito and Kami to stay at their hotel. Certainly another room could be found.

"Why not our room, Dad? We have two double beds," Toy asked.

"If that's okay with Mr. Kamito," Dr. Kerho responded.

"If it's not an inconvenience," Kamito said.

"I think we'd never hear the end of it if we didn't keep these two young'uns together," Dr. Kerho said with certainty and received a pleased smile from his own boy. He added, "We're staying at the Willard Hotel, just a few blocks from the White House."

At the Willard Hotel Kami hastily unpacked so he could go swimming with this Texas boy who had won his heart in one afternoon. Kami had been on a swim team for several years and always packed a racing suit. Should he or should he not? Kami reached in and placed his board shorts to the side to obtain the blue Speedo. What was so amazing, this Texas boy also had a swimming brief.

They each did a quick glance to catch the other changing, while the men chatted and waited for their sons to depart to the hotel swimming pool. Toy and Kami played like two neighborhood friends, diving for quarters and checking each other's physical beauty surrounding their sexy swimsuits.

Neither boy wanted to seem too eager, then Kami grabbed Toy's ankle to heave the peer back from capturing a coin in the deepest part of the pool. From ankle grabbing to pulling on legs and arms, finally a tug on a swimsuit hauled the Lycra fabric over a set of pearl white gluteus. Kami surfaced with a broad smile before sliding his suit back over his butt. Soon the play became more of a wrestling competition, twisting, pulling and pushing, then entwining each other's limbs in assorted knots, which brought their bodies in close contact. The physical effort contributed to obvious arousals and both boys were soon treading water a few inches apart. Their words became more automated with thoughts contrived while in a sexual daze. Not wanting to be too obvious, their eyes darted lower to stare into the water at each other's angled projections of sexual attraction, distorted by the blue liquid.

"You are so cute," Toy admitted when he was sure he wasn't being too forward. "Is that really all you?"

"It's a Marion curse," Kami said without bragging, only because he was tired of shower room remarks at school. He wanted desperately to lean forward and kiss this gorgeous figure a few inches away from his lips. His endowment was something he should have been proud of, but his focus was on how Toy's had filled up the front of his suit, as well.

"That's a periscope, dude. I want to take off your suit right now," Toy said.

"I'd like that but it might scare the guests."

The boys laughed and splashed each other; their frolic helped

drown their thoughts and concentration.   Instantly both boys felt their
bodies lifted up and thrown through the air.   A few guests caught this
frivolity by the hands of two adults who had tossed their sons in the air
to land with major splashes.   The enlarged crotches in these mini suits
were noticed by their fathers and a few bystanders on the pool deck,
which increased the gawking.   Kami and Toy collected their balance in
the pool and knew the previous romantic interlude was vanquished for
the time being.

"Let's get 'em!"   Toy yelled to his new friend.

Both boys dove underwater and tackled their fathers, or at least
tried to.   Not easily done by these two lightweights, Mr. Kamito lifted
Kami straight up, feet first, out of the water and would have de-shorted
the boy if this was in Japan.   Kamito didn't mind displaying his son's
erection bursting at the seams of this racing suit. Dr. Kerho gave his son
a monkey bite on the hamstring and then bench pressed the boy over his
head, before tossing him like a rag doll to the deepest end.

"He's not big enough to handle his old man, yet," the doctor said
with satisfaction.

"They'd like to think they are," Mr. Kamito agreed.

Kami wasn't one to give up, though he knew he was a mere
plaything for his sensei.   It was a way to have fun with the adult he
loved the most and, at the same time, enjoy being tossed and heaved
about.   He hadn't expected to be manhandled by both adults, but this
was okay also.   To have his new friend feel the strength and skills of his
sensei gave Kami a sense of satisfaction.   In the end the boys were no
match for either father, but that didn't mean surrender either. They sent
splashes of water at their tormentors, then sprinted away before they
could be caught.

Over a late dinner there were plans made for the following day and
a future date that the four of them could meet. It was daring of Toy to
ask his father if he and Kami could sleep together.

"Just what I want to hear.....sounds of love while Mr. Kamito and I
are trying to sleep.   Let's make a deal.   If you two want to shower
together I'm sure whatever hormones are heightened will be explored.
That way everyone gets a restful sleep and we'll not all be yawning

tomorrow."

The boys agreed but Toy had to make sure. "Kami can sleep in my room when he comes to Dallas, right?"

"If Kami agrees, who am I to stop two goofy kids from falling in love?" the doctor agreed and winked at Mr. Kamito.

"Hai! Dai-suki desu," Kamito said.

"What'd he say?" Toy asked Kami.

"My father is in perfect agreement," Kami said with a grin.

In the twilight hours of a most surprising day the two teens sprawled out on lounge chairs by the pool deck. They kidded each other about their monikers.

"Ka-mi-chan?" Toy questioned. "Can't I just call you Kam?"

"It's just a form of endearment in Japan. I might just call you Toy-chan. I guess if I can call you Toy, you can call me Kam. My friends did at school. There is something I should tell you."

Toy's expression turned worrisome with this statement. "You're going to tell me you already have a boyfriend."

"No, but I do have twin boys, not quite a year old. It was just an accident."

"You're kidding? Kam, getting a girl pregnant can't really be called an accident. I thought you weren't into girls."

"I'm not. Someday I'll explain, but it wasn't like that….I mean, romantic and all. My sons are with my mother right now."

"I have to confess something myself. If you can get a girl pregnant by accident, I made forty-five million school children homosexual by accident. I sort of sabotaged one of my father's experiments."

Kam roared with this thought. "That's so cool! You mean my brother and sister were actually gay?"

"For a few months. I've reversed the gene process since; at least turning the heteros back to being straight again. I figure someday I can do my thesis on the results."

"How does one simply reverse someone's sexual orientation?" Kami asked.

"Kind of easy, actually, but there are secrets within the scientific community that would be best to stay hidden. See, there are genetic

26

markers called methylation and these markers turn off certain sections of genetic code. So even though we inherit two copies of every gene—one from our mother, one from our father—whether the gene is methylated often determines which one of the two genes will be turned on. Methylation is inherited, just as DNA is. But unlike DNA, which has an enzyme that proofreads both the original and the copy to minimize errors, methylation has no built-in checks. Knowing which genes explain sexual orientation, which are approximately seven, helps in playing around with a person's attraction. My father doesn't appreciate this playing around though. I simply encrypted this methylation process and made a whole lot of kids gay, though I only meant to neutralize my dad's experiment. They had to find me to figure out the encryption."

The two teens laughed with their adolescent nuances and slapped each other's hand.

Hastily returning to their hotel room two very beautiful boys stepped out of their swim wear in preparation to share a shower together. Though they were on the fringe of six-footers, they both would readily admit they were 6 feet tall if asked. The flipping of two erections was a direct physical sign of their attractiveness to each other, attractiveness that didn't require 10 million genes to map out every base pair of their 46 chromosomes. Both were lean with narrow waists, nearly identical copies of finely tuned physiques: Toy's, from years of ballet training; Kami's, from hard physical exertion in the martial arts over the past year and a half. For any man, straight or gay, not to gawk at these specimens of youth would be fools in their own right.

Toy moved his hands over Kami's body as if he was examining a marble sculpture. "Whoa! You're hairless!" Toy said upon noticing the lack of pubic hair on Kami.

"A laser treatment in Tokyo. I wanted you to think I wasn't legal," Kami humored.

"It didn't work. With your length you couldn't fool anyone."

They pressed their groins together and felt the chemistry between their bodies in each other's arms. Holding this squeeze for almost a minute became a realization to each of them that they'd found someone far different than just a friend.   Though they each had experienced gay affairs over the past year they were each reluctant to express their knowledge or aspirations to each other by being too presumptuous. For this brief time they only wanted to explore, to relish the maleness that expressed a lust they had for each other.

Time didn't allow them the opportunity to totally envelope what they desired, but their hands swept, fingers brushed, and lips found tastes that each boy had learned was part of their makeup and soul. What they managed was mutual orgasm which kept them both happy and desirous of future opportunities.

"Will you teach me a martial art and Japanese?" Toy asked as he delighted himself in drying off his new friend.

Kami enclosed Toy chest against chest.   A peck on the forehead and he was as hard as rock again.   "That's a given. I'm interested in ballet and I want to hear you play the cello.   I'm a bit remiss in the musical department," Kami admitted.

"It's a deal.   Have you thought about where you might want to go to college?"   Toy asked, as he swept his hands over a bubble butt and squeezed the mounds to press their loins even tighter together.

"I was thinking about returning to Japan, but someone really close to me has possibly changed my mind.   I mean really close."

"Who?"

"You, Bozu!   Oops, that's a Japanese expression for squirt."   The boys playfully tongue wrestled which ended up in another prolonged kiss.

At bed time the boys stared at each other across the separation of beds.   What they wouldn't have given to get rid of these adults and have a night of sexual pleasure.   Kami was seriously considering skipping his remaining years of high school, only because Miyato, his half-sister, had said he was advanced enough to pass the SATs and enroll in the college of his choice.

Love at first sight was something he had scoffed at before, now he believed totally that he'd found his life mate. Kami wondered if Toy was just way too brilliant to consider him a viable lover.   In many ways Kami felt inferior to such a handsome, intelligent boy. Though he

28

couldn't read Toy's mind from four feet away, the boy in the other bed had also fallen head over heels for this North Carolina boy with a Southern accent and a thorough knowledge of Japanese culture. To Toy, Kami was the most amazing teenager he'd ever met. Such an accidental meeting had to be a miracle and his recent prayers had been answered. In a teenager's search for God and eternal life, meeting Kami wasn't really an accident—it was a godsend.

Toy reached out across the open space between the beds. His fingers soon found the touch of someone he already loved very much. Hands caressed, tickled, and soothed through caring and expression. They each fell asleep with their fingers still grasped and their dreams of tomorrow that this love wasn't a dream, but a future.

Chapter Four

Ralph Stevenson opened up a certified letter from the Caribbean. A check for another fifty-thousand dollars was made out in his name. A smile beamed from ear to ear. He'd almost forgotten his stress from months before, but it had all been worth it in the end.

On that memorable, but nearly disastrous evening, the ground had only recently thawed in Springfield, Illinois. With each scoop of the earth the sound of a diesel engine had rumbled in the night air throughout this vast cemetery called Oak Ridge. Located 2 miles outside of town its location didn't always draw a large amount of tourists or interference at this time of the night.

Ralph had been the man behind the wheel, doing what he'd been paid $50,000 to do, but neither his temperament or his patience were considering the windfall on that frigid evening. Ralph had worked at Oak Ridge Cemetery for the last thirty-three years, the last fifteen as the head funeral director. He valued his sleep and knew he'd rather be in bed at 2:30 in the morning than digging a ten-foot ditch beside one of America's favorite sons.

Stevenson cursed the men over a century earlier that had caused this debacle. A man who had been imprisoned in Chicago had hired two of his friends to simply steal the remains of the sixteenth President of the United States, Abraham Lincoln. The coffin, at this time, had simply been placed above ground so that anyone could view the burial place of such a beloved man who guided the United States through a most difficult war between the North and the South. Unfortunately these friends sought help from another, only to later discover they'd hired a member of the local police department. Though the plan never succeeded, Robert Lincoln, the last remaining son of Abraham, decided it was best that there never be another attempt to kidnap his father's body for whatever reason.

The 500-pound cedar and lead coffin was placed inside a steel cage, lowered into a 10-foot deep vault, and buried under two tons of wet concrete. It was now Ralph Stevenson's job to break through this vault and secure a few remnants of a highly valued skeleton. Halfway through this job Ralph had already decided to ask for more money.

When the ten-foot excavation was finished, Ralph collected the tools given to him by this "supposed" doctor. Only when he was inside this dirt abyss did it decide to rain, and rain it did. Within minutes the dirt began to slide back into the hole. Ralph was stubborn enough not to give in to nature, finally penetrating the lead vault to the dilapidated cedar coffin. Securing the prescribed remnants he wrapped them in a bag and stuffed them in his backpack. His accomplishment had taken over two hours and it would be light soon. A ladder had been placed on the other side of the pit, but there was one problem. The mud had become so thick Ralph found his legs stuck with more dirt gradually filling the freshly dug hole.

Ralph fought panic, then began to dig madly with his hands to move the mud from enclosing his knees. With his hands full, another avalanche of dirt collapsed on what he'd just dug out. Like a swimmer, Ralph moved his arms quickly, scooping mud to and fro until he felt his toes wiggle. With one giant effort he was able to lift one leg out without a shoe then forced his other leg out, minus the other shoe. He crawled on his hands and knees to the ladder, felt one secure wooden bar and began to climb upward. On top he lay out gasping for breath and thankfulness that he wasn't buried ten feet under. There was no way he was accepting just $50,000 for all this trouble.

Having the samples in hand had made the renegotiating a whole lot easier. He offered Dr. Jolson an extra bonus for more money, considering his efforts and the anxiety for nearly losing his life. Ralph's father had been a close friend of Robert Lincoln, having worked for Robert in the railroad office in Chicago. Upon Robert's death he left Mr. Stevenson with a collection of paperwork about his father. Inside this collection were an archive of letters written by the president to his lovers and letters written in return. It just so happened that these lovers weren't women but men like Joshua Steen and Col. Elsworth, amongst others. Robert had reasons to keep such personal and private letters away from the press and historians.

Ralph had no use for this remarkable collection, so, as a bonus, he was quite willing to grant this geneticist with this discovery. The odd thing, Ralph had never taken the time to read or ascertain the letters' significance to mankind. They were simply to him old correspondence. A hundred-thousand dollar retirement far exceeded any personal pride for possessing something from a long-dead president.

Chapter Five
(One year later)

Toby Kerho didn't bounce or stride, he gracefully glided as he walked down the student path to the Perry-Castaneda Library. This observation of gait was humorously noted by his roommate and significant other who was watching from inside the library.

Toby was just finishing his freshman year at the University of Texas at Austin. He planned on taking summer classes, then another eighteen units for each of the following semesters. Toby was determined to finish his undergraduate degree in three years before moving on toward his graduate degree in biological science. His goal was to become a doctor in genetics like his father.

Entering this massive library on campus it didn't take long until he spotted a smiling face with a wave that was meant to direct him to a table in the far corner. Toby's favorite person was studiously browsing the laptop in front of him.

"About time, wannabe doctor!" Kami said without glancing up again.

"Waga koi wa! (My love!) Long line at the snack bar. Of course a person I love has to have his onions removed from his hamburger," Toby sarcastically informed.

Kami stuck his tongue out and opened his hand to receive his version of what a hamburger should look like and received another kiss from his lover.

"Your father called while you were at dance practice. He says it's important. I told him I could take the message, but your father said no matter how much he loved and respected me this information was strictly for your ears only."

"It must be really important if he won't tell you. I'll call him when we get back to the apartment. Need any help with your homework? What is that, microbiology?"

"Yes to both. Do you know anyone that has the slightest clue about all this stuff?"

A punch on the shoulder had Kami giggling. He knew Toby could ace this course without ever attending a class. It was the lab that was the killer.

Thanks to Dr. Kerho's financial assistance and seminars to the university, he managed to secure entrance for both his son and his son-in-law, as he liked to refer to Kami. The boys had quickly rented their own apartment away from the university and adapted quickly to their independence away from adults.

Across town in the rustic section of Austin the boys shared a one-bedroom condo, finely decorated with Japanese furniture, art, and assorted marital arts swords and paraphernalia. As with every afternoon they each took turns showering so as not to spoil the anticipation before their daily midday romp in bed. It was usually one of three daily rounds of sex. On this afternoon, as soon as the futon was laid down, Kami rested his head on Toy's chest while Toy dialed home.

"Konnichi-wa, Dad-san. Are you mad at my Kam that you can't communicate?"

"You know I love Kam, unless he gets you pregnant. I had a question only you can answer. Do you know a Dr. Jacob Jolson? Apparently he knows you quite well."

"First of all Kam is on the verge of trying to get me pregnant after I hang up. Secondly, isn't he the geneticist from Israel? You know, the one who you've often referred to as a maverick in the field of genetics."

"That's far more information than your father needs about your sex life. And yes, Dr. Jolson is considered somewhat of a rogue geneticist. He wishes to speak with you at your convenience. I'd like to know what this is all about."

"I personally think a man could get pregnant, Dad. Not the sperm in the egg thing, but actually carrying a child."

"You have what, a twenty-eight inch waist, if that? A guy is not made to carry a child, Son. Leave that to mothers with wide hips and a God-given trait for pain. Now where were we?"

"Koi wo shi koiba." (If we really love our love)

"What did you say and why is Kam laughing?"

"I'd do it for Kam."

"Son, I paid for a surrogate for your first two. Now the both of you have four kids. Don't you think that's enough? And don't you two have sex while I'm still on the phone."

"We're not. Kam is lying on me, giggling. How are my babies?"

"They're doing great. Now back to this Dr. Jolson. I recommend you two come home for the weekend and we'll call him together. I'm not sure what business he wants with a seventeen-year old."

"Almost eighteen, Dad."

"Yes, I forgot, my son is beyond his father's expertise."

"You're learning. I almost have you trained."

"Tell Kam hi and to ride you rough."

"Dad! I can't believe you said that. We film our best episodes, you know."

"Why am I not surprised? Bye!"

Kami took the phone and threw it on the floor. A kiss prepared the main event.

"Taihen utsukushii desu," Kam whispered in his lover's ear. (You are very beautiful)

"You're just saying that because you want my ass." Toy smirked.

"That, too.

****

The drive to the Kerho home in Dallas was usually done when the boys wanted a home-cooked meal. There had been distinctive changes to the Kerho home since the family had spent a month in Japan the past summer due to the invitation from Mr. Kamito. Mrs. Kerho and her daughter Brandy loved the Japanese culture and soon adopted the more relaxed dress code and mannerisms of their hosts. Miyato, Kami's half-sister and Mr. Kamito's niece, was the ultimate and gracious hostess throughout the stay.

They had stayed at the Kamito home in Tokyo for a week before traveling down the coast to the vast resort near Kyoto, owned by this same distinguished Japanese statesman.

Toy and Kami had grown madly in love over the first few months they'd known each other. They had decided to spend their college years together to see if this love could last. As the boys readily admitted, their love appeared to grow stronger each and every day.

Though he wouldn't admit it, Dr. Kerho also was smitten by the Japanese way of life. He quietly installed his own roja (garden path) and took up ikebana (flower arranging) as a hobby. While his significant other became adept at cha-no-yu (tea ceremony), Brandy took up origami and designed kakemono (scrolls) for the household. Soon no one was allowed in the house with shoes on and all the family developed the custom of wearing a kimono or yukata, the smaller version of a kimono.

Throughout the house were shoji (wooden grids) and even a tokonoma—an alcove of honor. Toy had a strong vocabulary of Japanese; plus, he loved to eat with hashi, or wooden chopsticks. He wore the traditional fundoshi, as did Kami, when the two were around the apartment or at home in comfort. Dr. Kerho hadn't become that brave to wear a loincloth when his daughter was around.

Having sisters two years older than they were and with the same sarcastic manner was very coincidental. What the boys found out and feared, their sisters quickly became best friends and remained the ever present torment of years past, never passing up a chance to find fault or heckle their younger siblings. No more than the two young men had walked in the house when Brandy nearly bumped into them on her way outside.

"Hey, it's the Mutant and Gorgeous Two," she said.

Toy turned to Kami. "She's still upset that I have a penis and she doesn't. Don't you have something better to do, like play with your Chucky doll?"

"Let me play with Kam, and he'll forget all about you," Brandy said with her eyes locked on a boy she fell in love with the first time she saw him. Sister like brother.

"I have a brother you might be interested in," Kami suggested.

"Don't encourage her, Kam. Let's go see my father."

The boys swung into Dr. Kerho's library. A haiku poem decorated a central space just behind his desk.

"Good to see you guys," the doctor greeted, stood up and hugged them both. "Might as well call this character from Israel and see what he wants."

The three men sat around the desk as Toy's father dialed the phone number given to him by Dr. Jolson. When the Doctor Jolson answered he had his digital camera directed on his chair in a study with a view of the ocean to his rear.

"Thank you for returning my call, Dr. Kerho. I assume your son is present as I requested."

Dr. Kerho flipped on their own video cam so that the doctor could view everyone present. He began by introducing his son and Kami. "Your view is either a great photograph on your wall or you're not in Israel."

Dr. Jolson chuckled. "It's real but it's not Israel. I'm in my home above St. Barts. I love the ocean and would rather spend my remaining days in solitude. The reason I have decided to reach out to you and your son is not something I had previously planned. See, a few months ago I was diagnosed with pancreatic cancer. Unfortunately I was remiss in seeing a doctor earlier. I'm afraid there is little hope for recovery. Isn't it a tragedy that I dedicated my life to gene therapy and stem cell research, only to fall victim to the one thing I wished to prevent?"

"I'm sorry to hear this," Dr. Kerho interjected. "I've admired your work and your dedication to the field of genetics."

Toy rolled his eyes at his father, especially after his the comments about this geneticist being such a rogue in his field. He received a wink right back from his dad.

"Thank you, Doctor. Obviously I have taken notice of your great success with the Genome Project. This brings me to your son, Toby. His courage a year ago in an effort to counter such a nefarious mission from your former president was truly impressive. When I was thinking about someone who could step into my shoes and further my dreams your son was my first choice."

"And what dream is that, Doctor?" Dr. Kerho asked, eyeing this old man who didn't appear to be on his last legs.

"I've been in the midst of an astounding project of immense

magnitude to the world's future. Without going into detail I'm asking from the bottom of my heart for your son to dedicate himself to fulfilling my prophecy and goals. I realize this is an unusual request, and rest assured I have spent a great deal of time researching and determining that your son has this call in life." The doctor redirected his vision on Kami. "Mr. Marion, you are not a stranger to me and your history is commendable. What you have accomplished with Mr. Kamito is unbelievable in its scope. I'm so pleased that two beautiful souls found each other in love."

Toy and Kami glanced at each other in amazement. Dr. Kerho was not so amused.

"If I may ask, sir, how did you acquire such information about my son and his friend?"

"My reach is extensive and finances have no limit. Rest assured I mean no harm or intrusion into these boys' lives. I only wanted to assure myself of their orientation and commitment. I should have been so lucky to find such beauty in my day."

"Could you explain your project more, Dr. Jolson. I mean, what exactly are you expecting from my son?"

"I neither have the time nor the patience to offer a full explanation. What I do have are twenty-nine boys who need a home, a future, and intelligent supervision to make their lives successful."

Dr. Kerho might have laughed if not for the absurdity of this request. "Surely you jest, Doctor."

"My name is not Shirley, Dr. Kerho."

This certainly got a laugh from Toy and Kami. From this old man across the screen this was quite a comical effort.

Steve smiled but was intent on finding the truth. "I highly doubt if we'd be interested in all this, Dr. Jolson. Have you pursued any other avenues? Certainly an orphanage or boys' home would fit your needs."

Jacob leaned forward with serious eyes. "Dr. Kerho, these boys are not your average, run-of-the-mill street kids. I can promise that all finances will be taken care of if that is a reason of your cautiousness. What I'm offering your son and Kami Marion is the greatest opportunity to advance their education and create a vision the world has never known."

"Let me blunt, Doctor. You've been suspected of dealing with cloning. We're not talking about something really bizarre here, are we?"

"Bizarre? You make me sound like some type of monster, Dr. Kerho. Many of us profess this, that, or the other, but how many of us act in a manner in keeping with what we profess to believe? Simply the realization that we are that which we are, and part of the eternal scheme of life, that we are everlasting and perpetual and will never die, is tremendous."

"Are we talking reincarnation here, Dr. Jolson, or cloning?"

Toy and Kam sat rigid in their seats, amused at this tete a tete debate between two doctors.

"Is there really a difference? The soul does not pass forward if I cloned you into another identical human being. A clone is a new man whose brain is essentially a blank slate. I can assure you I have not taken a live human and cloned him into some project for eugenics."

"Then what do these 29 children represent?"

"Gifts to our world, Doctor. Children I had hoped to raise, educate, give and develop talents to their full potential. Yes, it was an experiment in human behavior, but its inventor and creator is on his last breath. I'm asking for a reprieve here, sir."

"My boys sitting next to me are in college, sir, and their own future...."

"Yes, I want them to remain in their studies. Their education, paid for by my endowment, will allow them the knowledge to fulfill the dream. Experience will come with these boys living through 29 other creations. Dr. Kerho, I beg your assistance. Your life will never be the same, nor will you ever forget this decision in the name of science."

Steve sighed with his frustration. "Dr. Jolson, the enormity of this request is far more extensive than my son or I can formulate at this time. I'm requesting a short time to discuss this before such a decision is made."

"I understand completely, Dr. Kerho. For my appreciation and you taking the time to listen to an old man's problems, the boys' schooling is already paid through the year. A college education isn't cheap by any means."

"Dr. Jolson, you didn't have to do that….." Dr. Kerho watched the screen go blank in front of him while the two young men to his side chuckled at the abrupt cutoff of the screen.

"The guy hung up on me." Steve had the most bemused expression.

"Nothing you're not used to, Dad."

"Very funny. Look it, guys, this is a no-brainer. The guy is an eccentric old fool who is probably imagining this whole thing. Twenty-nine children…give me a break!"

"I think it would be pretty cool. What do you think, Kam?"

"Hey, I'm neglecting my own two boys. What would I do with twenty-nine more? And you have a son and daughter now, Toy. I suppose another twenty-nine wouldn't make much of a difference. Where would we put them?"

"You boys can't be serious?" Dr. Kerho questioned with his best perturbed look.

"Maybe these boys are our age. Can you imagine all that testosterone in one place?" Toy joked.

Dr. Kerho wrestled his son over his lap and spanked Toy once on his bottom. "You guys are incorrigible. Let's find your mother and go out to dinner.

"Shouldn't we fill Brandy's dog bowl before we leave. She might get hungry."

"Don't you guys ever stop fighting?"

Chapter Six
(Two-days later)

Toy and Kami had returned to campus and discovered a refund of their tuition from Texas University. They celebrated with fish and chips at Austin City Limits, enjoying a jazz band they both loved. There were decisions to be made: buy Kami a car; upgrade their apartment; buy a complete media editing system to improve their own movies of love making; or buy tickets for a European summer vacation.

"Greece!" Kami shouted above the music.

Toy smiled. "Yeah! Beaches. Nude beaches. Men and nude beaches."

They opened a savings account for the $22,000 the next morning.

In Dallas, Dr. Kerho received a surprise phone call from the Dallas-Ft. Worth Airport. There was a plane due in three hours with its cargo to be picked up by one Dr. Steven Kerho.

"I haven't ordered anything," Dr. Kerho verified.

"That's not our call, Doctor. We're simply relaying the information given us," the shipping clerk said.

At the terminal Dr. Kerho was escorted to a receiving station. The huge 777 taxied up to an expanded   VIP area where the doctor waited in anticipation of what package could possibly be delivered that would cause this much attention. A woman dressed in a white smock was the first to exit the plane; she then asked an official to point out Dr. Kerho. She smiled as she approached.

"Dr. Kerho, I'm Dr. Jolson's assistant, Dr. Jane Meadows. I'm sorry for this late notice but Dr. Jolson died last night in his sleep. He left precise instructions and, of course, a check to cover expenses."

"Expenses for what?" Dr. Kerho asked.

"I assumed you two had made arrangements," Dr. Meadows said more rhetorically than as a question. She passed the envelope to Dr. Kerho and awaited his response. Inside the envelope was a cashier's check for two billion U.S. dollars.

"There must be some mistake. What am I supposed to do with two billion dollars?"

The answer was already in full view when twenty-nine women began to bring off the 777 carriages with twenty-nine babies in toll.

"This can't be happening!" Dr. Kerho stated in complete exhaustion of what he was looking at.

"Is there a problem, Dr. Kerho?" Dr. Jolson's long-time assistant asked.

"Well, yes! I never agreed to all this."

"Apparently Dr. Jolson thought you did and we wouldn't want to talk ill of the deceased now, would we?"

Flabbergasted, paperwork began to be passed to Steve for his signature: adoption papers, immigration paperwork, shot records, custodial agreements. In haste the doctor signed off only because it didn't matter what protests he instituted. The last thing Steve needed was an international incident.

"I have no way to transport these children, nor do they have a place to stay," Dr. Kerho stated rather emphatically.

"That's taken care of, Dr. Kerho. There's a bus parked outside the terminal and Dr. Jolson has bought a 15,000 square foot home a few blocks from your own residence. It's awaiting your arrival as we speak. And, oh yes, six of our nurses will be staying with the children until you find other arrangements."

"Yes, of course, other arrangements." Steve watched several cargo bins being taken off the plane. "And what pray tell are those?"

"Dr. Jolson's lifetime work. They're a gift for you and your son. I'm sure you'll find the information very interesting."

"One has to have time to read such literature to enjoy them," Dr. Kerho stated rather sarcastically.

The sound of crying babies began to fill the air. Steve thought sure he'd long surpassed this stage in his life, and this was before his own son had talked him into paying a surrogate so Toy could have children of his own like Kami. Nine months later fraternal twins, one boy and one girl became an addition to the Kerho household. Mrs. Kerho and Brandy adored having the newborns. There was no way he could put this on his wife and daughter.   In the heated moment he called his son.

"I don't know what you've done but the world has just tilted and your ass is mine!" Dr. Kerho's first words were to Toy.

"I swear I haven't done anything wrong this time, Dad.   Kam and I have decided we're going to Europe on vacation with the money we saved on tuition."

"Oh, how wrong you two whippers are!   How does your class schedule look for the morning?"

"We both have morning classes. We should be back by noon. Why?"

"You'll both go to the stadium parking lot and wait."

"Like inside?   The gate is closed."

"That's only so I can't get to you earlier. You'll do as I have instructed."

"Wow, Dad, you're being a bit over-dramatic here, aren't you?" Toy waited for a response but there was a silence with a click.   He figured his father had learned this from Dr. Jolson.

Following the instructions to the letter the boys waited in the abandoned parking lot for several minutes until the sound of a helicopter approached from far away.   They were taken back when the chopper landed and the pilot waved the boys over, then flew off toward Dallas.

"Are you sure you were to pick us up?"   Toy asked the pilot over the whirling of blades.

"You fit the profile," the pilot said with a smile.

The landing on the street in front of their home was quite an impression to the community.   The first person they met was Brandy, and she got her say in.

"You guys are in big doo-doo. I'd sign up for a prison sentence if I was in your shoes or the French Foreign Legion. They probably don't have a don't ask, don't tell policy."

"You've been in my underwear enough times, sis, so my shoes wouldn't be that big of step."

"Creep!   I don't wear boys' underwear with tire stains."

Toy was ready to do battle when his mother walked in.   "Ah, my favorite sperm donors.   Your father is waiting for you at another home.

43

Let's take a walk and I'll explain."

The two block saunter was filled with information that had Toy and Kami with gloomy faces and heightened expectations. Twenty-nine children to take care of waited their arrival. At the front door of this massive estate that had been on the market for 12 million dollars for over a year was the Texas State Protection Services. By the time Dr. Kerho was finished with them he had his sights on two boys.

"You're the college students, what are we going to do about all this?" The doctor's first question had all the devastation of a college professor's exam.

"What makes it our problem, Dad?"

Dr. Kerho held out a check made out to Toby A. Kerho and Kami A. Marion. Toy and Kami counted the zeros three times. "Whoa, we're like zillionaires," Toy announced.

"That's it? We're rich?! Boys, this Dr. Jolson decided to leave you guys most of his estate. What in the world did you do to this guy? He wasn't in one of your motion pictures was he?"

"Steven! That's not fair to accuse the boys of corrupting senior citizens," Mrs. Kerho said.

"Thanks, Mom. For this much money we would've helped the guy out, but that's kind of weird, considering his age."

"Thank you, my son the porn star! All I'm saying here is that you can't expect your mother and me to bring up 29 children, especially after surviving you and Brandy."

"Gee, thanks, Dad. We weren't that bad, were we?"

"What your father is saying, we need to come up with a solution," Mrs. Kerho suggested. "I'm sure we can survive until a home will accept the boys."

"Mom! We can't just accept all this money and not do what Dr. Jolson wanted. Kam and I will come up with a plan. Didn't you and dad always tell me to make my life a gift? 'Remember always, you are the gift!' you said. 'Be a gift to everyone who enters your life, and to everyone whose life you enter. Be careful not to enter another's life if you cannot be a gift. When someone enters your life unexpectedly, look for the gift that person has come to receive from you.' That's what

44

you said, and now you expect us to turn our backs on this special gift."

"Great! We raised a kid who remembers what his parents told him. As if we really meant it," Dr. Kerho teased with a straight face but then busted up laughing.

"Okay, let's take this one step at a time," the doctor said after a deep breath. "You and Kam better be prepared to spend your weekends here in Dallas. We'll make a helicopter available for you so you won't have to drive. With this much money we have some flexibility. I want a devised plan by the weekend on how we proceed. For now, you two are going to be experts at diaper changing and feeding."

Dr. Kerho had the boys sign the check, then he endorsed the amount because the boys were underage. All the documents needed his signature due to the minors being the beneficiaries and sole guardians.

"I suppose this means an increase in my allowance," Toy said with a straight face.

"Only if you prove worthy of changing diapers and taking out the trash. You two wanted kids—now you have them. You shot the sperm; you do the term. It lasts eighteen years, if you're lucky. I'm beginning to think raising children is a lifetime project."

"We didn't contribute to these twenty-nine, though, Dad," Toy informed.

"No, but you're now their daddies according to these documents. Good luck. Mother, shall we leave?"

Mrs. Kerho played along, winked, and then left with her husband. In the background there were enough needy cries and whimpers to convince the boys they had far more than they could chew.

It was actually fun for the first few hours playing with children nearing the age of one. Kami could relate to all this because he had two sons barely older than these boys. For a few minutes the boys thought of all the things they could buy these tots, more toys and Winnie the Pooh characters than they ever imagined when they were kids. To spoil these toddlers would be a blast.

Returning to college was a relief, though they both spent the late night hours tossing ideas back and forth, plans for an academy, or ways to give the boys everything they needed to be happy and spoiled. A

McDonald's restaurant on campus sounded great, but this idea was finally ruled out.

There was a sign of maturity when the boys used their tuition rebate by hiring an architect, Joe Bradley, a young gay architect who they had specifically searched out through the Gay Yellow Pages. At first this architect was convinced he was dealing with a fantasy that the boys had put together for some future project or a high school assignment. There were rich kids who could play with their father's money and design elaborate settings, Joe thought and hadn't yet been told about the large endowment available for this academy or about the prospective students. The boys' envisioned of what they would want if they had to relive their childhood all over began. They considered sports, play and school, before deciding on making these boys far more advanced than the average Texas boy. Joe was sure he was wasting his time.

"Fellas, I'd love to stay and chat but I have to make a living."

Toy wasn't sure what the guy meant. "Oops! Dad gave me this check for your retainer. I almost forgot all about it." Toy reached in his pocket and pulled out the cashier's check for $250,000. "Just let us know how much this is going to cost and I'll cover all your expenses. We want everything done in gold. Just kidding."

Finally this young and aspiring architect saw this adolescent vision as a professional architect and not a game played with two teenagers.

"Sorry about that. I wasn't sure whether you guys had any money. What exactly are you thinking here?"

Kami spoke right up. "Well, we'd like you to find us some land north of Dallas, hundred acres, or so. We want a large building, like a pyramid in glass, but it has to have an observatory on top. A fifty-meter Olympic pool with a restaurant overlooking this venue. And of course a main basketball arena with a huge scoreboard, just like the pros. A synthetic track, awesome baseball diamond, soccer field, and three paths around the facility: one with grass for running; one with a cork surface for bikes; and one on the outskirts for horseback riding. If there's trouble with water or sewage, we'll build our own."

"Yeah, we can do that," Toy verified. "We want to be environmentally friendly so have a lot of solar power and maybe wind power, too. We want an eighteen-hole golf course on one corner of the property—make it over seven-thousand yards; my dad likes a long course. And a monorail system, mostly underground; plus, a lake with a water slide park, lots of sand. Fill it with bass and bluegill, but nothing that will bite male appendages, if you get my meaning. We'll need an auditorium with an orchestra pit and a great dance floor for ballet. I just hope we can get one of our students interested."

"Don't forget the dorms and a church, Toy," Kami mentioned.

"Yes, a dormitory overlooking the facility with fifty-inch LCDs in each of their rooms, no beds, but futons, and their rooms have to be in glass with a skylight that can have digitally transferred pictures from the telescope. The dorm will have an indoor pool, but not like a normal swimming pool. It has to have white-sand beaches with boulders and a pirate ship that has a slide. Make it have palm trees around the beach and no more than four-feet deep. A theater with a digital projector and heated floors which are sanitary. The boys will be in bare feet most of the time. Put in a beach volleyball court next to the pool. We'll need an arcade with the most modern virtual reality games, and bathrooms that have all the new appliances like the Oral-cleanser to clean teeth which removes the need for kids to brush them. Of course a library and two gymnastic centers: one for training with pits, cameras, full safety straps—all the works, two cafeterias that aren't gaudy and zoo furniture for the dorm."

"What's zoo furniture?" Joe asked.

"Actual size animals that are all soft and fury that the boys can crawl on, sleep, recline and play with. I had a stuffed giraffe when I was a kid. It had a neck as long as Kam's.... Oh yeah, make sure the floors are safe and soft. Oh, six tennis courts: two clay, two grass, and two hard court."

"Do you boys have any idea how much this is going to cost? We're talking about two-hundred million here."

"That's a lot lower than Kam and I figured. We're looking more at three-hundred million with what we want," Toy responded. "Our labs and classrooms have to be state-of-the-art. I want a MRI machine and the best scientific lab money can buy. If a college has it, we want it."

Joe smiled and kept taking notes. "I'm figuring this academy is for about one-thousand to fifteen-hundred kids, right?"

"Not exactly," Kami said. "Let's see, I have two and Toy has two; plus our other boys. That gives us thirty-three. With staff and coaches we'll be looking at sixty, or so, but only make the dorm for, say thirty-five."

"But you're only teens! How did you manage to have four children already?"

Kami glanced at Toy. "How did we do that?" They laughed. "We're gay and ambitious. Why do you think we choose you, Joe? We want to see gay men successful."

"There's a lot more but we haven't finished our wish list yet," Toy said. "We did want several houses built, so make sure you get a residential permit, as well. These have to have indoor pools and offices that overlook the grounds. I would like our main building to have an entrance that makes people think they're entering the Getty Museum, and our library should look like the Huntington Library, but we don't need originals or anything that expensive."

"I like your style boys. I think you've just made my career," Joe said.

After numerous sketches and mutual agreement between Toy and Kami, a large size model was done to show Dr. Kerho in Dallas. Joe was elated to discover that the boys had an adult in their life who also captured this project with the same enthusiasm of his clients. Dr. Kerho was impressed with the boys' effort, then turned to this starving architect and asked, "When can we start?"

"You mean you really want to build this academy?" Joe asked with wide eyes.

"You are capable of handling such a project, aren't you?"

"Yes, sir, but the expense.......will be something an oil tycoon might not be able to afford. I'm not sure kids these days can appreciate such a complex."

"I have a feeling these boys will be different," Dr. Kerho said with a premonition he hadn't figured out yet.

The boys picked out a piece of property about an hour's drive north of Dallas. All totaled there was about five-hundred and thirty-five acres within a bowl type valley. It was pristine with hills surrounding its setting that would be perfect for security.

"I believe all of this will fit nicely within 260 acres," Dr. Kerho added, but we'll use the rest for future plans. A manmade lake will have to suffice."

Joe grinned from ear to ear. "Who are you guys?" he asked of the two boys.

"They're our future, Joe, just like you are. These two boys are your bosses. I want no short-cuts or shady construction. In addition I expect you to have a full report and updates weekly given to my sons here and one to me."

"Yes, sir."

"Since you're dealing with minors I expect they will be treated with the same respect you would give an adult. They write the checks, not me. If one of them gets too big for his britches, call me. If you're wrong, I'll settle with you first."

****

Within weeks permits and licenses were issued along with further layouts and plans to commence construction for an academy that any educational and athletic facility in the United States would envy for its magnificence and creative design. The initial name of the academy was simply, **BOYS ACADEMY**. This would soon change.

Toy and Kami had sacrificed a great deal in their pursuit of developing an academy to their dreams and visions. No longer did they partake of concerts or athletic games at the university, though Kami still found time to work out and be an assistant sensei at the Aikido school in downtown Austin. Toy had joined a dance group to maintain his ballet skills. At night they relished their time in bed, then instead of finding comical relief with television, they began to read through Dr. Jolson's multitude of research and volumes upon volumes of studies. When their finals were finally finished, it allowed them even more time to scan the ton of literature now stored in the Kerho basement and at a local storage building.

Kami had been the fortunate one to open a box and discover the magical letter with the boys' names written on the envelope. Through this disclosure Dr. Jolson's secret was finally revealed.

Toby and Kami:

Forgive an old man's visionary utopia and tampering with the world's sense of humor. You are about ready to embark on a project I'd give my life to—actually I already have because you're reading this letter. The boys you now have in your possession represent the history of homosexuality in its finest hour. I expect you two to carry on this tradition of excellence and pride and invoke in this group of boys the spirit of what it really means to love another male.

I've chosen you not because I see perfection, but because I have captured your ability to love another with openness and fervor. May your sons express in complete freedom the essence of the human body and all it has to reveal in their discovery of finding out who they really are in God's Plan. Enclosed is a list of my efforts to give you the best minds and bodies of generations past. I pray these boys never experience the shame and homophobia that their predecessors did.

Love and prayers for your future,

Dr. Jacob Jolson

Chapter Seven

There was a difference of opinion on how long they would keep this information to themselves. In this one large box contained a philosophy that the boys consumed and embraced as their own. Quotes from the ancient Greeks had its message of endearment, but this alone hadn't kept the boys up to the wee hours every night devouring the words of this now deceased geneticist. This was a message from the grave, a futuristic inspiration that the boys now saw they were living out for a man who had been born a hundred years too early.

Dr. Kerho also noticed an immense dedication within these two. He had sat down and gave his best father-to-sons talk to present a reality lesson to these two new entrepreneurs.

As these fresh, vibrant faces sat before his desk Steve had to smile at this innocence of taking care of so many young boys.

"Have you two boys ever considered what you'd do if you decided to break up and date other boys? I mean, guys, gay people love variety. We, speaking as a heterosexual, date a lot of girls because it's fun before selecting who we finally think is our soul mate."

Toy winked at Kami. He was going to have fun with this one. "Father, dearest, why do heterosexuals feel compelled to seduce others into their lifestyle? Would you really want both your children to be heterosexual, knowing the problems they'd face? Look at Brandy, per se. Heterosexuals are notorious for assigning themselves and one another rigid, stereotyped sex roles. Why must you cling to such unhealthy role-playing?"

Dr. Kerho leaned back on his desk with a bemused smile. He was prepared to reply, but Kami beat him to it.

"I agree with my partner. How can you, adopted father, enjoy an emotionally fulfilling experience with a person of the other sex when there are such vast differences between you? Does a man truly know what pleases a woman sexually or vice-versa? How can you hope to actualize your God-given homosexual potential if you limit yourself to exclusive, compulsive heterosexuality?"

51

Dr. Kerho held up his arms in surrender. "Okay! My bad. Didn't mean to step on any toes. Sorry." He hugged both of them.

"Apology accepted," Toy said.

His sister wandered by and her sibling couldn't help but comment, "There does seem to be very few happy heterosexuals. Techniques have been developed that might enable, say, my sister, to change if they really wanted to. After all, you never deliberately chose to be a heterosexual, did you, Brandy? Have you considered aversion therapy or Heterosexuals Anonymous?"

Brandy wondered how she got caught in a cross-fire here. "And to think you're a father," she eyed her brother.

"Maybe someday a guy will take pity on you and buy you an appointment at a sperm bank," Toy countered.

"That's enough from you two! Go your separate ways before Kami thinks you're not lovable brother and sister."

Kami smiled. He knew how to play neutral.

After Brandy departed with sworn vengeance, Toy offered his final solution. "Dad, Kam and I have talked about this problem of a zillion cute boys who offer nothing more than the two of us have already. We were meant to be together and I don't want anyone else but Kam in my life. Why would I? He's beautiful, intelligent, and you should see his penis." The three of them laughed.

"I have. Your mother and I came home early last week and saw you two skinny dipping in the pool. Needless to say your mother asked why we weren't born so lucky. We also noticed you two have abandoned your pubic hair."

"We've removed it permanently with laser treatments," Toy verified.

"Tired of growing up, or what?"

"Dad, have you ever got a pubic hair in your mouth. It totally destroys the moment."

Dr. Kerho laughed. "I believe the missus and I have shaved a few times down there. That's our secret—not to go beyond this little colloquy."

"Was Brandy with you when you saw us in the pool?" Toy asked.

"To be perfectly honest she appeared a little overwhelmed and went straight to her room. I hope she doesn't think every guy is built like Kami."

"He's mine and we're getting married as soon as possible," Toy admitted.

"The State of Texas might have something to say about that, yet I know your mother and me are very supportive of your decisions."

Toy and Kami glanced at each other and nodded. "Dad, there's something we think you should know."

"Don't tell me you're pregnant."

"Very funny. Look, we've found out some things about all our boys. I believe they're all gay."

"It's kind of early to get that type of impression. They're barely one year old."

"No, really. We've been reading Dr. Jolson's work. His writings center on homosexuality as the ultimate gift of humanity. He liked to dismiss psychiatrists en masse as moral conservatives who based their teachings and mindless traditions on society's mores and not on science. He says no true scientist would endorse definitions of perversions that rested on moral prescription. Dr. Jolson even quoted Plato in his Symposium that love between homosexuals is nobler and more spiritual than love between heterosexuals."

"That hardly means he had somehow found a way to make twenty-nine babies all gay."

"I did, but not on purpose. I think he did. He talks about the education of Cyrus the Great, about making the boy a healthy, able, and honorable man; the youth learns the virile sports, the arts of war, the habit of silent obedience, and the capacity for effective and persuasive command over subordinates. As to love, Dr. Jolson said any young man should be permitted to indulge in it without prejudice of gender for obedience and affection."

"That's still vague. Who really knows if Cyrus was gay?"

"Kam and I did our research—he was. We think these boys are clones but we're not sure. I know you are against cloning, but I just wanted you to know."

"Is it fair for these boys to be programmed for homosexuality, boys?"

"Why not? I don' know why he would do this."

"Have you two considered that this doctor might have been a pedophile and he literally arranged to have a harem of young boys for his delight?"

Kami spoke up. "I don't think that was his aim, I really don't. I think by his words that he envisioned this Rubric of Athletics—the worship of health, beauty and strength. Sure, I believe Dr. Jolson held youth as this phallic symbol of the ideal. Don't we all. I've been hit on by more professors and older men at the university than in all my life. I think he was all wrapped up in reliving youth and creating his own destiny within the boys. He talks about beauty, like virtue, as lying in fitness, symmetry, and order. A work of art should be a living creature, with head, trunk, and limbs all vitalized and unified by one idea. Love is the pursuit of beauty."

"Yeah, I agree with Kam. Who says twenty-nine boys can't live happily as gays. Dr. Jolson quoted Aristippus: "'Whatever we do is done through the hope of pleasure or fear of pain. Pleasure is the ultimate good, and everything else, including virtue and philosophy, must be judged according to its capacity to bring us pleasure. What we know best are our feelings; wisdom, then, lies in the pursuit not of abstract truth but of pleasurable sensations. The keenest pleasures are not intellectual or moral, they are physical or sensual; therefore, the wise man will seek physical delights above all. The use of philosophy is that it may guide us not away from pleasures, but to the most pleasant choice and use of them. It is not the ascetic who abstains that is pleasure's master, but rather the man who enjoys pleasures without being their slave, and can prudently distinguish between those that endanger him and those that do not; hence, the wise man will show a discriminating respect for public opinion and the laws, and will seek as far as possible to be neither the master or slave of any man. The art of life lies in plucking pleasures as they pass, and making the most of what the moment gives. The present is probably as good as the future, if not better.'"

"Impressive speech. You two have really been examining this guy," Dr. Kerho realized. "Can you imagine the flak from the community if it's not only discovered these boys are clones, but that they've been made gay? Imagine the religious reaction, boys."

"Dad, religion also once objected to autopsies, anesthesia, artificial insemination, and the entire genetic revolution of our day—yet enormous benefits have accrued from each of these developments."

"Let me think about this. I'm not sure I need any more surprises."

Chapter Eight
(Two Years Later)

Dr. Kerho had come to love all these new children in his life. The boys called him Grandpa, and this he could relate to. His empathy for these boys grew day by day, and he began to wonder even more what was in Dr. Jolson's mind when he created these new human beings. Were they just 29 gay men who he cloned? Lovers? Men who had died of AIDS and never had a chance; thus, he felt sympathy for them? Did these boys deserve a lifetime of harassment and uphill struggle in their quest to be accepted by society?

He'd long taken blood samples of every boy and discovered that the gay gene, as he called it, was never tampered with. In other words these boys weren't heterosexual at one time, then changed to fit the doctor's protocol or agenda. Steve did find evidence of chloroplast-to-nuclear gene transfers, movement of nuclear genes among chromosomes, or movements of genes between hosts and vertically transmitted endosymbionts. What this proved was a form of eugenics; human hybrids indoctrinated against diseases and inherited mutations. Many scientists would conclude that these children were designing babies, enhanced genomes. To Dr. Kerho this type of genetic use was the future in preventing human beings from inheriting the same physical and mental problems of their predecessors.

The doctor was less concerned with this type of germ line engineering than about sexual orientation. He was very tempted to make half the boys heterosexual and keep the remaining gay. This would offer an interesting contrast and balance with the academy ready to open within a week. His little secret was one that gave him a lot of anxiety.

Toy and Kami graduated with their Bachelor of Science degrees in three years: Biology and Kinesiology, respectfully. Their only break for the summer was to open this academy with very little fanfare. They loved returning to Dallas every weekend to play and enjoy the company of their own children and 29 bubbly boys of enormous energy. To all these children Toy and Kami were called Daddy, a name that was hard to deal with at such a young age. To two teenagers on the verge of leaving adolescence it was too much fun not to involve themselves in every aspect of the boys' lives, and one girl—Toy's daughter.

Given a few days between the ribbon cutting ceremony, which would include just family members and those professionals involved in the academy's construction, Toy and Kami decided to investigate the last few remaining boxes of material they had yet gone through. One was marked with a radiation symbol, which they didn't touch.

Though the boys had never considered that these 29 children were ever given last names, Kami discovered an array of birth certificates in the box he was browsing through. Not considering any significance he spanned the list and was intrigued by a few last names suggested by Dr. Jolson. He showed the list to Toy who dropped what he was doing and ran for his computer.

With giddy expressions the two new college graduates spent the rest of the afternoon in deep research and an imagination that went beyond science fiction mentality. They came to an astounding conclusion, then decided to dismiss it because no one would believe them anyway.

* * * * * * * * * * *

A beautiful summer day greeted the Kerho and Marion family when they collected on the grounds of what was then known as the **Boys Academy**. Dr. Kerho had recently sold the large mansion from which the boys had grown up in the last three years. A bus arrived and children ran with joyous expressions on their faces toward their new home. The academy was built on what was once cotton and corn fields on flat farmland. A few petroleum fields were miles away, though the smell of Texas oil never reached this far north.

In front of the boys' dormitory a blue ribbon was spanned across the walkway. A huge plaque had two posts in the ground and a tarp over the gold lettering would announce the name of this magnificent facility. The final tab on such a masterpiece exceeded three-hundred and fifty million dollars. Their architect had one of the biggest smiles on his face that day.

In one arm Toy had captured a happy Alex, a boy with reddish hair and brown eyes and was as charismatic as any of the boys.

"I never thought this day would come," Dr. Kerho told his son and son-in-law.

Dr. Kerho invited both boys to his sides to shake their hands. Kami had a boy named Bobby on his shoulders. Steve had grown to love Kami like a son and would always consider him a member of their family. He kissed both boys, then the antsy tots who were delighted to be in their fathers' arms.

Toy thought there would never be a better time. "Dad, I want you to meet Alex."

"Son, I know who Alex is. I changed his diapers and have fed him for the last three years."

"Yes, I suppose that's the case. But did you know this was Alexander the Second?"

"What are we talking about here?"

"I also want you to meet Bobby Newton. His progenitor was a famous scientist you should recognize."

Dr. Kerho stood spellbound, blood drained from his face. His legs felt rubbery before he simply melted to the lush grass below him.

There were a few moments of panic. So many people flocked to the doctor that the boys' pediatrician, Dr. Mark, had to pause just to keep people from crowding too close.

"A little water, please," Dr. Mark asked for and received instantly. As Dr. Kerho sat up he regained his composure to the relief of the guests.

"I'm afraid I lost it there for a second," Dr. Kerho apologized. "Must be the heat."

"We're going to have to keep an eye on you," Dr. Mark humored, but was more than attentive to a man he respected very highly.

Toy helped his father stand, relieved that it had been only a fainting spell. "Dad, I didn't mean to shock you like that."

"I almost made a terrible mistake. Excuse me a second, Toy, I have to make a phone call." Dr. Kerho stepped away from a puzzled son and dialed through his cell phone to his office. "Stan, cancel that little experiment on the GGS computer.   I won't need those genetic samples after all." Steve felt like a child whose hand was caught in the cookie jar.   In this case he had ample warning to resolve the problem. There was no way he was going to change anyone's sexual orientation now.

As the ceremony reached its climax Toy and Kami cut the ribbon together in front of an exuberant crowd that included their mothers; sisters;   Mikki and Rikki, Kami's four-year old twin sons; Mr. Kamito; Oshi—his friend in Tokyo—; his brother Reece; and Ryan, his first love. Ryan was living in Japan now, acting in the Japanese theater and in love with another actor.

Mrs. Kerho had her own biological grandchildren in each hand, three-year olds' Wendy and Lance. Two sets of grandparents for Toy and Brandy were also in attendance. They were still puzzled on where all this money came from.

The tarp covering the sign of the academy was released with the sound of firecrackers and a musical number through a set of speakers. What the crowd had expected wasn't forthcoming. **KINGS ACADEMY** was in gold leaf letters below a crown. Olive wreaths were on each side of the sign. Everyone clapped and were amazed at the beauty of this entrance to a true architectural wonder in the enormous glass structure before them.

To Dr. Kerho the sign made obvious sense.   He winked at his boys and reassured them that this was far more appropriate and fitting than Boys Academy.   The two adjacent glass buildings, a library and science hall, gave a balance to the main structure of a massive pyramid shaped cone that had a bubble top.   The boys thought it resembled a Hershey Kiss when the telescope was fully extended from its tip; otherwise it was more like a pyramid with a gumdrop on top.   The three buildings had a glass tint that resembled the colors of the rainbow; a vision that caught the constant attention of so many children.   Just below the observatory, the academy's church spanned the grounds below in its entire splendor.

"Now for something really special," Kami announced and had everyone look to the hillside.   The black drape was dropped, revealing another glass structure with a triangle on top extending above the hill.

This was the boys' dormitory, with all thirty-five rooms overlooking the complex and valley.

The congregation was directed a half-mile away on a monorail installed to zip students, staff, and guests around the campus. Barely hidden behind a slight knoll, the man-made lake had the children giddy with expectation. There on the five-acre lake stood a water slide park, a wave pool for surfing, kayaks, rafts, sailing craft and an old pirate's ship on the far side by an island that had more slides and an obstacle course.

"You're not selling tickets to this, are you, Toy?" Mrs. Kerho asked.

"It's just for the academy only, Mom."

The boys were used to swimming in the pool at their last residence. As soon as they all exited it was a free-for-all on how fast the kids could take off their clothes and run for the water slides. Toy and Kami hadn't quite expected this reception and quickly asked for assistance in being lifeguards for an hour.

While the adults relaxed and had refreshments served in the shade by the lake, the boys and one girl played as they'd never played before. One hour turned into two, then Dr. Kerho managed to take control and round all the children up to view the insides of the main assembly building and sports complex. It was decided that they had thirty-three tired children on their hands, so Brandy and Brittany took the kids to the dormitory while the adults viewed the interior of the academy.

The classrooms were state of the art with upgraded labs and the finest of equipment available for professional science buildings. From the classrooms to the sports complex, even Mr. Kamito was impressed with the massive dojo, capable of handling numerous martial arts. The Olympic size swimming pool was one of the fastest and finest built in America. On a second and third level a viewing area curved along the length of the pool where people could dine while viewing practice sessions or competition. A separate diving pool complimented this venue. Below the pool's surface a glass enclosure assisted coaches in analyzing their swimmer's progress.

On the other side of the restaurant, offices, and ground level conditioning room, a huge arena encased the splendor of a basketball court.   The court could be quickly changed to accommodate gymnastics, volleyball and numerous other sports.   Overlooking the arena on the opposing side were more offices for the assistant coaches. As the sports complex spread outward, a gymnastics center was adjacent to volleyball courts, split by a locker room and a sports medicine lab.

Kami's brother, Reece, was soon tugging on his older sibling's sleeve.

"Why couldn't you do this when we were kids?"   He teased.

"You like it, huh?"   Kami asked.

"I can't believe this is all for four year-olds.   I want to use this facility."

"Four year-olds grow in to fourteen year-olds, Reece. Look, you're almost seventeen.   Finish your college degree and I guarantee you I want you to work here.   You're an exceptional athlete and the boys can learn so much from your experience."

"You really mean that, Reece?"

"Absolutely!   I've approached Brittany about working here as well.   She loves the idea of being around Brandy."

As the group elevated to the top floors, the men congregated into Kami's and Toy's executive offices overlooking the entire complex. There were televisions that could be channeled into every arena and classroom.   Dr. Kerho sat in one of the plush chairs and viewed out over such beauty.

"We've outdone ourselves, boys. This is more fitting for a university or an Olympic training facility. I foresee a few problems already, if you boys would like to hear them." He received a nod from both of them.

"Toy, your own boy, Lance, asked me why he doesn't have a penis like the other boys.   It's obvious to everyone out there on the lake that these twenty-nine boys each have a looping flower that's beyond a four-year olds normal physique.   There's no mistaken that Dr. Jolson did some designing with these boys and I've examined the evidence under the microscope."   Steve didn't want to admit his near tampering.   He was ashamed of that thought and still had trouble forgiving himself for this errant idea.

Kami spoke up to confirm this reality. "My mother hinted to me that this is what Reece and I looked like at this age. It's going to be easier for these boys because they will all have the same size."

"I agree," Toy said. "Look, Dad, we both know there have been some improvements to these boys' genetics. I can do a little designing myself for my own son. I've already given it consideration, but I didn't want to discuss this with you until you knew what we knew. Since you like ballet, I think you'll like Dr. Jolson's mind-set. The man definitely loved ballet and the best dancers the world has ever known. We now have their genetic makeup at the academy."

"I'm from the old school, boys, but I'm game for change. I knew something was up when Brandy brought home some of her college friends to teach the boys elementary math, colors, and shapes. Bobby had walked over and pulled out one of the girl's algebra books, then immediately wanted to know what the crazy numbers and letters were. She thought the boy was just being cute, but, when she came again, Bobby remembered what an exponent was, a virgule and vinculum, though he had trouble pronouncing them. He'd go, "Let's do Al…gee….ba today, Suzy. Do you know the boy can now do beginning algebra and is teaching several of the boys? Tell you what we must hire; we need artists, dance instructors, musicians, and the best teachers money can buy for these boys."

"Dad," Toy interrupted. "Kami has already pursued this line of reasoning. The staff arrives tomorrow for their tour and seminar on what we expect."

"I should have known you guys were on top of this. You've done a terrific job designing this campus. The idea of putting astronomy at the apex of this building was brilliant. The boys will gain so much by viewing constellations and the heavens."

"Actually we'll able to transmit the telescope's images over to the boys' rooms so they can download the images on their walls if they want," Kami said.

"If I was only a boy again," Dr. Kerho admitted. "Human advancement is going to come from many angles and this academy is on the threshold of a world we are now just witnessing. What these boys are about is beyond anyone's expectations or dreams. To even think the public will accept this experiment, if that's what we wish to call it, is naïve and foolish." Steve glanced around the office at the faces of both youth and one of a proud Japanese. Even the intelligence of a prominent executive as Mr. Kamito would not understand the narrow mindedness of an American Christian in the heart of Texas.

"Let me be blunt," Steve started with all seriousness. "Growing up I was not prone to be empathic toward boys I thought of as gay. I've never told my son this, but having a gay son has been an education, and my love for him has grown leaps and bounds. I couldn't imagine anyone better in his life than Kam. When my boy was honest with his mother and me, as a twelve year old, our lives took on a new twist. Even the church we once attended had to be set aside so that Toy could mature without shame or mistrust.

"Our previous church, as is common in the south, relies on submission to church authority; after all, this is a potent form of emasculation. It entails a surrendering of conscience and personal control and deadens emotions and feelings. Glorified acts of force, judgment, and violence against outsiders, like gain, nonbelievers, compensate for this unquestioning submission. Those who do not fit, who are not subservient to dominant Christian males, must be proselytized, converted and "cured", like gay boys, through an exorcism that is moronic, at best. Our earlier church sought to banish mystery, the very essence of faith with their fundamentalism dogma. We cannot allow judgment on our children. I guarantee if the public knew what we possessed their moral permissiveness would become exciting and seductive and empower their followers to carry out acts of violence, often with a clean conscience."

"What's the answer, Dad?" Toy asked.

"It's the responsibility of anyone connected with this academy to condition these children to welcome nonconformity, not blindly obey adults like they are God-like authority figures. They must take their own responsibility to make moral choices, how to accept personal responsibility, how to deal with the chaos of human life, and learn empathy for their brothers. They will learn how to build mature, loving relationships, which will build their self-esteem through their intrinsic qualities, by their actions of capacity for self-sacrifice and compassion, not by the rigidity of their obedience."

"Love," Mr. Kamito said with softness.

"Yes, love, sir. Our academy might well represent a fight for survival. Don't be complacent with the radical Christian Right. It is a sworn and potent enemy of the open society. Eventually this problem is one we'll have to battle, but teach that tolerance is a virtue, but tolerance coupled with passivity is a vice. We must all promise to keep our academy's heart and soul a secret for many years to come, if not forever. Are we in agreement?"

Everyone nodded. Mr. Kamito was even more pensive with this thoughts. He knew that human history was not the battle of good struggling to overcome evil. It has been and always will be a battle fought by a great evil struggling to destroy the simple expressions of human kindness. The best thing, what is human in human beings has not been destroyed even now, then evil will never conquer.

Chapter Nine
(A decade later)

At 88 years of age Yuri Borzov was on his death bed. The proud father of five children, twelve grandchildren and six great-grandchildren, Yuri could be quite proud of his role as the family patriarch. He'd lived a satisfying life in a very difficult environment of St. Petersburg. To him it was the Paris of the north; others felt it was a decaying city. His vocation and responsibilities never allowed time to appreciate the various cultures of the world and its beauty.

The Borzov family had owned the same residence over a hundred-and-forty years, not unusual for a generational family structure growing up and growing old in the same city. Consequently, three guests from America, two years previous, had very little problem in locating the Borzov residence, thanks to an address left by Dr. Jolson.

Just two years earlier, Yuri wasn't used to knocks on his door, since his family often would come right in. His first glimpse at the two young men in their twenties brought a curious frown on his wrinkled face. He took his hand and swept back the few strands of gray hair left around his temples. It was when his eyes drew to this child of twelve when he froze a prolonged and intense stare. This moment of connection between this young boy and an old man was a dream almost forgotten and a promise he thought would never come. Yuri knew immediately he was looking into the eyes of Peter the Great. He began to cry.

Kami had reached out and cupped the elderly man's elbow out of concern he might not comprehend a visit from strangers.

"Sir, you don't know us, but…"

"Oh, but I do," Yuri replied in broken English, a language he'd had to learn around the Cathedral to communicate with tourists from the West. His voice cracked with the gesture to please come in. Tears began to slide down his cheek. He turned to Trevor and said, "You're Peter, right? I've waited almost fourteen years for this day. Where's Dr. Jolson?"

Toy glanced at Kami on who should take the lead here. "Mr.

Borzov, Dr. Jolson is deceased. We're fulfilling a promise made to you fifteen-years ago. This is Trevor."

"You're Peter to me," Yuri said very adamantly. "Oh, you're such a beautiful boy. Just want I'd pictured for a Romanov."

Trevor sat up straight on the sofa between his fathers. His royal blue blazer had a navy crown on his breast with a peach polo shirt underneath. It was a perfect ensemble for the 60 degree weather on a balmy St. Petersburg day. He had waited for this day for two years after his fathers had told him that, upon the academy's vacation to Europe, this would be one of their stops. Trevor rattled off a greeting in Russian that delighted this St. Petersburg resident.

The two communicated for several minutes in Russian, while Kami and Toy smiled at the connection the two of them had made. Mr. Borzov took on a heightened spirit and addressed the adults.

"You've done a good job with the boy, gentlemen. I compliment you. Peter is as intelligent and polite as the leader we'd expect him to be."

"Thank you, Mr. Borzov. May I ask how you knew Trevor when you saw him?"

"I knew. Just look at the eyes and mouth. The boy is already tall and lean just like Peter was at his age. You know the man grew to be nearly seven-feet. Yes, he's a Romanov all the way through. Oh, I wished I hadn't made that damn promise to Dr. Jolson to keep this all a secret. Russia needs young men like Peter, a future to move us forward. I'm afraid St. Petersburg has seen better days."

Toy had no intention of correcting his host. Peter the Great was tall but more likely around 6'8", and Trevor was well aware of his potential height. Mrs. Borzov was startled by the guests in her living area. Smiling and polite she approached everyone with the utmost curiosity.

"Mama, these are friends of mine from America," Yuri explained.

"Papa, I didn't realize you knew anyone from America. You never told me about your worldly travels."

The males laughed. Kami offered an explanation which was basically true. "You might say we're old friends from our interest in the Cathedral of Peter and Paul."

"I'm just glad someone takes an interest in an old man who did his best for so many years. God knows our country seems to forget its citizens at our age," Mrs. Borzov said with some bitterness. "How would the boy here like some of my freshly baked cookies?"

Trevor looked to his fathers.

"This is a special occasion so I don't see why not," Kami said and received a pleased smile from one of his twenty-nine adopted sons.

Trevor devoured his first cookie with a glass of cold milk while the men talked. He was seen staring at the piano by Mr. Borzov.

"Do you play, Peter?"

"Yes, sir. May I play you something?"

"I'd be honored," Yuri said and didn't expect much in the way of talent from someone so young.

Trevor wandered over to the old Steinway, touched a few keys, and then picked up the cover to twist a couple of wires until he was satisfied with the sound. He sat down and began with some soft chords before speaking to Mr. Borzov.

"This is from all my brothers to those who helped make our existence a reality."

Yuri had trouble believing such a voice could come from someone so young, as the boy sang to him in Russian: Michael Jackson's, **I Just Can't Stop Loving You**. Halfway through Yuri was sobbing and absorbing this moment as one he would remember the rest of his life. Trevor finished and began to retreat to the safety of his parents, only to be intercepted by a weeping old man.

Toy and Kami were in awe of this affection. They were equally impressed by the way Trevor handled this emotion. When the embrace was broken, Trevor was in tears himself. The men had an extreme warm feeling that maybe they'd done a few things right bringing these boys up.

"Sir," Toy spoke up, "we had no intention of intruding for too long. Trevor wanted to visit your ballet company. He loves to dance."

Yuri stood and wiped the wetness from his face. "I must be your guide. I'm not too old to show old friends around St. Petersburg. I just happen to know one of the teachers at the ballet school."

The men didn't want to be rude and accepted Mr. Borzov's offer. With Yuri as their tour guide the men walked the main pedestrian artery, the Nevsky Prospect, during rush hour to see the bustle of street life. Quite impressive was the St. Isaac's Cathedral, with a dome visible from the Gulf of Finland.

They spent the afternoon at the Hermitage Museum, and Trevor was the first of his brothers to visit the tomb of his progenitor. It was very surreal to take a twelve-year old to view his beginnings and the creator of this great city, not to mention the once czar of Russia. The three adults watched for a reaction, but Trevor's deep thoughts remained buried within a mind so young and innocent of such a history. Each boy at the academy was required to study the life of their progenitor. Trevor knew Peter-the-Great as if he was reincarnated. This was somewhat true, Trevor ruminated.

Trevor observed the opulent ballrooms and artwork inside, including the excellent explanation of and artifacts from Peter the Great's victory over Sweden's Charles XII in 1709 in the Battle of Poltava, a turning point in the histories of Russia and Sweden. Several glances occurred while these three tourists stood before paintings of a young Peter the Great. The resemblance between this American youth and the boy in the painting was uncanny. When Yuri saw the interest that Trevor invested in Peter's possessions, the elderly man invited his guests to the Naval Museum, which featured more than 800,000 items, including Peter the Great's rowing boat.

As the evening unfolded Toy decided on an upscale, Tsar restaurant, created in a 19th-century feeling of dining at the emperor's table. Trevor gave a boyish "Yick" at the Russian caviar, but he enjoyed the Russian dishes. Trevor was quite excited when he returned from the bathroom, where each toilet is built into a high-backed, wooden throne.

Their night stay was in a three-thousand dollar deluxe suite in the Kempinski Hotel. Being in the company of his two fathers without any of his brothers around was like Christmas to Trevor.

Yuri had promised to come to the hotel the next morning and escort his new friends to the ballet school.   In Russia boys and girls studied to be accepted as part of the St. Petersburg ballet, as American children dream of being an Olympic gold medalist. Trevor had been practicing ballet since the age of six, though Brandy and Toy had worked with the children in dance since the age of two.   Brandy, with her college roommates, used to put on the Bee Gees and Michael Jackson just to watch 29 boys jump and bounce around to the music.   The initial idea was for the children to express themselves in rhythm and wild abandonment.   This grew to a daily session of music and movement, then to dance.   Brandy's and Toy's ballet experience was soon implemented on all the boys as a daily class.

Trevor had naïve aspirations of practicing with the other kids of this dance school.   This would never have happened but for the input of Mr. Borzov.   Even Toy and Kami had told their son not to get his hopes up until the school's director, a spry and earnest woman named, Inga Stravotski, walked toward them like a woman ready to scorn their very appearance in her school.

"Mr. Borzov tells me you can dance, young man. How long have you taken ballet?"

"Six years, ma'am," Trevor answered.   "I'd be pleased if you allowed me to work out with your company."

"For a few minutes, but that's all I can allow.   Have you brought your own tights?"

"Yes, ma'am."

"Change quickly so you don't delay the other children."

Trevor reached in his pocket and pulled out his ballet jock and white tights.   He stuffed the jock back into his coat pocket.   His ballet shoes were in his other pocket.   He didn't bother to ask where the dressing room was but changed right then and there, as Ms. Stravotski grunted and turned the other way.   Practicing at the academy the boys didn't usually bother with a ballet jock.   There were no secrets to those who sneaked a glance that this twelve-year old was gifted in more ways than being able to dance.   A few girls giggled, which received a verbal reprimand from their teacher.

The initial stretching exercises were an easy warm-up for a boy used to strenuous exercise and demands of a pedantic and no nonsense ballet teacher at the academy. The other children watched suspiciously this American boy, yet were quite surprised when the twelve-year old spoke as good of Russian as they did. They had no idea they were looking at a twelve-year old copy of Peter the Great.

Through a variety of dance maneuvers Trevor showed a skill quite advanced for an American. Ms. Stravotski didn't excuse this boy as quickly as she had thought she would. Another 12-year old Russian boy took a particular strong interest in this handsome American boy with very long legs below a high rounded butt, supporting a sleek, thin torso. A long neck into a narrow face, sculptured to cuteness by blond hair cascading over the boy's forehead.

Sergei had also been dancing since the age of six. He had a mother who had religiously prepared her son for a life in dance and to pass the gruesome tests to be accepted for a lifetime of training; a privilege much like being accepted for higher education.

Sergei was impressed with Trevor's leaps and turns, but his eyes had also gravitated to the boy's groin. Trevor's posture was so refined, his jumps, precise and energetic. There are many types of dance philosophies and St. Petersburg had always been known for the more classic teaching methods. To be this athletic was the true essence of Ms. Stravotski's methods.

Trevor survived the strenuous practice and had gained a devoted friend in Sergei. They were soon whispering and smiling with each other, which gained a hastily noted dissent from their instructor who admonished their lack of attention. She demanded constant diligence and attention, though the boys were acting just like twelve-year old boys were supposed to act.

Ms. Stravotski might have been stern and pompous, but she was also very observant for those with talent. After the two-hour practice she approached Trevor and asked if he'd prepared his skills for modern dance as part of his repertoire.

"Yes, ma'am. I'd like to show you if you have the appropriate music," Trevor responded.

The proper woman scoffed, as if Russia wasn't westernized

enough.   She asked Trevor what type of music he desired.

"Michael Jackson," Trevor blurted out and had Kami choking so as not to laugh.

A smirk came across Ms. Stravotski's face.   She snapped her fingers and quickly had one of her assistants at her beckon call.   A CD was quickly brought and installed in a player.   Trevor was at his best with the freedom to dance, selecting the song, **Billy Jean**, to do his choreography to.   His jumps and leaps were years ahead of his age, so athletic and high.   Trevor used a variety of movements to his dance: gymnastics, martial arts, ice skating, Michael Jackson moves, and break dancing to add variety and flare to the rhythm.   This was something his brothers had done for years, sometimes on a nightly level in seeing who could outdo the other in moves.   Trevor would have loved to tell this group about his brothers, especially Dane and Trent.   He also relished this opportunity to exhibit his love for expression and had the awe of the many Russian dancers who had been trained specifically in the classical rendition of ballet.   This was very typical of the Kirov-Mariinsky school of Ballet.

Mr. Borzov and Sergei laughed at the boy's antics, while Kami and Toy were more afraid that their son had embarrassed or insulted this old institution of dance.   Trevor's penis was overly defined, which had an effect of making this the center of attention.   With 28 of his brothers having this same genital blessing Trevor was not conscious of its profound focus by others.

When finished Sergei was the first to applaud and come forward to kiss his friend on the lips.

"Sergei!   Enough!   You are way too presumptuous in your behavior," Ms. Stravotski spoke.

"That's okay, ma'am.   I like Sergei, too," Trevor responded and settled this refute.

"Your compliance is no excuse," the director replied.

Ms. Stravotski marched over to Kami and Toy as if they were responsible for all this.   She wasn't about to admit that this boy had a talent above and beyond anything she had ever witnessed in a twelve year old.

71

"Am I to presume that one of you is the father of this protégé?" She asked with the sternness of a school principal.

"We both are, ma'am," Toy informed.

"Interesting. I've never heard of a boy having two fathers. Nonetheless, I am offering your child a chance to train at this school. Perhaps he could wear a supporter when he returns. He appears to be of age that support might lessen my students' eyes on another obvious gift and more on the art of dancing. We have our children living away from their parents, so I can assure you he will be properly taken care of."

"Ma'am we appreciate your offer but Trevor is part of an academy in America. I have great respect for your school and your reputation, but he belongs to something beyond just the scope of ballet. I would hope you can come and visit our academy someday."

"Yes, I would like that, though I have yet to find an American company to the likes of what we offer."

"And your reputation proves this out," Toy said as not to be argumentative.

Trevor was beside Kami with a puzzled expression on his face. "Dad, why is she mad at Sergei for kissing me?"

Kami and Toy had insulated their sons from societal homophobia and reactionary bias. This had to be handled with kid gloves or the subject could have a major effect. "Let's ask Ms. Stravotski, shall we?"

"Ma'am," Kami started, "Trevor was concerned why you disciplined the boys for kissing each other."

The surprised director, who had recently replaced the aging Valery Gergiev as the school's chief conductor, was taken back a bit by the question.

"There's a time and place for such antics. I don't wish to give the impression that we condone such behavior."

"Behavior, meaning what?" Toy asked.

"I only meant……..not that I have anything against anyone's orientation, but we do have a reputation to uphold and I don't want anyone to think that only those kind are wanted here."

"Those kind meaning homosexuals, I presume. My son here is homosexual and your words have sent the wrong message to him," Kami

72

said bluntly.

The woman stuttered as someone not used to be corrected. Mr. Borzov had his own smile at the courage of these men.

"I apologize if that's the impression I put forth. There is nothing wrong with your sexuality, young man. I wasn't sure how you'd feel about being kissed."

"That's okay, ma'am. My brothers kiss all the time," Trevor said. "It seems to me that more boys like me love to dance and express themselves because it's in our nature. If straight people would embrace our love they would enjoy dancing, as well."

"Very wise words from a twelve-year old," the director said. "I believe you're right."

A trifle frazzled, Ms. Stravotski thanked her old friend for bringing the boy, but Yuri whispered a most interesting comment that had this woman jerk back in reaction.

"This can't be," she replied softly.

Yuri nodded, which had Ms. Stravotski march right back to where Trevor was conversing with Sergei. She curtsied and kissed the boy's hand. Trevor was too speechless to know what had just transpired.

"Your presence would be appreciated at any time, young man," Ms. Stravotski said and smiled for the first time that day. "Sergei, you have chosen a good friend and an honest one. You have my blessings."

"Wow, what did you do?" Sergei asked his new friend after his instructor's departure.

Trevor shrugged his shoulders. Grabbing Sergei's hand they all but ran toward Toy's and Kami's side.

"Dad, what's our address at the academy? Sergei wants to write. He gave me his."

Kami and Toy had never had this problem, but Kami didn't really expect the Russian boy to write, so he wrote down the PO Box number in Dallas. With Mr. Borzov smiling the whole time the three men watched the exchange of addresses, followed by a rather intimate kiss between the two boys, which wasn't that unusual at the academy, but it was the first time an academy boy had shown an interest in the another boy outside of school.

****

Yuri Borzov reflected back with fond memories on those precious days, two years earlier.   He was now extremely ill with prostate cancer which had him bedridden. Yuri broke a slight smile when he remembered he had had to explain to these guardians of Peter that he hadn't told the dancing teacher that this American boy was Peter the Great.   What he had said was the truth, the boy was a Romanov, a true heir to Russia.

Mrs. Borzov had questioned her husband why these American friends never wrote or called after this surprise visit.   Yuri knew these answers and maybe it was because he had been adamant in calling the boy Peter--a right he thought he had, considering what the boy truly represented.

A shaky hand reached under his pillow and pulled out a picture that reminded him of that special day.   He'd never admit it to his family but he had been the happiest with the hours he'd spend with this twelve-year-old boy. Yuri left a note for his wife that he wanted this picture placed on his chest in his casket. Tired and feeling he'd done his best and that he had explored life to its fullest Yuri Borzov died in his sleep that very night.

In the morning, as the family grieved, Mrs. Borzov read the note her husband had left and examined the photo that puzzled her tremendously.   She questioned whether her beloved had fathered another child, yet the age of the boy didn't work out to her husband's impotence in later years.   Neither did the boy look like him.   The woman was embarrassed to discuss this picture or show it to anyone, only because her husband had had the boy dress in the actual blue-green uniform of the Preobrazhensky Regiment, which Peter the Great had worn at Poltava.   It was almost humorous to see a boy in these giant's clothes, but it was a strange resemblance of someone so young to the features of a long-dead ruler.

All these historic clothes: Peter's hat, an enormous pair of black boots, a sword, and an ivory headed cane, were temporarily borrowed from the museum by one of Yuri's old friends for the few minutes of pleasure. The clothing items and artifacts needed dusted and polished anyway, so told this curator to the police a few weeks later.

The writing on the back of the picture was more troubling: **Peter the Great, at age twelve**, her husband had written. Yuri must have been delusional during his finals days, she considered. She didn't abide by her husband's wishes and kept the photo. By the following week she had a private conversation with a museum curator which led to an investigator. It wasn't uncommon for offices of Russian administrators to be bugged; thus, her conversation with this curator had been overheard. A complete search of the Borzov residence led to more turmoil followed by the woman expelled from her home. Mrs. Borzov had been smart enough to copy the photo and kept the original. She refused to give up this picture to these investigators and had secretly passed it to her daughter.

The investigation led to a curious search at the Cathedral of St. James and Paul, and evidence of a break-in was discovered. It was difficult to keep this news from spreading, even within government and political ears. A dropped comment to the press and an alert newspaper was quick to jump on front page headlines, even they thought was hyperbole: **Peter the Great's Remains Stolen!** This was hastily refuted by the government as not to alarm the public. Their reply was to minimize the reaction, simply to state that the tomb of Peter the Great had mysteriously been tampered with but there was no proof that any remains were extracted.

Mrs. Borzov had her own resentments by losing her husband's government pension and the home of her ancestors. She was an old lady who had only wanted to do right and find the truth. Now she was at the mercy of one of her daughters and being ostracized by those who knew of her husband's involvement. Their family reputation was under harsh criticism and ostracized by the neighbors. At the assistance of her daughter she called the German magazine, Stern, and offered a copy of the picture and her husband's final account of his involvement in stealing Peter the Great's remains for money.

Stern Magazine thought this story made for a good hoax on the past reputations of an archaic Europe and its leaders. They gladly paid the

$10,000, only to discover with their own investigation that there might be more to the story than just a ruse. Being that the cathedral had been closed for a few days for a "renovation" appeared too coincidental to dismiss. The editor slanted the article to the possibility that someone had discovered a means to actually clone the dead and copied the DNA of one of Russia's national treasures.

The story broke to mass hysteria within the week: ***ON DEATH BED, CURATOR SAYS HE MET PETER THE GREAT!*** There was a photo in combination with an art work from centuries before showing a teenage Peter. There were striking similarities. The lengthy article was meant to create the illusion that this was indeed the fact based on the gullibility of the public. Stern hadn't expected quite a community outrage. The hysterical reaction had overwhelmed the offices of one of Europe's most honored tabloids.

Mrs. Borzov had revealed a lot more than her husband would have ever desired. The talk about two men who had escorted this boy brought forth a vision of boys and men with a navy crown on their chests. Throughout Europe people remembered seeing boys and adults with this insignia a few years earlier. What exactly did this mean? Was Europe crawling with little boys who were once the great leaders of Europe? The chief editor of Stern loved that idea and ran with it in subsequent issues. All it did was stir up the anxiety and imagination throughout the European continent.

The rumors and investigative path led to a ballet director at the Kirov-Mariinsky Theatre. She pleaded ignorance but was accused of not reporting this so-called American boy to the authorities, let alone, having him dance with such a select company. Sergei remained faithful to his friend and never did reveal the address he possessed and communicated with quite often. Ms. Stravotsky was fired from her position amidst protests and a brief strike at the school.

Such a topic was soon the talk of Europe. Curious men in position of protecting historical tombs were cautious, if not paranoid of being investigated themselves. A custodian at Westminster Abbey hinted that he'd noted a chipped wing of a cherub above Isaac Newton's tomb, a fact that probably didn't mean anything but to his own allegiance to the

76

Abbey's security. The director of the Abbey didn't take this too seriously at first, but then a routine inspection revealed a slight off-color repair work no more than a 3/8 inch hole into the marble. It was enough to bring in Scotland Yard. The further inspection proved that, indeed, the hole had gone into the sarcophagus and the remains had been disturbed.

The English press was soon privy to this news which made the Stern publication appear amateurish. Massive sales progressed to where reports of more grave robberies were reported, many without foundation. It soon appeared that anybody who was anybody had had their bodies stolen. It made quick work for jokes and insinuations of horror movies gone to the extreme. There was a fact that many countries began their own routine examinations, inquiries, interviews, and downright interrogations of all churches, cemeteries, cathedrals, and monuments throughout Europe. Museums were often closed for inspection and even grave sites were exhumed to accurately assess any damage or stolen remains. For the most part the discovered invasive techniques into caskets revealed very little was taken, if anything. Only after a thorough search was it noticed that a bone was missing or a nail was amiss.

Within weeks Stern Magazine had raised the ante by posting cash rewards of several hundred thousand dollars for information and proof that a body had been tampered with. The magazine hit the newsstands with the headlines: *EUROPE ROBBED!* They sold 10 million copies the first day and spent three days supplying needed inventory.

There were as many attempts to scam reporters as there were legitimate reports. People speculated that someone was holding the great leaders and scholars ransom, an amount to be determined for each country to be paid at a later date. The Middle Eastern countries were accused of terrorism, supplying themselves with dynamic power figures and brains to develop nuclear weapons or use against the West. To bring Europe down to its knees all these Islamic extremists had to do was threaten the lives of past icons. Soon no one knew what fact from fiction was.

As weeks went by names began to surface: Michelangelo,

Leonardo de Vinci, Caravaggio, Oscar Wilde, Peter Tsychikowski, Richard the Lion-Hearted, Isaac Newton, Lord Byron, Lawrence of Arabia, Herman Melville, and Peter the Great.   An archeologist who had discovered the tomb of Alexander the Great at Siwa wasn't sure if someone had managed to procure a specimen from the site.

The European Community was cautious in their reaction.   Some were concerned, others humored by the events.   No new leads came forth to the identity or location of this boy in the picture and no one pursued the report of seeing dozens of youngsters a few years earlier with the same blazer like the one Mrs. Borzov had identified.   As quickly as the hysteria started, it began to dwindle as weeks melted away.

To keep the interest alive, Stern allowed a tabloid in America to run the story for only two hundred thousand dollars.   When all else failed Stern went to the West to continue this sensationalism.   Anyway, if the Jihad couldn't be blamed, it must be the Americans because they were usually behind every political plot or debacle depending on what side of the fence one was standing at the time.

Scotland Yard was the ingenuous investigative unit to take a look at the Abbey's employees from 15 years earlier.   The retiring of a security guard appeared to be coincidental, but Mr. Alfred Janson was found in an upgraded apartment with more soccer memorabilia than the officers had ever seen.   Despite the man's pension he had certainly lived above his means; plus, this senior citizen was extremely nervous during his interview.

The refusal of a polygraph had the Yard very concerned.   It was the impression of those in authority that incarceration might be the answer for a while that had broken the man's spirit and confidence. The best part, he had long forgotten the name of this doctor who had given him all that money.   One thing Alfred did remember, the doctor was from Israel and did something with designing family trees.

Chapter Ten

The fishing village was just a few miles from the capitol of Iceland,

Reykjavik. A sculpture stood majestically at the village entrance, its nakedness, by Western standards, was more than explicit to solicit the wrath of an American mother. To the children of this village the nudity wasn't anything different than they would see in a communal sauna or sometimes on the shoreline during the summer months.

For most of the boys in this community the age of sixteen couldn't come quick enough so they could quit school and become fishermen, much like their fathers and grandfathers and great-grandfathers. Ask any Icelandic boy, and, chances are, they would tell you they think of school as purgatory on the way to a future of finding riches at sea. For girls, school represented their ticket out of town, an education that would send them to a university, hopefully in the States.

Two fourteen-year olds sat beneath this sculpture on this balmy, late summer day. They were aware of this naked structure above them and both had glanced up and cracked a joke that their seating arrangement made them vulnerable to being peed on by the groin above their heads. Leif and Gisli had been friends for many years, living their mundane existence often in pure boredom. Out of lack of anything to do, this was often one of their camping spots to talk about school or their peers.

"My father returned yesterday from sixty days at sea off the coast of Norway. He had one-point-five million krona, almost twenty-five thousand dollars!" Gisli told his best friend.

"Cool," Leif answered but wasn't all that impressed. Fishing was not something he wanted to pursue.

Gisli continued, "My brother has been out for over forty-five days and he's driving a Porsche Boxer and getting all the women he wants. I can't wait until I turn sixteen and can get the hell away from this place. How about you, Leif?"

"I don't know, my mother wants me to go to college and so does my teacher, Mr. Eldjarn. It's tough keeping up with the girls; all they want to do is study because they there isn't much else to do in this town. My teacher said we're the only country in the world where girls are brighter than boys in mathematics and science."

"So what?! I know enough science to know where my dick goes between their legs. We can't even count trees because there aren't any.

Who cares that the girls are brighter than we are? They eventually give it up so we can have our fun."

Leif laughed, only because it was expected of him. "I'd rather concentrate on other things than girls."

"You're such a faggot at times, Leif!" Gishi hit his friend on the shoulder, mooned him, and then dashed off in a sprint.

Leif caught up with his friend with his lightning fast speed and the boys wrestled in the sand, until Gishi felt the arousal in his peer's shorts. He pushed Leif off of him and leaped backwards.

"Do I look like some freakin' girl, Leif? God! You really freak me out sometimes. You're so weird."

Leif was embarrassed and totally ashamed. He'd mistaken his friend's display of dropping his shorts for something he wanted to be sexual. If Gishi knew what Leif truly thought about him and the fantasies he had of the two of them together, he'd really freak out and abandon him as a friend. Leif's secret was something he couldn't reveal, not even to himself for fear that it would consume him completely. This passionate longing to share himself with another boy was felt deep inside, far from the light of awareness where secrets lived--down beneath the layers of public façade, personal myth, and fantasy. If Leif peeled away the well-crafted layers, only then would this teenager see the secret clearly for what it was: his own self-contempt.

Leif stayed sitting on his feet, his arms draped over the obvious protrusion in his shorts. "Come on, man, I was just practicing. It's not like you don't practice," Leif tried to convince his best friend.

"Not on a boy. I use my bed for that. You're so lame!"

"I gotta get home. I'm supposed to help my mother at the supermarket, and then Mr. Eldjarn is picking me up to go surfing."

"That guy is such a fag. You never see him with a girl. I bet you two do more than just surf."

This time Leif didn't care how his shorts looked. He leaped up and hit Gishi in the stomach a whole lot harder than was playful. "You ever say that again I'm going to beat the shit out of ya!"

Gishi was scared of the rage in Leif's eyes. He backed off and held up his hands. "Dude, you're one scary weirdo. I didn't mean to come between you and your surfing buddy." Gishi began to walk backwards until he felt safe enough to turn and head home without further words.

****

Leif often teased his mother about her tastes in paperbacks and soap opera rags that read more like romance novels. He'd grown accustomed to escorting her to the grocery store, where the assignment for her only son was to push the cart behind her while she selected food items, the latest assortment of tabloids, and juicy books with covers of bare-chested, muscled men making love to glamorous sex goddesses; some book covers were even more daring. The one great aspect about bringing up the rear, Leif tossed his favorite bicycle and surfing magazines in the basket.

Leif was quite used to being the only child. His mother had told him that his father had died on a fishing expedition, which is one reason Leif never cared for this vocation. He noticed his friends' fathers and was sometimes envious of these bonds. He wasn't sure this lifestyle was right for him, all the excessive drinking, probably because there wasn't a whole lot to do in a country where, for six weeks, the sun never sets.

Given the colder climate and the lack of other pursuits Iceland is well-known for its hedonistic qualities and appreciation for all things sexual. Course, there again, few cultures accepted the gay culture with open arms, and Leif was well aware how cruel kids his own age could be with other boys who weren't like they were.

Leif's world revolved around his creative ingenuity and bicycling. He was an adept surfer, a love he learned from his teacher at school, Mr. Eldjarn. What Gisli had said hurt him deeply; though, deep down, he wanted Mr. Eldjarn to be gay, like he felt his own attraction was aimed. The problem was, his teacher had never so much had touched him, outside of helping him put on his wetsuit. Leif could barely glance at his teacher when they stripped their clothes off in preparation to put on their wetsuits at the beach to go surfing. He was too afraid of his body's reaction to look at another nude male.

81

The teenager had the typical Icelandic and Nordic appearance: blond hair and blue eyes; a somewhat longer face, white and unblemished complexion; thin and lithe—a body type conducive to a future Tour de France rider than a fisherman's strength. If the women of Iceland were considered the most beautiful in the world, Leif had as much to give the males of Iceland for the same reputation. To anyone gay in this remote island this young man represented the quintessential model of gay pride and beauty in youth.

Leif began to recognize his attraction to males at the age of twelve. With male beauty so prominent and a cultural virtue, Leif's excursions to the local swimming pool gave him many an opportunity to admire his peers and men in the skimpiest of swimming attire. Only lately had his own growth made him overly-conscious of his endowment in a racing suit. Boys had begun their lewd comments, which he considered pure jealousy. It was when he received the eyes of men that heightened his sense of sexuality. There was no secret that Iceland has one of the lowest ages of consent at fourteen. Often times now, to hide his protrusion, he would wear a board short.

At fourteen Leif developed an enormous crush on his history teacher, which brought forward several after-class discussions, then an invitation to go surfing. He liked when Mr. Eldjarn mentioned in class that certain historical figures were gay and had contributed to the many advancements in science and scholarly pursuit. Leif knew the comments weren't to demean or humor those men as it was to educate the class that gay people were to be respected. The more Leif thought about this he had to wonder if Gisli was really right: Maybe he was some type of freak or weirdo. To Leif the mention of any gay figure reinforced his own self-esteem and admiration for people like himself. His goal was still to find a peer at school that shared the same interest.

With this infatuation grew a discovery that his teacher not only surfed but loved to bicycle as well. Leif often saw a racing bike on the roof of Mr. Eldjarn's Toyota. After school Leif had waited at the top of the knoll overlooking the school, then followed his teacher home to a single residence where, it appeared, no one else lived. The boy celebrated this discovery with a sense of satisfaction. He felt like an adolescent stalker, but this obsession grew more to a day-to-day surveillance.

Surfing became the one aspect they could share and be together. Leif relished this opportunity and kept secret from his teacher the manipulation it took to find something they could share together. He was tempted to tell his teacher about his web site on MySpace, an adolescent creation where he communicated with other kids around the world, but Leif was more expressive than others. His short films started as a thirteen-year old, sometimes in his pajamas; other times in his sweats. He often rolled on his mother's exercise ball, a huge blue rubber ball that allowed him to lay his body over it and do exercises. As he became braver he stripped down to his underwear and pretended to be a model in front of the camera, all the time thinking that there might be one gay boy his age who was watching.

Little did Leif know, but there were thousands who had found this web site to their liking. When the mood was right he leaped upon his bed, flew off his clothes and began dancing in his underwear to his favorite group, White Stripes. His web site, **ICELANDIC BOY** became immensely popular by the number of hits on his site. There were times he did a striptease, faking the removal of his last garment, then teasing those who wished to play voyeur with his grinding and pulsating movements of his hips. Leif wasn't quite sure he was doing it right, but the dancing invigorated his need to express his emotions and love of feeling sexual. He had made sure to purchase a dozen pairs of assorted briefs, the smaller the better. Often times he forgot he was doing this in front of a camera and became aroused to the point of masturbating through the garment. When his erection grew he kept this slightly hidden, but then allowed it to bounce up and down in his underwear to the amusement of those watching. Near the end he'd jump toward the camera to where no one could see his genitals, then removed his underwear until the only thing the viewer saw besides his legs tightening and swaying to the masturbation was the white liquid when it splattered on the video cam. The worst thing for Leif, it was a bitch cleaning up. Dozens of responses came back saying, 'That bulge can't be real! Show us!' Leif wasn't that daring, yet.

On this day Mr. Eldjarn was waiting for the mother and son to return home. Leif flew from the car, grocery bags in both arms. He hurriedly grabbed his wetsuit, booties, and hood before throwing them into his teacher's car. The actual surfing spot was several miles outside of town, off of Ring Road—which runs around the entire perimeter of the island. The area, known as Reykjanes Peninsula, was surrounded by a barrier of reed-covered sand dunes on a long, crescent-shaped beach. The trip had given the two males time to tell jokes and how their day had gone. Leif never said what he really wanted to say.

There was a solid high-pressure delivering brilliant sunshine, a light offshore breeze, and temperatures in the low 60s. The surf was a vibrant six to eight feet, a prime condition that had both Sveinn and Leif smiling with delight. With the beach to themselves Sveinn had no reluctance to strip down in preparation to put on his wetsuit, a must even with the air temperature so mild. Leif kept his eyes glued to this recent college graduate a little too long. By the time he was aware of his reaction he knew he had to turn his back. With a frantic sense of panic he began to hurry. Leif's haste caught his wetsuit on one foot, just before his body began to tumble to the sand.

Laughter from behind him wouldn't have been so bad but for the attempt from Leif in covering his erection. He didn't want the help, but he also knew resistance was futile. His secret was up, literally, he had no doubt, but Mr. Eldjarn didn't tease him nor make comment. There was one other thing Leif didn't fully understand and was somewhat discomforted—he had no pubic hair to speak of, yet he had a full set of adult genitals. By the time he had tucked what he needed to tuck into the wetsuit Sveinn was zipping up his back.

"Glad you surf better than you dress, boy," Sveinn kidded.

"I should have put this on before I got in the car," Leif admitted.

"That would've spoiled all the fun," Sveinn said with a wink and got a giggle from Leif.

Whatever had happened between them and the embarrassment, the way his teacher had reacted put a new determination and uplift in Leif's demeanor. He hadn't wanted his teacher to see him like that; now he was glad it happened.

The water was an indefinable, lucid, grayish blue, and nippy, but not quite numbing. Powerful head-high sets broke top to bottom and spun along the reef. Leif was getting more efficient with each outing and now could actually do turns and an aerial that resulted in a crash landing. Being out in the ocean with just his teacher made Leif giddy with joy, no matter how often he ate it with his wipeouts.

After surfing they warmed themselves in the nearby Blue Lagoon spa, Iceland's most famous tourist attraction known for its milky-blue waters rich in blue-green algae and mineral salts. Leif sensed happiness in his teacher, but then the question was asked that verified his teacher had seen his groin.

"I noticed you haven't hit puberty, Leif. I find that hard to believe, given what I saw."

"I can't explain it. My voice has lowered but no hair below my head has ever grown. Maybe I'm a late developer."

"You do spank the monkey, right?"

"Huh?"

Sveinn laughed and imitated the obvious. Leif giggled and replied, "It's what I'm best at. I probably shoot further than I can stand on a surfboard."

Leif felt he could probably tell his teacher anything now, yet he hesitated, still not sure of how his sexuality would be perceived or accepted.

Driving toward his home Leif eyed his teacher for his Viking heritage. Mr. Eldjarn was a proud and independent person; yet, Leif also knew the man could be a hard disciplinarian and expect a diligent learning behavior in the classroom. Whatever was the nature between them had been dissolved by their physical awareness. All Leif knew for sure was that he was in love for the first time in his young life. In bed that night he put aside the pleasure he received from dancing to the invisible faces on the other side of his web cam. Instead he wrote how his heart felt: 'Today I viewed the naked splendor of his beautiful body and was rewarded far beyond my expectations of what love is all about. He would think my thoughts are boyish and silly, but I see them as if I saw God for the first time today. My imagination could not have come close to the vivid reality of desire and lust I have for this Adonis of my dreams. He took hold of me by a thousand threads of sensation, in which my manhood will never recover. I think my blood is permanently fixed inside this throbbing animal between my legs and every burst of semen is a reminder of his love for me.'

Leif reread his writing in his diary, a diary that often contained frustration and hope, more than it did joy and contentment. With this day, he smiled and giggled with the realization that he'd struck gold. His hardened member had long protruded from the fly of his pajamas and was staring at him as it throbbed to be caressed. Leif whipped off his pajama top and bottoms, then rested back and thought of this man who had captured his heart.

Mrs. Bjornsson often waited until Leif was in bed before she strolled in to kiss her son goodnight. Sometimes she knocked; other times she didn't feel a boy's privacy was all that important. Her entry was neither heard nor acknowledged, but her eyes took in the adolescent growth of her son as he was arched upward from the bed in full erotic splendor of masturbation. If Leif's eyes had been open she would have apologized and dismissed herself; instead, her stare glued on her son's hand, milking his penis up and down to the rising of his hips.

Mrs. Bjornsson remembered that Leif, as a child, didn't have a normal size penis for a child, and now this growth had her awestruck. She'd never known what sexual union felt like between a man and a woman, her imagination had never prepared her for a sight of such magnificence in a male's body. Beside his torso lay her son's diary, and she wondered what a boy could possibly write about that would excite him so. Leif had read his daily notes to her as a young boy, and they both had laughed at a child's insight of the world. She was tempted again to glance at his private thoughts, sure that a young lady was at the heart of this sexual escapade.

She had to first catch her breath and then realize she had invaded on a most private moment. The door was barely closed when the quickened moans made her eyes dart back upon the erection. Leif was whispering a name she didn't recognize in his last few squeezes as he reached orgasm. The orgasm was artistic and an athletic event by itself, as Mrs. Bjornsson found this threshold too chaotic and beautiful to ignore. Only when she saw her son's surrender to the sensation and relax with his last massage to bring the last jolt of pleasure to his loins, did she quietly close the door and leave no trace of a mother's interference.

Leif was up bright and early for school. The teenager was used to preparing his own breakfast. With the assumption that his mother was already at work, he was doing an imitation of Tony the Tiger from a recent commercial he saw on TV. His growl was animated as he emptied the last contents from the box of Frosted Flakes into a bowl. Being nude wasn't that uncommon with the house just to himself.

Mrs. Bjornsson strolled into the kitchen in her bathrobe and tried to subdue her giggle to this comical mimic, but it was too late. She wasn't about to explain a night where sleep escaped her with a body that needed pleasured for an hour. The viewing of her son had sent her hormones where they hadn't gone in years. A quick call to her workplace and she had taken the day off.

"Mom?!" Leif drew back onto the counter with the realization that he was stark naked. His hands went to his groin.

Mrs. Bjornsson might have been embarrassed for both of them if the previous night hadn't happened. She had no intention to scorn her youngster for his lack of dress.

"I'm sorry, Leif. I should have gotten up earlier and told you I was taking the day off. Please don't think I'm mad at you; I often traipse around the house nude when you're at school or surfing."

"Really?" Leif asked as if he couldn't picture his mother doing such a risqué thing.

"Really," his mother replied. "Sit down and let me make you some bacon and eggs." She walked over and kissed her boy on the forehead and pretended to examine his body for the first time in years. "My, how you've grown! Sexy, if I don't say so myself." She patted him on the butt and said, "You do a funny Tony the Tiger."

"Thanks, Mom. I'll just go put some shorts on."

"Why? Do I embarrass you?"

"No. I just figured you'd be all sorts of ticked off."

"Trust me, I'm not. Now sit down and let your mother feed you something more nutritious than sugar and flakes. There is one thing that puzzles your mother, why no pubic hair?"

Leif sat on a bar stool by the counter and eyed his own groin for an answer. He somehow expected an answer that had never occurred to him. "I don't know. All the other boys at school have pubic hair and I don't even have peach fuzz anywhere. Was my father like this?"

Mrs. Bjornsson was caught for words. "I believe all men eventually have pubic hair, Leif. Maybe it will all come out at once. Do you want to see a doctor?"

"No. He'll probably think I'm a freak, too?"

"Now why would you say that? When girls find out what you have, I'm going to have to chase them away from our door."

"Did my father have one of these, too?" Leif laughed at his question. "I mean, as long?"

Leif's mother had always taught integrity above all. This was a difficult inquisition that she had brought on herself. "Let's just say he was gifted in his own right," she replied in an off-handed way without answering the question. "You can take your time this morning because I'll be able to drive you to school."

"I'd rather ride my bike, Mom. I like to ride after school."

"You don't have a girlfriend you're not telling me about, do you?"

Leif smiled. "None that I know of, Mom. Bicycling trips help me clear my head."

"Just remember, honey, your mother understands boys' hormones at your age. If you're getting serious about a girl, I can bring home a few condoms from the hospital. Better be safe than sorry."

"If that happens, Mom, you'll be my supplier," Leif joked. "Would anyone I choose to love meet with your approval?"

Mrs. Bjornsson brought over a plate of eggs and sausage and sat down next to her son. She checked underneath his arms and on his lip for the slightest trace of fuzz. Her glance to his groin was more for admiration than checking hair growth. His penis was twice the length of the sausage. "I'm sure my boy has great tastes, so why would I complain?"

"Just asking. You never know when I might run into a stranger who I fall head over heels for."

"Maybe I should stay home more often if this is what I have to look forward to. I can't remember the last time we had a mother and son talk. Being that your mother is a nurse, is there any question about sex I can help you with?"

Leif thought for a few seconds and had several, but they would reveal too much of his secret. "If someone swallowed sperm, would it make them sick?"

"Interesting question. So are kids at school doing this with their dates?"

"Mom! You can't ask a question to my question. You said..."

"Yes, so I did. No, sperm has several nutrients and would be absorbed as any other food. I suppose it might taste a bit salty, though."

"Oh. Thanks, Mom. I can't think of anything else."

Mrs. Bjornsson wasn't sure she wanted anymore questions about adolescent behavior. "Well, you better eat up and head for school. I haven't seen my boy in his birthday suit since you were ten when I bought you that cute swimsuit. You had to try it on as soon as we made it through the front door. You were as cute then as you are now." She made her boy blush.

Throughout the school day Leif couldn't help but smile every time his eyes caught his teacher in the hallway or at lunch hour. Sveinn often did the same thing. The two had struck a chord, that's all Leif knew. The morning events kept flashing back in his mind. He had no idea that his mother wouldn't trip on his nudity, and he was tempted to tell her that he felt he was gay and in love. Leif knew he'd have to build his courage up for that one.

Leif was hesitant on asking Mr. Eldjarn if he could bicycle with him after classes were done for the day. He was afraid of rejection, so he waited patiently for the man to remove his bike from the car before heading away from town. Leif stayed a soccer field's length behind as they rode by clusters of brightly painted homes. Some of the older dwellings were made from salvaged Siberian driftwood. It also seemed isolation had forced Icelanders to make the most of their surroundings.

The landscape transformed this coastline from the barren lava fields of the Reykjanes Peninsula to vast lava fields covered in moss and sprinkled with red and yellow lichen and spongy tundra that sinks a person up to his waist.   The land changed again to green pastures with grazing Icelandic ponies, fish hanging out to dry on wooden racks, and old, crumbling stone homesteads at the foot of massive, craggy hillsides interspersed with the tallest waterfalls in Europe.   The coast was lined with sheer, vertigo-inducing cliffs, while, just off it, giant sea stacks— those huge pieces of rock isolated from the land—were being hammered by the elements.

About ten miles out Sveinn had stopped at one of the many shell sand beaches near an abandoned fishing village.   A rest area offered the in-season tourists, usually in discovery of caves, bizarre rock formations and nesting cliff birds, a lavatory and picnic location.

Leif also parked his bike and watched the adult hustle across an old lava field to a concrete structure with a stick figure of a male on the outside of the door.   The teen assumed this was a momentary bathroom break and waited, and waited, and waited.   Finally Leif decided he would just have to make this appear as an accidental running-into-your-teacher happening.

Leif zigzagged around the partition that led into the restroom and then stood at the open doorway with complete astonishment of the absence of people.   He scanned underneath the toilet stalls and saw two sets of shoes pointed at each other in the last two stalls, one of them with a pair of shorts lowered to the ankles.   A divider separated the two stalls and Leif was confused by the positions until a light clicked on in his mind.   He crept up silently to the fourth stall and slipped into its confines, then climbed upon the toilet seat.   On his toes Leif peeked up and over the edge of the next divider.   The fellow with his back to him wasn't all that much older than Leif, and the guy's bare rear had tightened, his head arched backwards, and it was quite obvious that this guy wasn't screwing the divider.   Leif began to rub his own member which had quickly risen in his shorts.

Instantly this stranger in the second stall backed away, raised his shorts and all but ran from the restroom.   Leif squatted down on the toilet seat and waited to see the shoes of his teacher pass a few moments later. He hustled to the stall where his teacher had vacated and verified the hole in the divider, a hole that was too tempting not to experiment with. The mere fact that his penis was inside this hole and that he now knew his teacher was gay brought him to orgasm without even touching himself.

Another two days went by with Leif's testosterone running open throttle. Again the same thing happened, but, this time, Leif tiptoed in and watched another man getting jacked off, so he figured. Once this guy was serviced and vanished, Leif listened and waited, only to hear the sound of a hand slapping skin. Leif knew what that was all about. He slipped into the next stall, his heart pounded through his chest, knowing this was the most daring thing he'd ever done in his life. He found the courage to slip his hardened member through the hole. His knees shook, a fear of not knowing what to expect, though he figured a warm hand. Instead a warm mouth cradled his organ which sent a pleasure over Leif he'd never experienced or known to exist. Up on his toes his head jolted backwards until he had no choice but to put his fingers up high on the divider for balance. The instantaneous orgasm was swallowed with never a release, and Leif tried his best to mum his moans while his body wanted to melt within the divider. He fantasized that his teacher knew it was him.

Taking a clue from the previous recipients Leif zipped up his shorts and ran from the stall. It was another two days of absolute joy and fantasy; several times he was within a second of telling his mother how his heart felt, only to have the phone ring or his mother would divert the conversation to something that happened at work. Leif had never seen his mother so animated and alive. She talked about men at work and how one of the doctors had taken an interest in her, and what did Leif feel about his mother dating?

The next fair-weathered day Leif waited and followed his mentor again out of town. The sequence was identical and Leif couldn't wait until his teacher was in the restroom before he sprinted into the empty stall next to this man of his dreams.

Sveinn Eldjarn was not a gullible man by any means. He'd

guarded his orientation from many, especially his peers in the teaching field. His incidental meetings of young, available men like himself were mere luck and word of mouth. Often times he was able to be the receiver and the recipient of this restroom meeting place. He was pleasantly surprised when a rather long penis sought his assistance the other day. A second such occurrence was very rewarding but the thought came to him of another such young man with a penis length that well exceeded the norm. A lack of pubic hair and then the glance under the partition gave his curiosity a mind-awakening. The mysterious boy was Leif!

Sveinn wasn't positive of the boy's motives. There were students who would play cruel tricks on teachers, then brag to their peers to be popular. He knew Leif well enough not to think the worse of the boy, and even though the age of consent in Iceland was fourteen, his role as a teacher precluded such a relationship.

The following day Sveinn tested his protégé as he approached this rest area. There was a curve in the road which he was sure Leif was far enough behind not to see him veer off behind a rock. When Leif breezed by, Sveinn waited and came in behind his student for a change. There was a surprise on Leif's face when he didn't see his teacher's bike parked in its usual spot, but that didn't stop him from traipsing to the bathroom.

Sveinn came in a moment later and saw the pair of sneakers with the shorts already lowered. He simply opened the stall door and saw Leif hard and ready.

"Well, odd meeting you here, Leif."

"Mr. Eldjarn, I was just,….er, well,….," Leif raised his shorts quickly, blood rushed to his face and embarrassment flashed through a guilty grin.

"You could have just told me and this secrecy needed taken place."

"I….I wasn't sure if you'd still like me."

"Leif, I'm gay, too, but I guess you know that. I'm also your teacher and such a relationship is frowned upon. What do you say you follow me to a hot springs nearby and we'll talk?"

They rode together for a change, pensively quiet, occasionally exchanging glances and smiles. At the secluded hot springs they stripped and sat near each other, both as nervous as the other.

"So, how long have you known?" Sveinn asked.

"I just suspected, that's all, and then when I followed you, I saw."

Sveinn laughed, but noticed how this confused the young man. "No, that's not what I meant. When did you discover you were gay?"

"Oh, that. I think when I was more attracted to those hunky guys on my mom's paperbacks. I also find several guys at school attractive, but they're not gay."

"You'd be surprised. So, what do we do now?"

"You can teach me everything you know. I mean, this place is so boring, I've never met another gay person in my life."

"You have, you just don't know it. The guy at the pool who's always joking with you would love to get in your pants."

"His jokes are so bad. He's really gay? He's kind of dorky looking though. I want a handsome guy like you."

"Well, thank you. Does your mother know?"

"She couldn't handle it. I mean she's into all these romance novels and asking me when I'm going to bring home a girl. She's real curious why I don't have any hair under my armpits or any peach fuzz. Women are so nosey."

"That is strange. Almost like you have a new genetic breakthrough. I wished I didn't have to shave every day. I do find a hairless person sexy."

"Great! Then you'll like me."

"We'll have to talk about this. Are you willing to give me up as your teacher just to have sex?"

"Sex is more important than school. Can't we do both?"

"If you weren't already an A student I'd think you were trying to get in my shorts just to get a better grade. I'm afraid other students would soon pick up on us two devouring each other with our eyes all day."

"I do that anyway. I have wanted to ask you a question."

"Fire away," Sveinn said and put his arms around this gorgeous admirer, now sitting between his legs."

93

"My mom gets these newspapers with stories on movie stars and famous people. Well, this German magazine has this article on missing bodies; you know, like someone is taking some hair or bone from a casket. They think someone is cloning famous, historical people. See, you've mentioned a lot of these people and I've come up with my own conclusion."

"And what is the secret, my boy? I have also seen this bit of gossip. I think it's all about selling magazines, if you ask me."

"Well, sir, I've noticed that all the names are people who might be gay."

Leif's teacher thought about this with a pensive expression. "Interesting insight. I'll have to check the names again, though you may be correct. I'd advise you to send in your hypothesis and make them put up or shut up on their offer for information and a financial reward."

Leif twisted his face toward his teacher and kissed him on the lips, then submerged in his first attempt at returning favors.

The next day in school Leif was bubbly with anticipation of all that was new in his life. He knew something no one else in his classroom knew. More than that, he'd enjoyed his first sexual episode and knew what his sexual appetite desired. Just thinking these thoughts had Leif praying that Mr. Eldjarn wouldn't call on him this day to come to the blackboard to answer one of the ten questions asked. No more had he wished this then his teacher gave Leif the first question.

Leif hesitated and reached down his pants to flatten his erection beneath his underwear band. His classmates were beginning to stare, already assuming that their peer didn't know the answer and was stalling. When Leif stood up, within the first two steps his erection had slipped out and was pointing his pants out and to the right. It was the first time he was glad he sat in the front row. He immediately shoved his hands in his pockets, though, by this time, Sveinn had seen the difficulty.

"I see your mind is thinking very hard about this, Leif. I'll stand right behind you while you list the events for the start of World War One. The rest of the class can compare their answers to your own."

The two of them laughed about this incident on their after-school bike ride. Sveinn made sure they left the school grounds at separate times to cover any suspicion. They rode again to the hot springs, and though the weather turned to a warm rain, their bodies found comfort in the warmth of the other. For Leif he was sure he was in love, a feeling he'd never possessed before. There was something profoundly exciting about kissing someone of his own sex. That first experience of feeling romantic love blended with erotic surges that burned itself into Leif's brain. The joy of finally having touched the innermost secrets with the first feeling of completeness it had brought was monumental in Leif's life.

Sveinn was cautious in his role, always empathic to how Leif would perceive this in years to come. He knew this boy idolized him by the words spoken to appease. After their first orgasm Sveinn took the time to speak of this interlude.

"First and foremost I'm your teacher, obviously in more ways than one. A great teacher has no interest in being followed or worshipped or even believed. I invite skepticism, questioning, challenge, and provocation. I want only to wake you up and suggest possibilities. I won't trespass upon my own discovery, only because I might step into a sanctuary of my own heart. It's important you learn to love and respect another's boundaries at the same time."

This was above a fourteen-year old's head. "I love you, Sveinn," Leif blurted out and hoped it didn't sound way too corny.

"Okay, so much for boundaries. Look, Leif, romantic love is often most exciting when its object is new, not well known, not familiar, and ripe for conquest. Friendship thrives on time, on knowing its object very well. I suspect that in the long run, friendship partakes of divine love more often and fully than does romance."

"Fine with me. Let's be friends then," Leif considered.

Sveinn shook his head and gave up. Having this body in his arms was overshadowing even his common sense.

A week later Leif's life had taken even a higher level of excitement. Stern Magazine sent the boy $1,000 in a cashier's check as a reward for his investigative efforts. Stern wasn't sure about the gay angle but it added another bit of sensationalism to each upcoming article. Leif couldn't wait to tell Sveinn and invited his teacher to dinner with the approval of his mother.

Leif's life was floating on a cloud. Even Gisli called up and invited Leif to a street hockey game, which must mean Gisli had forgiven Leif for being so blatantly gay during the last time the two had done anything together. Leif accepted quickly and began to ponder if his friend had thought about being closer than they were.

Five other boys were already warming up by slapping a puck on their roller blades into a portable net that was rapidly slid sideways when a car approached. Leif waved and whipped off his jacket to join whichever team who chose him. There were smiling faces, but way too suspicious for Leif's perusal. Gisli went first and dropped his pants to his knees, showing off blue briefs below his American Boston Bruins sweatshirt. The other four boys followed instantly, and every one of them had on briefs. They gyrated their hips while Gisli hummed a way too familiar tune to Leif's ears.

"Bet our Icelandic Boy could do this better than we can," Gisli said and laughed.

Blood drained from Leif's face. He knew what his mother had often said, 'Don't take things so personal, Leif. A smile quickly disarms a peer's attempt to belittle you.' All Leif could think about was, "They know I strip and jack off in front of my web cam." He wanted to laugh, to join in on their caustic humor. Instead he flipped them the bird and stormed off. He figured that they'd be calling him fag and queer before the night was out.

Humiliation felt like a wet shirt hanging from his body. Word would spread quickly through school. Mr. Eldjarn would find out and probably not even talk to him anymore in fear of other students making but of a teacher who sided with an exhibitionist. The sky had fallen and a middle finger hadn't exactly brightened the prospects for a brighter future.

Mrs. Bjornsson was delighted that her son had someone he could look up to as a male mentor. She had always felt guilty of not giving her son a male's point of view and lack of male company. When the discussion came to one of her favorite magazines she was even more pleased with the conversation. Though she saw this sullen look on her son's face when he walked in the door, she quickly reminded him that his teacher was due for dinner in fifteen minutes.

Leif could have kicked himself for forgetting this important night. The mere fact that, if his peers had not made fun of him, he would have been playing street hockey for the next two hours.

Leif had showered, splashed on aftershave--for whatever reason-- and dressed in his best shirt and slacks. The prior incident with his peers had almost been forgotten.

Though Leif's mom had no idea the magnitude of this invitation, she was the most gracious host. She was convinced that Mr. Eldjarn was her son's favorite teacher without any more suspicion. After homemade ice cream and cake Leif ask Sveinn if he'd like to see his room while his mother did the dishes. Of course Mr. Eldjarn offered his assistance in the kitchen which was quickly refused. No more had Sveinn entered the domain of a boy's quarters than Leif wrapped his arms around his teacher and found the man's lips. They fell on top of the bed and Sveinn's mind went from nervous to pure panic on the repercussions of a mother walking in. Sveinn broke the embrace and collected his composure.

"Leif, we have to be careful," he warned.

"My mother doesn't intrude on my bedroom without knocking. Anyway, those dishes will take an hour." Leif whipped off his shirt and shacks. It didn't surprise Sveinn that the boy hadn't worn underwear. Leif had an agenda and didn't wait for his teacher's response. His hands undressed Sveinn as fast as possible before anymore words could be spoken.

There were new things Leif wanted to try and what better place than on his own bed. He rested on his back and raised his knees to his chest to feel Sveinn's hot tongue penetrating his anus. This sensation exceeded Leif's expectations and he began to plead to have his teacher enter him. With fingers full of saliva Sveinn put his erection at the entrance and lubricated the tip of his penis.

Minutes of absolute ecstasy with a complete lapse of time. Their bodies glistened with sweat; Leif motivating his mentor with sexual needs of total abandonment. It wasn't the knock that didn't occur, but the immediate entrance of his mother, curious that she hadn't seen her son and guest in several minutes. The entwined males had only time to break their bond, an immense stare came from Mrs. Bjornsson with a shocked reaction that didn't need words.

"Shit!" Sveinn's word spoke his displeasure as Leif's bedroom door closed abruptly.

Leif was less panicked. "I'll talk to her. I should have told my mother I was gay a long time ago. I'm her only son, what's she going to do, kick me out?"

"It's not always about you," Sveinn countered Leif's lack of panic.

Mrs. Bjornsson was sitting on the sofa when they presented themselves for this inquisition. Sveinn had one hope, that Leif's premise would be understood. The lady of the house beat both of them to the punch.

"Mr. Eldjarn, apparently you have influenced my son in more ways than one. The boy is too young to recognize when he's been manipulated and groomed to your needs. I believe you are in the wrong field of study and vocation."

Sveinn stood frozen with the words that penetrated deeply. His dreams of teaching for years felt like an exploding meteor in his mind. Leif wasn't going to take this distortion as fact. He hadn't even bothered to put on many clothes to face this crisis.

"No way, Mom! Sveinn…..Mr. Eldjarn is not at fault here. I am. I'm gay! Don't you get it? Why don't I ever bring home any girls or appease your constant badgering for me to have a girlfriend? I stripped the person I love the most and wanted to be loved."

"Leif, you have no idea who you are. By God, you're only fourteen. This man is an adult who should find men his own age. I think its best that Mr. Eldjarn leave this house immediately before I call the authorities. As with you, young man, you will be changing teachers next week, whether you like it or not."

"No I won't!" Tears began to flow and his face heated with disdain. "And I thought you were accepting me, telling me that I was sexy and cute. You were lying!" Leif fought for words from which he felt betrayed by his own mother. Instead, he dashed for his room.

Sveinn was caught speechless for several seconds. He had no retort to this mess. He knew this was always a possibility but he had successfully beaten this common sense to a pulp because of a single image in his mind--Leif's body.

"I'm sorry, Mrs. Bjornsson. I do love your son immensely, but I've violated your trust and your expectations of my behavior." Sveinn moved toward the door as quickly as he could.

"Mr. Eldjarn." His name caught him in mid-stride.

"I've done my best with this boy, but I didn't raise my son to be a faggot. The boy is confused about his sexuality, and you're not helping. I will have to think whether I will report this to your school. In the meantime I'd appreciate it if you transferred Leif immediately from your class. You will have no further contact with him."

"Yes, ma'am."

****

Leif had stayed in his room only long enough to wait for his mother to go to bed. He was glad she hadn't approached him further with her judgment of him or her outdated opinions. Climbing out his bedroom window he was at his teacher's residence in fifteen minutes on his bike. Sveinn was still frustrated, nervous, and immensely upset when he opened the door.

"Leif, my career is ruined and I'm looking at prison time if you pursue this. Is this what you want?"

"I just want you." Leif begged for recognition.

"No, you want the sex. Any boy would do. You have to be brave enough to put yourself out there and find another person like yourself. Shit! If your mother knows you're here I might as well kill myself or leave the country."

"Okay. Let's do it. There's nothing keeping me here. My mother said so herself; she doesn't want a faggot living in her home."

"She was being irrational given the moment. She'll come around to accept you."

"Oh sure, after she makes my life miserable for months! Remember, I'm her little boy who doesn't know who I am. Fuck that! I know I like dick, to suck it and be sucked."

"Leif, you're getting carried away. I've never heard you talk like this and I won't tolerate it in my house. Now go home and get some rest. We'll settle this next week, but know what's done is done. There won't be any more bike rides or hot springs."

"Fuck you!" Leif said with tears running from his eyes. He was practically out the door when he turned back. "You're such a hypocrite! You told me that until I lovingly embraced my own existence, I have no context in which to love another person or the world or indeed life itself. Well I know who I am and loved you because I thought you loved me. You're a liar!"

Sveinn was a thought away from running after the boy, to hug and cradle this fragile ego and calm a heart he loved very much. Leif hadn't walked but ran from the premises. Tomorrow the boy would come to his senses, Sveinn thought, but even he didn't really believe it.

Leif walked his bike for two hours. His world had tumbled down and his emotions were spinning with accusations and resentments he couldn't deal with. For a boy this age the messages he'd received from his friends, his mother, and even his teacher were too insidious and seductive, way too overwhelming for a young ego, already frightened of being different. He wanted to take a stand against, to reject this abuse. His healthy self-esteem from the week before had been attacked and reduced into a self-fulfilling prophecy that being gay was bad.

Tired and worn, Leif returned to his room through the window, packed his clothes and pocketed a check made out to himself from Stern Magazine. Inside his dresser he found his passport, a document he'd obtained when his mother took him to Holland a few years earlier to meet some of his mother's relatives. One more glance around the room and he spotted a hobby from his boyhood days of setting up wooden and metallic soldiers in mock battle. It all seemed so stupid to him now. He hated war and people who judged. To him this boy's room belonged to someone who had hidden for way too long. This was about to change.

100

A half-hour to the Blue Lagoon he slept by the warm water, occasionally bathing his tired body in the liquid. He didn't need anyone in his life, he decided. The world was his oyster to be found away from this hell hole. Slowly convincing himself of his only possible solutions, Leif arrived at the bank the next morning and cashed the check for currency. A bicycle ride to the airport and he soon found himself on a plane to New York City. Sveinn had told him that any man would find him irresistible. All he had to do was smile. A few days short of his fifteenth birthday this was something Leif was quite willing to do for love and freedom of expression.

Chapter Eleven

Senator Robert Franco sat in the front of a school bus and felt like a teenager again. Such a position was condescending for a Texas Senator but it was a minor inconvenience to make for his three sons, all athletes at this prestigious Catholic school in San Antonio. St. Bonaventure was considered an athletic powerhouse throughout Texas in the field of high school sports. Robert was as proud as a father could be to have three sons wearing the gold and blue colors of this elite institution.

Robert had cleared his schedule in Washington and at his Austin office just so he could partake of this first-time competition between St. Bonaventure and this new academy from the Dallas area, Kings Academy. Though the senator had asked one of his many subordinates for information on Kings Academy, there wasn't a whole lot known by the general public about this unique school: its curriculum, philosophy, goals, or even the application process to be accepted as part of their student body. There was one thing very apparent to anyone in the educational process, Kings Academy scored higher than any public or private school in the State of Texas on achievement tests, especially the National Examinations in math and science.

A smile creased across the senator's face. How could a school compete in athletics with powerhouses like St. Bonaventure, using a bunch of geeks and eggheads for students? For a second the senator was pondering that this trip might be a waste of time for these rambunctious and talented teenagers. St. Bonaventure had a reputation of being in the top five in every sport and highly respected by even the top powerhouses out of Houston and Dallas. Many a school avoided scheduling this Catholic institution if they wanted to maintain a winning reputation.

Now only a few miles away from their destination Robert pulled out of his briefcase the state's Superintendent of Schools' file on Kings. Whoever had built this facility must have had a great deal of money, Robert realized. Kings Academy spanned over a three-hundred acre site; even an eighteen hole golf course for its staff and students had been recently installed beyond its northern boundary. There were rumors that the athletic venues were finer than many of the state's universities, but the senator had trouble believing this since St. Bonaventure was the first high school to visit this academy.

The names of the headmaster and other school officials weren't recognized as prominent administrators with Ivy League backgrounds or long term reputations as Texas university types with athletic resumes. Just as the senator turned the page of his report a paper airplane went sailing by his ear. One of his sons leaped from his seat and snatched the airfoil from his dad's lap.

"Sorry, Pop!" The boy remarked and guided the dynamic paper plane to the rear of the bus.

Coach Venegas was sitting next to the senator and was empathic for anyone enduring a group of rowdy teenagers. "Sorry about the noise, Senator."

"Reminds me of old times, Jess," Robert replied and was already making a reminder to himself to travel by limo to their next away competition. His eyes went back to page two of his report. What struck him the most was the total enrollment of Kings Academy—thirty-three students. This must have been a typo in his report. To have such educational achievements and promote an athletic program of any magnitude would require a student body of at least five hundred. The report certainly gave the senator more questions than he could answer; in particular, what he was told by his own sons on the number of sports the two schools were competing for that day.

"Coach!   How many boys do we have on the two buses? Fifty or sixty?" Robert asked.

"Yes, sir, fifty-six! We're competing against the Kings in gymnastics, wrestling, volleyball, tennis, and swimming and diving," Coach Venegas yelled over the constant noise. "Varsity only, though. Their numbers are too weak to include junior varsity. I believe they don't even have grades nine through twelve, yet."

Robert checked his stats on his lap, then asserted, "Says here that they've been in existence for eleven years. I dare say this is a mistake but the report given me has the Kings with a total of thirty-three students."

"Must be a mistake, Senator. This same academy visited Sam Houston High last year and shut their volleyball team out in three straight. Did the same thing with their tennis team. Their talent must be tremendous. We'll soon find out. Look! There's the academy, sitting there in the valley."

The entire left side of the school bus leaned against their windows, a vision of beauty equated their interest.   The academy resembled more of a Caribbean resort than an academy for teenagers.   Only three primary buildings stood out from this massive estate, glass structures with a rainbow effect rose like silver pyramids. The afternoon sun gave off reflecting streaks of light from the panes like silver laser beams. Surrounded by lush green hills a crystal clear lake with white sand glistened adjacent to outdoor athletic fields, groomed to perfection. Gardens and Greek arches adorned pathways leading to a terrace carved into the hillside where half-a-dozen tennis courts were installed. One St. Bonaventure athlete pointed out a sand volleyball court near the beach, while another admired the multi-colored sails on a combination of docked sailboats and miniature catamarans.

"Cool! Are we here for a vacation or competition, Coach?"

"Cool your jets fellas! Let's take care of business, then we'll take over their school," Coach Venegas humored. The boys wanted to believe him.

"Man, do you see that, Dad? They even have a water slide park. What kind of school is this?"

"Son, you can go to the water slide park in San Antonio anytime. Just kick some ass and don't worry about play time," the senator lectured Jay, his youngest.

Coach Venegas stood up and motioned everyone back to their seats because the bus was actually leaning to the left. "Miller! We're not here for slip and slide! Focus on your event, gentlemen, and let's show these nerds what we're all about."

The bus rocked with the shouts of approval and enthusiasm as the vehicle slowed in front of an iron-gated entrance. Two security men, both Japanese, approached the bus and had the passengers disembark. Fifty-six athletes, coaches, trainers, five female cheerleaders, and one State Senator exited the bus and were taken down a few steps to an underground tram. As each person moved through a surveillance platform they each had to place a hand on a heat-sensitive pad for identification purposes. An orange band was snapped to one ankle before they boarded.

"You'd think we were at an amusement park or earmarked as terrorists," one of the coaches commented.

The tram was soon underway, accelerating amidst a labyrinth of internal lights and sensors.   Above ground the tram swooped around the lake to give these visitors a pleasant look of the water, its sand of almost pure alabaster, white and soft, making a powerful contrast with the monochromatic turquoise sea characteristics of the lake.   The beach resembled a deep-pile carpet of whiteness with a bouncy quality that invited even these disciplined athletes to frolic in fun rather than compete for the glory of their school.

"Oh, man!   Did you see that?!" a boy yelled.

"What?" a few others asked, their necks jerking toward the pointed finger.

"I swear it was a dolphin. Look! There are three more."

"Hey, look! There's a kid riding a dolphin! And he's naked!"

Several boys laughed until they saw a young boy in the lake with his hand gripping a dolphin's fin. Two more naked children, a girl about twelve and a boy around the same age were surfing and waved to the tram as it moved around the lake. A smaller boy went splashing from one of the larger slides into the lake.

"Better not tell Sister Anne about this school. She'll restrict us from ever coming again," a St. Bonaventure swimmer said very seriously.

Another boy yelled. "Yeah, but tell Father Flannigan and he'll be riding that dolphin with the boy!"

"No, he'll be riding the boy!" another teen shouted.

"Watch your language!   We have adults on board," Coach Venegas warned.

"Don't dolphins live in the ocean?" one of the cheerleaders asked.

Travis Franco, the oldest son of the senator, clarified this distortion. "Not all dolphins are salt-water creatures."

Below another hillside a structure that wasn't recognized earlier came into view.   The sun had managed to blend it into the hillside. The facility's semblance was accurately stated as a dormitory. Whatever the tour accomplished it took the reason for being there off of the athletes' minds.

Underneath the main pyramid the tram came to an abrupt stop. Met by more security all the athletes were taken by an elevator to their dressing area.   Spectators and cheerleaders were escorted to a main hall where signs pointed them to the various arenas. Robert had joined the main entourage rather than have to smell the aroma of a boys' dressing area. He stood by a glass pane overlooking the facility and awed the magnificence of this architectural wonder.

There was a sense of being in another century, far removed from what was to be expected in an educational institute. He had traveled much of the world and this facility was more like what he'd seen in Germany or a Scandinavian country, but even they didn't measure up to this wonder. Spaces within the complex flowed together easily and the structures were entirely unobtrusive. To scan the other glass buildings nearby was like looking at shimmering lakes in a field. Robert chuckled with the reality that, whoever designed this, if their goal was to make the architecture disappear, they did a bang up job.

A momentary spark of recognition and Robert realized he'd lost contact with the rest of the spectators and parents who had accompanied the team in cars. He moved around a corridor and saw the gold-lettered signs pointing to a restaurant, fitness center, boutique shop and botanical gardens. Paintings lined the hallway and the senator moved forward to where several cheerleaders had stagnated and whispering in their perusal of a naked youth with wings. Robert well knew the painting as one he'd seen in Venice, Italy. It was a Caravaggio masterpiece, though the likeness was too uncanny for it to be merely a duplicate. Robert knew even this academy couldn't afford an original.

There were other paintings, all by renowned artists, and their resemblance to the originals were spellbinding. This display was way too graphic and expensive to be in a high school setting, Robert thought. With his attention drawn he didn't notice that the five cheerleaders were directed to a women's locker area. Though Robert wanted to see what else this hallway led to he decided to join the team in the locker room.

What the boys had already found amazing, the senator walked in and was equally awestruck that the walls were painted gold and blue, St. Bonaventure's school colors. The more he examined this design he discovered that the colors were somehow a colorized, electronic version that could be switched to other colors. What a remarkable gesture to put the opposing team's color on display in the visitor's locker room, he considered.

"Hey, Dad, check this out," his youngest son, Jay, said.

Jay demonstrated that a mere touch on a thumb pad became a lock-unlock mechanism for each locker. No one else had access as long as the boys were there. Even the locker area had a variety of bright colors, colors that any elementary youngster would find attractive. A fluorescent inscription above the lockers had a **WELCOME ST. BONAVENTURE banner**, which impressed even the senator.

Across the lockers were creative designs on the walls, images depicting the physicality of human beings, from blood vessels to nerves, to muscles and fingerprints, to eyes and faces—anatomically correct and nothing left for the imagination.

"Ouch!" An athlete shouted next to the senator.

"What's wrong there, son?" The senator asked.

"I tried to take this ankle loop off and it shocked me. I'm not touching that thing again."

"I'll ask one of their security men if that's safe," the senator replied.

It had been years since the senator had attended high school. Though he pretended to scan the various displays he surreptitiously eyed his sons in undress, taking a glimpse and comparisons of the wrestlers from the gymnasts, to the swimmers with their narrow hips and no ass. As the shorts, compression tops, or skintight swimsuits were slid on, Robert moved from his corner stare and traipsed toward the trainer's room. An odd device called a Physio-Scan was one of the first strange items of equipment the senator came across. In digital information it flashed: **Remove all clothes**. Robert wasn't about to do that.

"Hey, Arthur! Come over here."

A senior wrestler trotted over with just his jock on. "Yes, Senator Franco."

"Stand on that pad and see what happens."

The athlete did as instructed and the message flashed: **Remove all clothes**. The jock was slid down and the athlete resumed his position. In seconds all sorts of information began to flash on a screen: Stan Arthur, 6'1", 210 pounds, blond type, B+. The senator and the boy gawked at the screen as it continued to present data on the boy's health, his posture, balance and muscular deficiencies. It noted that the boy's musculature was in imbalance, the right side stronger than the left; the tilting of the spinal cord; the left thigh being 23% stronger than the right, and so on.

"Wow!" is all the boy said.

When this wrestler retreated in contemplation Robert called over his youngest son. "Jay, step on there, but first remove your wrestling outfit."

"Ah, Dad."

"Don't ah, Dad, me. All of it!" The senator was glad his son had reached puberty.

Jay did as he was told and the results were more impressive than the senior wrestler. At least Jay was almost in balance and had good muscle tone with equal strength. The physiological scan also noted deficiencies in nutrients and excessive uses of sugar and salt in food. It further suggested that Jay needed to remove excess in his bowels which embarrassed him for a second. The boy had no problem making his 115 lb. weight limit for the lightweight division, but he decided he should use the bathroom anyway.

Convinced that this academy had more advanced machines than an Olympic training center Robert wanted to continue his tour. Even in this sterile, white-walled area were 3D images of various body types, examples of white-twitch and fast-twitch muscle fiber, and how each type determines one's capabilities or predetermination for success in any given sport. This had already been relayed to the boys on the Physio-Scan. There was also a device similar to a video game which an athlete could play with to determine his muscular makeup and sport predilection. This was but another distraction these athletes didn't need, so Robert made sure none of his sons saw this until after they competed.

As the boys gathered in their respective groups with their coaches the senator saw that each room had a computer imbedded in the wall. Such titles as: Music, Temperature, Lighting, Ambience, Color and White noise were among the variety of selections. Robert's fingers couldn't resist and he moved an arrow to contemporary music. The locker area quickly filled with what one athlete immediately asserted as "elevator" music, and, after enough stares, Robert nullified this selection.

"Put on Stone Temple Pilot," a boy suggested.

"What?" The senator asked. "Am I that out of touch I've never even heard of this group?"

His middle son, Niqui, helped him out. "Dad, its alternative grunge."

"Alternative, what? Help me, Jesus!"

With the various athletes moving to their venues the senator found he had to make decisions he'd failed to consider. With a son in different sports he figured each boy's time period for participation and knew wrestling was his best choice since his youngest boy was in the third weight group. Except for the pool the rest of the venues were below ground level as the escalator moved the athletes sub-surface past displays of art and science, enclaves more befitting a museum than an academy. And all this time the senator is thinking, 'This is for just thirty teenagers?'

The wrestling hall was as majestic as the rest of the facility. Two massive blue mats had a white crown in the center with a capital K. Senator Franco wasn't all that impressed with this semblance of the school's emblem, but he was very pleased with the plush seating in blue corduroy chairs with armrests. They were almost as comfortable as his office chair. He felt like he'd bought season tickets in an executive suite for Houston Astro's games. In the middle of this arena hung a large banner that said: ***KINGS ACADEMY WILD WEST HEROES.*** Robert thought that was a strange nickname for a school.

Robert's eyes went to the surrounding motifs and designs on the venue's walls. Greek prints and someone's creative imagination of how the Parthenon might have looked in its day decorated the massive interior. Naked wrestlers grabbled with one another which gave this great hall a feeling of importance and energy for the effort and discipline of these young wrestlers. He'd almost forgotten about other parents and spectators who had now begun to enter, their smiling faces and constant chatter were about their first impressions of this academy. The senator returned a dozen greetings with the same flare as if he was campaigning for the next election. He proceeded to take his comfortable seating arrangement with some air of delight and his status of importance.

Ten Kings Academy athletes entered from a far door. Their blue and white sweats, if that's what they could be called, were tailor made to their sleek bodies, faces of focus, and their eyes darted toward their competition like any young athlete's inquisitiveness to see if their competitors were built any different than they were. Robert was pleased when he saw these athletes with a short, military cut hairstyle; he'd expected a long-haired radical look, given the nature of their test results. Children's smiles were the same everywhere, he pondered and relaxed with an increased confidence that his son would win his match.

The Kings' kids appeared young, yet in excellent shape. They tumbled around the mat for their warm up, exchanging wrestling moves with their teammates, but not doing any kind of stretching or exhaustive workout. The first bout was an exciting display of wrestling skill. Lightweights, light and fast, the teens reminded the senator of two twelve-year olds in play, but these teens were skilled and moved in and out of reversals and near falls until the Kings' boy pinned his opponent with thirty seconds left in the second period.

While the St. Bonaventure boy had on a typical wrestling singlet, Robert found the uniform for the Kings' boy quite interesting. The garment appeared odd, at first, as if it was painted on. There were no seams or edges and there were no signs of underwear or a jock. The boy's rear end was graphic with apparent no covering but it was a bright red in color. Stomach muscles were easily seen through this type of top and it puzzled the senator why the boy never adjusted his straps or saw the garment actually move.

On the second mat each boy had a Dutch-boy haircut that could have made them twins but for the difference in uniforms. The St. Bonaventure athlete was long-legged and skinny, a factor in why the boy weighed around 120 pounds. The Kings' athlete was shorter, stockier, with a bubble butt of firm muscle. It was still an excellent contest of leverage of limbs versus quickness and muscle.

With as hard as these two boys were working on pinning each other, Robert enjoyed himself immensely. Wrestling offered its moments of humor, but, in the end, it was another Kings' victory. The boy had swept in underneath his taller opponent and caught the lad in a fireman's carry, then succeeded with a quick pin.

His youngest son, Jay, was next on the first mat. Undefeated in the first 10 matches of the year, Jay had a strong work ethic, quiet in his demeanor, yet the boy felt in the shadows of his two older brothers. This certainly didn't stop Robert's son from his desire to win every match. Winning was Jay's way of getting attention from his father; unfortunately, his brothers were champions, as well, and this still left Jay with a mere nod of approval.

The other wrestler from the Kings Academy was a reversal from the teammate of the previous match. Taller, slim, long legs and a sleek body, the athlete almost appeared too cute with a bundle of hair tucked under the helmet. For the first minute there were a lot of fakes, attempts to capture a single leg or attempted arm rolls. The referee soon had the wrestlers into a "par terre" position (on the ground), and the Kings' wrestler secured a cross-face hold with one arm while bringing the other through Jay's crotch. It was a near-pin, until Jay managed to wiggle out of bounds.

Senator Franco yelled his lungs out in a frantic effort to support his boy. Once more he wondered about the Kings' outfit and was puzzled that this athlete had very little crotch. A moment of hope vanquished when Jay's opponent shot in for a souplesse, a belly to belly maneuver. Up and over Jay rolled and, a few seconds later, the match was over. No more had the referee raised the winner's arm, than the Kings' wrestler undid the helmet, allowing blond hair to cascade down to her shoulders. Her teammates swarmed all over her in celebration and Jay realized he'd been beaten by a girl.

It was not the senator's finest moment. The only thing remotely uplifting was when the young lady was scolded by her coach, before she ran quickly over to Jay and not only shook his hand but hugged him. At least he had something his teammates wouldn't receive from as pretty of a girl as they had at their own school.

As the matches progressed, St. Bonaventure was totally outmatched. Robert tucked his head down and quietly slipped out in the middle of another lopsided bout. He ventured to the gymnastics' venue to watch his middle son, Niqui. Niqui was considered the finest all-around gymnast in the State of Texas. Only a sophomore, he was being watched by several major universities. By the time his father had arrived, Niqui had already competed in the pommel horse and high bar.

The five Kings' gymnasts were dressed in practically the same style of short as their wrestling teammates, only shorter up the thigh. Once again the boys' rear ends appeared to have no cover and the suit was more like a ballet tight. It certainly didn't have the typical outer short over a leotard. They all had an interesting definition of the crotch as it appeared the penis espaliered across the top toward their thigh. The clothing was so anatomically revealing, only a well-toned, linear male could wear such a thing.

If the outfits were meant to distract the competition it served its purpose. Robert was not amused that their five cheerleaders were gawking at these gymnasts with hands across their mouths like giggling teenyboppers.

"Senator!" The voice caught Robert's stare and distracted his concentration.

"Excuse me, Phil. I didn't see you approach," the senator greeted and welcomed the appearance of their own school headmaster. "I was just admiring, huh, well, a most interesting design in gymnastic wear."

Phil Masters chuckled. "I have been in awe since the second I arrived. Can you believe this place? Our Olympic team should have it this nice. I doubt if I could get our boys in outfits like those, but when you perform like these athletes against our own, I can only sit back and admire such skill. If I didn't know better myself I'd think this was one of those prestigious East German sports schools."

"They're that good, huh?" Robert asked but had convinced himself of this fact already.

"All their boys did three release moves on the high bar with a triple dismount. I have to wonder if our Olympic team knows about these kids. Your son is doing a bang-up job, but these gymnasts are in another league. Have you toured the other floors and seen the science experiments? Sara and the girls are still examining the exhibits."

Robert scanned the dozens of spectators and wondered if his own wife and daughter had arrived. "I noticed quite an art collection, Phil. I hope they're not kidding themselves by thinking those are originals just because they have the signatures of the masters on them. Someone must have sold them a bill of goods. How does the science exhibits compare with our own?"

"They're above our level of expertise, I'm afraid. One student did an experiment with lasers: beam circularization which employs anamorphic prism to correct for the naturally divergent, elliptical nature of beams produced by laser diodes."

"Whoa! I'm just a U.S. Senator, Phil."

"Sorry 'bout that, Senator. A little love of mine; something I studied in college. I'm only red with envy that these boys are studying lasers in high school. I was impressed with their research on creation versus evolution. One of their students calculated by mathematical equation that it would be easier for the earth to accidentally composite a space shuttle than for the human body to evolve from cellular multiplication. See that blond-haired kid on the rings? His name is Bobby Newton, the same boy who did the research. I know, I asked his coach."

"God! The boy is a little dynamo in the floor exercise and pommel horse. Now you're telling me he's a genius, as well. What kind of academy is this, Phil?"

"One I'd give my right arm to teach here," Mr. Masters admitted and added, "and I'd be real happy to be back in the classroom. Hey! There's my lady. The girls must be still wandering the hallways. Here comes your youngest. Bet he's real disappointed after his loss."

"Hey, Dad," Jay said with a smile that didn't show a sullen face.

Robert didn't embrace his son but went right to the jugular for some explanation. "Didn't you know she was a girl?"

"No! Not right off. I mean, I kind of wondered why she didn't have this bulge in her shorts, but after we wrestled for a while I saw a crease between her legs and I knew it was her lips."

"Vagina, Jay. Can't you even use appropriate terminology?" his father scolded.

"Not with Mr. Masters here. It's not like we can use words like penis and vagina in front of the nuns or priests," Jay explained.

"What's our school system come to when the boys can't even discuss normal sexuality?" Phil Masters asked. "I'd blame the church but I'd hear it from all our self-righteous parents. No wonder an academy like this is so far more advanced."

Robert listened but didn't respond. "So, after you recognized she was a girl you were distracted and maybe went easy on her."

"Heavens no, Dad! I didn't want a girl to beat me, but she was really good and really, really, nice—much nicer than the girls we have at school."

"Why's everybody taking a liking to our opponents?" Senator Franco asked his youngest.

"I don't know. Guess, because they're fun to be around. Many of them are in two to three sports today. See that one boy there? He was their first wrestler. Really talented! I'm impressed that their school requires them all to be very versatile."

"What do you think of their uniform, Jay?" His father asked.

"It's not material, Dad. It's like paint and real slippery. There's nothing to grab but it sure is revealing."

"I knew you were distracted. Phil, is that legal, the Kings' uniforms?"

"I believe they are. They cover up all the parts that need covered. Hey, I better get moving. See you guys at the banquet," Phil noted upon walking away.

"Banquet?" Jay asked.

"Something this academy is putting on to shove our losses down our throats again," the senator said. Jay didn't believe that one.

After watching two of his sons meet defeat Robert knew his eldest had more talent in swimming than any Kings' swimmer. Travis was considering scholarship offers from Stanford and the University of Texas, both swimming powerhouses in the NCAA. Only a California boy had a faster time in the 100m freestyle. It was obvious the St.

114

Bonaventure swimmers were in awe of their competition by the time the senator had sat down at this 50-meter Olympic pool, with its own diving complex. Above where he was seated a few parents waved down at his presence. Robert again witnessed a one-two finish by the Kings Academy in the 800M freestyle, then kissing and hugging again. He'd seen this on television from the Russians and Romanians but America kids were usually far more discreet, especially boys this age.

There was also a noted absence of Kings' supporters compared to the number of spectators from San Antonio. There was a man in a blue blazer with a crown on his breast pocket, sitting with four children and a woman. Robert moved across two aisles and introduced himself.

"Don't mean to intrude, but I'm Senator Franco from St. Bonaventure."

"Pleasure to meet you, Senator. I'm Dr. Marion; my sister, Brittany, and her four children. Do you have a son in the swimming competition here?"

"Yes, I do. He's the tall, brown-haired boy talking to one of your Kings' swimmers there."

"A fine looking young man. You should be proud. My boy is in the 200m butterfly. My other boy participates in gymnastics--the rings, actually."

"I saw him, I think. He was so steady and defiant I was very impressed with his handstand and double somersault with a no-step landing," Robert remembered and knew he'd impressed this gentleman. His memory of the Japanese-American boy was vivid in his mind. The boy's muscles strained from young shoulders, a rear end so taut and exhibited it appeared the boy was competing nude. Even the St. Bonaventure fans gave him a standing ovation.

"Thank you for your compliments. I'm sure Mikki will be pleased to hear you thought highly of him," Kami said.

"Are you a physician here at the academy, Doctor?" Robert asked.

"Yes, I'm the boys' pediatrician."

"These aren't the same youngsters I saw enjoying the lake earlier, are they?"

"They probably were," Brittany replied. "They rarely get a

chance to enjoy the lake by themselves without dozens of the older boys around them."

"To be young and so carefree. Were those actual dolphins? I didn't know there were fresh water ones."

"They're dolphins. We brought them in for the boys' fifth birthday, then decided to keep them after the boys loved them to death," Kami said.

Robert wanted to ask which children the doctor was referring to but he had other questions. "I'm interested in the kids' outfits. Fascinating."

"Designed by one of our boys a few years ago. There are still a few kinks left in the design but, overall, the advantages are tremendous," Kami elaborated, then added, "The material is an organic silicone coating, not a fabric. It's all derived form a method of polymerization."

"Beyond me," the senator joked. "Rather revealing, but functional."

"The boys call it the STREC suit, meaning State of Texas Required External Covering."

The men laughed but Robert found this absolutely interesting. "You mean it's painted on?"

"No. It's put on like a heat transfer without getting too technical. There's a P.A.F., which means Penal Airfoil, a small pad to keep the penis from becoming drag. It also has an I.R.S. implant. Don't worry, it has nothing to do with the agency we love to hate, though the boys thought it was funny to name it as such. It stands for Internal Rectal Surveillance; a digital probe that is a miniature computer and sensor. It helps regulate the body's temperature during competition. The Russians tried such a thing years ago with a diathermic device by raising the body temperature through the rectum by a few degrees and discovered a ten percent increase in performance. Only problem, the duration was short and fatigue was optimal. We've improved upon it."

"You mean the kids have, excuse my words, a rod up their rectum?"

"Not a rod but a sensor, per se. Don't worry, it's hardly noticeable and is regulated under very safe guidelines. The results we get on a physiological scale are priceless for each athlete."

"Isn't that cheating?" Robert asked with concern.

"It's modern medicine. The athletes still have to be in top condition and develop skills that exceed their opponents. We're improving the body's ability to withstand stress, yet cope with diverse adjustments during competition. Senator, when you have the time and at your convenience, I'd love to show you our research lab and how we're improving athletic efficiency through modern science."

"I might just take you up on that, Doctor," Robert replied and couldn't wait to tell everyone that these kids were practically guinea pigs.

The senator began to think of himself as an investigator. "Don't suppose the outfits they wear over their suits have any benefit. Robert kept fishing.

"They are called ET gear, or External Thermals, and, yes, they serve to keep the body at the correct temperature to assist the warm up and preparedness."

The senator was caught off-guard when Brittany celebrated her nephew's victory in the butterfly. Other Kings' boys affectionately kissed and hugged the boy.

"They all love each other quite a bit," Robert expressed.

"They've grown up with each other and this is the way they show support and love," Brittany answered.

Senator Franco smiled for many reasons. He found the affection to be way too effeminate and gay. His desire to ask a hundred questions was only delayed because he didn't want to seem too much of a nuisance. Robert thanked the doctor and wished he could spend several hours with the man just to pick his brain. Dismissing himself back to his original seat the senator began rooting for his eldest son who was preparing to race his favorite stroke, the 100M freestyle.

"Okay, boy, show them what you got!" Robert yelled. Travis gave his dad a nervous smile and climbed upon the starting block.

While Travis had a half body suit on, which cost his father $300, the Kings' swimmers wore the same style of suit as their gymnastics and wrestling teammates. The only difference, the swimming version was extremely skimpy; the boys' hip area appeared to be painted red, with something undetectable keeping the penis and testicles flat against the skin. Robert was just waiting for the water to dissolve the paint.

Travis had lane four in the middle of two Kings' athletes, and on the outside was another St. Bonaventure swimmer. When the swimmers took their starting stance, Robert was sure the spectators could see the boys' anuses. He'd have to protest this blatant display of nudity. The race was tight after the first 50-meters, both Kings' swimmers were right with Travis. With ten meters to go Travis passed one of the Kings' boys to come within an arm's length of the leader. When they touched the wall, it was close, but Travis lost by a foot. The times were impressive on the electronic scoreboard above the pool: 49.90 for the winner; 49.97 for Travis. Their times were the top efforts in the 100-meters for all high school swimmers in the nation and well under the Olympic Trials' qualifying mark of 51.59.

Robert felt so numb he couldn't sit down, and then had to watch his son being hugged by his opponent. The teens' groins actually touched. Travis smiled and congratulated this handsome, reddish blond boy from the academy. The scene made the senator clinch his hands into a fist. No wonder his kid lost, the boy didn't have a competitive bone in his entire body, as Robert ruminated in his rage of anger. He moved to the railing and, with his voice rising, called out to his son to come over.

Though the St. Bonaventure swim coach had immense praise for his swimmer's exceptionally fast time Travis lost his smile when he saw his father's face. Robert didn't reach out to his son but withdrew to a towering position.

"What the hell do you think you're doing hugging that guy? Maybe you should have kissed him, too!"

Travis's face turned to embarrassment. "It was a great race, Dad. Can't two guys hug?"

"You'll lose your focus, boy. How's it going to be the next time you race this kid? You've complimented him like he's your best bud. Do you think the heavyweight champion of the world wants to hug his opponents? Hell, the guy wants to beat their brains in. Here's what you're going to do—walk in that Kings' locker room and tell that kid you swallowed a little water and lost your stroke. The next time you're going to kick his ass, and if you have to put a finger in his face while you're doing it, so be it! Maybe that little probe up the boy's ass will record your words."

"Huh?"

"Yeah, ask him about that, as well."

Travis lowered his head and sighed, then nodded because he knew there was no refusing his father. He moseyed back to his team area and saw his opponent with the Kings' squad. Travis knew he'd be smiling too if he'd won, but that wasn't the point. Then he thought what reaction he would have had if he'd been victorious and it wasn't much different. He would have hugged every kid in sight with a time like that. With his emotions empty, Travis, all of a sudden, felt swimming had lost its fun.

A physical presence came over the boy's despair when he glanced back up and saw his opponent come toward him. A warm glow flashed through his body and made his loin stir.

"I'm Alex, by the way. No one has given me that much competition before." Alex swept his arm around Travis and got a smile in return.

Travis quickly jerked back and was relieved to see that his father had left. The skin to skin contact was something far deeper than just two guys talking. "You were awesome," he whispered in Alex's ear. For those seconds not winning didn't seem important until his father dictated otherwise.

"Relay, Franco!" His coach yelled and broke the spell between the two boys. They hugged and sensed the spongy growth between each of their legs that had pressed into one another. Before the protrusion became too obvious Travis dove into the pool, glad that he was swimming anchor.

****

Senator Franco desperately needed a distraction to cheer his disposition. Arriving on the third level Robert stopped in his tracks with a life size statue of King David. He laughed when he saw the signature of the sculptor—Michelangelo. Someone had a great sense of humor around here, he decided and began to tour the glass enclosed exhibits of the school's recent science fair. The projects varied from designed vaccines to vision tests. The health issues had more entries than the others. One boy had experimented with an umbilical cord and a substance called Hyaluronan. The boy showed how it prevented scar tissue after surgery, plumped wrinkles, and reduced inflammation in arthritic knees. Robert doubted if this was all true or his own doctor would have recommended this for his bad knees.

There were at least two experiments on double-stranded RNA, but this appeared to be boring for the senator. The next banner read: **SEX AND THE ADOLESCENT**. Robert expected an elementary assignment on birth control or stages of puberty. He wasn't ready for an explanation of masturbation, a clinical analysis of sperm, and the philosophical influences and social persuasions for the practice of such. No wonder those girls were giggling when they were coming from this location. A statue of David was revealing enough but real-life pictures of a rather long penis being moved to orgasm was quite graphic to cause a school board to make astronomical changes before hysterical parents fired the whole lot of them.

The senator swung his head to his left and right just to make sure someone didn't notice his focus. He began to read the student's research, which was rather convoluted, from simple to complex. The exhibit was blatant in its message: *THE PENIS IS AN ANATOMICAL MARVEL.*

(It can change size and shape, become rigid and flaccid, and expel semen and urine through the same collapsible tube. What other organ goes through so many changes or has so many various functions?)

Robert nodded in agreement and was quite proud of himself and his little toy for a few seconds. He read on: (Yes, an orgasm lasted about four seconds and it didn't stem from the testicles, but stored a combination of seminal fluid, sperm and other ingredients. The testicles alternated in filling this supply, like a double-barreled shotgun.)

This analogy brought a grin to the senator's face. Though he wouldn't admit it to anyone this was information he wished he'd known as a teenager. Robert bent from his waist and read the ingredients: Sperm contains a trace of creatine, protein, and carbohydrates. Robert remembered his own sons talking about the effects of creatine. His eyes bugged out in reading that a clitoris has six times more nerves than the tip of the penis. No wonder his wife went spastic when he orally copulated her. Once more he scanned the hallway. This was damn good research, Robert realized. Five gallons of sperm in a lifetime, the chart noted. Given the thousands of times Robert thought of masturbating, he agreed and chuckled. He re-examined the digital photos from the waist to the thighs and realized the model didn't have a stitch of hair. Possibly he had shaved it for the experiment. Those St. Bonaventure girls now knew a whole lot more about boys than when they arrived to watch a sports event.

Senator Franco continued to read about sea horses that become pregnant—male sea horses, that is. How a paper nautilus, whose idea of sex is to heave his manhood like a tiny torpedo into the female, had the senator's eyed glued. Once implanted, the male organ serves as a built-in sperm bank. There were several accounts of bonobos enjoying male to male sexual contact, penguins having monogamous homosexual relationships, and dolphins making it with sea turtles. He chuckled when he read that a chimpanzee usually has sex for about six seconds, much like an adolescent boy, Robert thought. The senator was surprised to see that this student only received a third place. At his age he didn't know half the material this kid knew. He checked the name of the researcher—Mikki Marion, the doctor's son.

Robert checked his watch and realized he'd lost all track of time. He began to look for a men's room, then passed a group of St. Bonaventure kids who had finished competing. A few cheerleaders were with these athletes.

"Hi, Senator Franco," one of the cheerleaders greeted.

"Afternoon, kids. I'm not sure this is the best display for your eyes. The swim team could use a little support and your appearance would be much appreciated."

"We've seen the display earlier," a boy remarked. "It's nothing we don't already know."

"You should see the nude statues by their track, Mr. Franco. Ah, I mean, Senator." The young lady giggled at her mistake.

"You didn't learn to be obsessed with sex at St. Bonaventure, young lady," Robert said very seriously. "Now run along."

There were a few grumbles as they strode off, then one of the wrestlers spoke loud enough for the senator's ears. "As if we don't do it." A few chuckles supported the offhand remark, but the senator let it pass.

The senator was intrigued with this vision of nude statues, so he moseyed around the curve of the hallway to the south point of the pyramid. On the grounds below lay a green all-weather track facility with gleaming white plinths scattered around the 400-meter track. Robert's first thought was his trip to Italy, where he visited the Foro Italico and Stadio dei Marmi. These statues before him were just as beautiful and revealing in all their splendor. It just appeared odd to have such art at a school for high school kids, and blatant art at that.

Again with the urge to urinate, he retraced his steps in haste of finding a restroom.

"Daddy!" Robert recognized his daughter's voice which jolted the senator from his concentration. His wife and daughter were quickly approaching from the other direction.

"Well, there you two are. I've been looking for you at the venues."

"We saw Travis's last race. Their relay team lost by three body lengths," Maria told her father, disappointedly.

Mrs. Franco comforted the man of the house. "Guess our sons didn't have the best day, but those other boys were really good."

"Too good, if you ask me," Robert snapped.

"Pretty neat science experiments, huh, Daddy?" Maria asked.

Robert didn't want to appear too impressed. "A few have merit. I wouldn't recommend a couple to my children."

"We've seen it, honey," his wife informed and pointed at the masturbation case. "You don't give your children enough credit if you think they're ignorant of their sexuality."

"Yeah, Daddy. You don't think I know that my brothers do that?"

The senator's eyebrows rose. "And how do you know this, Maria?"

"They walk right in to my room without knocking, so what's fair for the gander is fair for all us hens."

"Hmmm, I see I'll have to have a talk with my sons."

"Lighten up, Dad," Maria teased. "It's all in the family."

"Yeah, Dad, lighten up, you old fuddy-duddy," Mrs. Franco teased, which got a laugh from her daughter.

"You condone this behavior, my dear?"

"Better they learn accidentally and innocently, may I add, from their siblings, than have a complete stranger or horny teenager from school teach them," Mrs. Franco lectured and grabbed her husband's arm.

Maria changed the subject to reduce any friction. "I'd love to go to school here. There are so many cute boys."

"I hate to disappoint you, Maria, but they're not all girl crazy," Robert said.

"Yeah, we know. We read their poetry and short stories down the hall there. Very homoerotic, but as long as there's one cute boy available, I'd be happy."

Robert glared at his wife. "Since when does my daughter know about homoerotic?"

"You must think all girls are stupid, Daddy," Maria said with her best pout.

"D'oh!" Robert joked to make his daughter smile. "Only my baby girl do I want to keep innocent."

"I'm not as innocent as you think, Daddy." Maria strolled off with an accomplished gleam in her eye.

Robert put his arm around his wife. "See what our sons have done with their exhibitionisms? She's too young for boys. And then there are these athletes from this bizarre academy. Did you see their swimsuits? They left nothing for the imagination! Why you could practically see up their rectum."

Mrs. Franco tweaked her husband's cheek. "Please! We all have the same waste disposal equipment next to our recreation area. It's

123

God's way of humoring us. Your daughter is fourteen, Robert. All she thinks about is boys. It's amazing how quickly a male forgets his teenage years. Would you consider for one moment that maybe she was the voyeur? Those Dutch doors you installed for our children's rooms have excellent viewing cracks. I dare say they each have a novel of secrets of each other's habits. I've peeked myself," she hinted and laughed.

Robert wasn't giving up his own secrets, so he faked a shocked look and kissed his wife before departing for that restroom he had yet to find. Inside the sparkling laboratory with marble sinks and all the amenities Robert took out a cigarette and lit it up. Almost instantly an academy security man swung through the doorway.

"Sir, there's no smoking on the academy grounds."

"It's just one. I'll be done in a few minutes," the senator tried to persuade.

"No exceptions, sir. I must insist you put it out now."

"Do you know who I am?"

The security man was not to be intimidated. "I'm only aware you don't look like my supervisor. The policy is for everyone. Possibly I can escort you to the front gate, if you wish."

The senator threw down the cigarette and crushed it with his foot on the shiny Italian tile. It left a black streak of ash. As the senator stomped out of the restroom the security man grabbed a damp towel and cleaned the tile to its original state.

****

Two floors below Alex and Travis had their eyes and hearts attached to each other as the swimming competition came to a close. Travis wouldn't hurt this adorable teenager for the world.

"My father thinks I'm too nice of a loser. I'm supposed to be all sorts of revengeful so I can beat your butt next time."

"You're not a loser to me. Any time we can swim, we're winners. We're doing something we both love and had our best times today. I think we both won."

"Yeah, that's how I see it, too."

"You can shower in our locker room, if you like."

"I'd like that," Travis said with a broad smile.

Chapter Twelve

On this same afternoon, Leif Bjornsson was on his way to Tampa, Florida.   He'd spent two weeks in New York City, a visit that had both matured him and hardened his adolescent sense of the world.   He was now fifteen, though he had told no one on the day of his birthday.

Leif had arrived at the LaGuardia Airport, only to find himself facing customs.   He saw a family that he knew was from Iceland, as well.   Blending in behind this family of five he watched the father place all the passports down on the counter and wait to be seen.   Leif had walked up with one of his Nintendo magazines and asked the father if one of his children had left this on the plane.   As the man proceeded to ask his kids, Leif slipped his own passport underneath the others.

The Customs' clerk had his usual questions for the head of the household, then stamped all the passports without glancing anymore at the names inside.   Outside the gate, Leif quickly ran up to the man.

"Sir, I think that agent accidentally put my passport in with yours." Sure enough the man was gracious enough to return the mistaken paperwork.   Leif was now officially in America.

Taking a shuttle to downtown, Leif bought a New York magazine and scanned the back pages for an idea how to get to a gay district.   One advertisement directed him to a place called Greenwich Village.   A short walk to an underground subway and he was at a ticket window asking for directions to this village.

A tap on the shoulder distracted Leif's purchase.   "Excuse me.   I heard you were interested in going to Greenwich Village.   That's exactly where I'm headed," the older man spoke with a most pleasing smile, an overwhelming aroma of cologne, and a well-manicured goatee. Leif was pleased for such assistance and this grand stroke of luck.   He was already nodding when another man had made a hasty walk to intervene.

"Young man, if I could talk to you a second, I might save you a lot of grief," the second gentleman said, far more astute looking with a black suit and a white collar.

The first man with the goatee became abrupt and nasty. "You again! Why don't you mind your own business for a change?"

"I'm quite aware of your game, my friend. You've found another chicken here and your intentions are not in his best interest. Youngster, unless you're looking for that style of life, I need to speak with you."

Leif was stuck between two annoyed faces; yet they both smiled when they stared at him. "Okay, I'll hear you out, but I'm in a hurry," Leif lied.

By the time he'd listened to this well-dressed priest, who spent an inordinate amount of time in search of runaways, the other type of hunter had disappeared. No, he wasn't a runaway. No, he wasn't into prostitution. Leif answered all the intrusive questions with various degrees of deception.

"All right, but do you need a place to stay? My organization helps keep kids off the street and gives them three squares and a cot for as long as you need it."

Leif thought about this. This was a great way to get his bearings and settle in. Leif agreed.

Interesting enough this type of boarding place was in Greenwich Village. It had dozens of kids and he was put in the same room with a boy his own age from Dayton, Ohio. Though Leif had no idea where this city was, his new roommate introduced himself as Rodney Miller and pointed west to denote where Ohio was. He and Rodney didn't talk beyond a hello for 24 hours, then the other boy offered Leif a joint. It would change this relationship.

"We can't do drugs here," Leif said. "It's in the rules. They'll kick us out."

"No they won't. They're scared we'll go back out on the streets and hustle our asses. You've probably never had one anyway."

"I've had ass before," Leif humored and had both boys cracking up.

They smoked the marijuana and, for Leif, it was his first time. Dizzy and disoriented he talked for hours about Iceland and why he'd run away from home, only it was more about how bored he was with life and school than the truth. Rodney admitted he had run away because his father was a drunk and his mother hadn't given a damn about him.

127

His only memento from home was a rubber, toy football, with Fairmont Dragons written on the side—his high school. Admitting he didn't care for football or the kids at school, he'd learned about narcotics and drugs from his peers. He had arrived in New York City, only to have his shoes and jacket stolen the first day. A man had approached him and offered him a job and a place to stay. Within the hour he was playing bottom for some father of three kids who liked boys. So went the story of this teenager from Ohio.

Leif was amazed at this story and realized this could've been him. "You mean you did it with guys? Are you gay?" The excitement sounded in Leif's voice.

"I'm no faggot!" Rodney admitted. "I did it for the money, though I had to turn it all over to my pimp who barely gave me enough spending money for the arcade. I was like his property, man."

"Oh," Leif said solemnly and fell asleep while staring at the yellow ceiling.

Over the next few days Rodney and Leif became fast friends. He made sure Rodney called him Erik, to go with his story he gave the director of this boarding house. Rodney told him about the routine inspections of their quarters and made sure Leif hid his passport. The stories Rodney told were true, though he had a way of adding humor to them and a sense that he'd gotten over the trauma. He'd gone to this one apartment expecting a man, only to see four guys. Regardless of his refusal they'd spiked his coke with Ecstasy and, by morning when he had woken up, his butt was sore with four rubbers thrown on the floor. When Rodney laughed about his adventure, so did Leif.

"Check this one out," Rodney continued. "This dude had me stand naked on this pedestal that revolved in circles. He lives on the rich Manhattan side. So here I am whirling around, posing like Donatello's David, and he's over there on the couch telling me how beautiful I am and jacking off. I was glad he came 'cause I was getting dizzy."

The boys laughed and rolled on the bed together, then the kiss happened. It was mutual in that nanosecond when their eyes closed and fate brought their lips together. While Leif was in seventh heaven, Rodney began to cry.

"Whoa, man, I didn't mean to violate your boundaries," Leif said, remembering what Sveinn had told him.

"You haven't, man. You're just so gentle and kind. I'm not used to having someone love me. I've been through more than you'll ever want to hear."

"Don't think that, Rodney. I have the same thoughts." Leif revealed how he had seduced his own teacher and all but stalked the man to love him. Rodney smiled and confessed his own secrets.

"I used to have this paper route and collected from my customers once a month. There was one time when I stopped at this family's home down the block where I lived. The guy had four kids and invited me in while he rummaged through his wallet. "Can I put this in your shorts' pocket?" he asked. I told him the money was more than the bill was, but he didn't care. When he slid the cash in my pocket his hand stayed there for a few seconds and I became hard in less than that. Down went my shorts and I came in his mouth. It was my first time, and I didn't know what to think. The next day I knew where someone cared for me and it became a regular stop. Soon I was wearing the tightest shorts I could find, often without a shirt to let my shorts droop down to just above my pubic hair. I had one man who asked me if I wanted to cool off in his pool; another who offered me a Pepsi, only to drop my shorts and relish me with his tongue.

"Then my best friend's dad got the hots for me. I'd sometimes sleep over at Brent's house and have to endure sleeping next to him with only our underwear on. He didn't like to fool around, but his dad did. I knew where the floor creaked in front of his room before he'd meet me in the bathroom. So, while he's humping my rear and jerking me off, his eleven-year old daughter walks in, just as we're close to orgasm. Best thing, she didn't yell and he didn't stop. Shit! I had six married men in the community on my list for sex."

From that moment on the two runaways were inseparable. Rodney wasn't so sure he wasn't infected with some disease, so he wouldn't allow Leif to experiment with anal sex. They did everything else. Leif began to involve Rodney in all aspects of his life, including his daily trip to the library. Leif was researching historical figures, both in Europe

129

and America.   He had taken the role that, if he was to recreate men, as he called it, from the dead, who would he select?   Rodney had no interest in this research of dead people and stayed in the art section to look at naked pictures of men and women in paintings and sculptures. He looked forward every day to this venture with Leif and occasionally would slip over to see what Leif was doing with an arm around the boy's shoulders as if they were best friends now.

One of the characters Leif had looked up was Charles XII from Sweden.   This information, found in a biography of Peter the Great, was put aside for the time being, only to be picked up by Rodney.

"This kid sounds so cool and even looks like you, Leif.   See, you could be twins."

"Quit fooling, Rodney.   Can't you see I'm busy?"   Leif rebuked.

Rodney kept reading the biography of this king.   "It says here that he used to wreak havoc in his castle with his friends, and he was crowned when he was only fifteen.   That would be so cool to be ruler over all those Swedes at fifteen.   I'd pick the cutest boys, like you Leif, and have three in a bed every night.   Guess when you're a king that young you can do anything you want.   Why's this guy so important to you, Leif?"

"He's not all that important to me. They recently found that his tomb had been tampered with and his remains were taken from Riddarholm Church in Stockholm.   The guy was probably gay, that's all I know."

"Guess you can't have sex with him; he died of a bullet hole through the head while peering over a trench. That'll teach him to wanna see what's going on."

"Reminds me of you," Leif said.   "You seem to think you're immortal and can do anything you want without consequence."

Leif, for the first time, noticed the needle marks on the inside of Rodney's forearm.   "Rodney, please promise me you'll give that up."

"I don't know, man.   It's my only escape from this lousy life."

Leif glanced up and saw a boy like himself who was reaching out. "You're really cool and I really like you, but I want you to give up the drugs."   Leif bent forward and kissed the boy.

Rodney was stunned and gave a wide grin to someone he thought was the kindest person he'd ever met. Leif was the first person he'd truly fallen in love with, but he wasn't going to tell Leif that.

When the two boys returned to their quarters, a note was left for Rodney to see a staff member. Rodney returned fighting back tears and angry with the world.

"I have AIDS, Leif. Shit! I thought I was careful with the sex thing, but those needles weren't always clean. I had sex one time without a rubber; this guy wanted me to screw his young daughter as his means of getting off. One sick bastard; the girl was a runaway like me."

"Did you screw her?" Leif asked.

"Yeah. I'd never done a girl and she laid there like a dead fish. After we finished and this old fart shot his sperm on top of us, she left with me and said I was the nicest guy she'd met so far. It's a sick business, Leif. And here I was going to suggest we use rubbers tonight and get it on."

"Fine with me. You deserve a sex life as long as you protect everyone around you."

"Fuck, dude. You act so mature for your age. What are you doing in a dump like this?"

"I'm fucked up just like you are. Sveinn was right, we're not ready for games like this. I don't know what the hell I'm doing."

Rodney gained a new breath on life with this kid from Iceland. He decided to run down and get a few free condoms from the office before Leif changed his mind. A police officer was visiting the boarding house and passing a new wanted sheet for runaways, an up-to-date report on parents looking for their kids. Rodney let his eyes drift and saw a picture of Leif. He sprinted back up the steps.

"Hey, man, you've got to get out of here now! They're on to you. I just saw the latest most wanted list for gay boys."

"Don't joke with me, Rodney."

"Dude, I'm not joking. Look, I know somebody down the hall who counterfeits drivers' licenses and other documents. Let's go visit him and get your passport to say you're eighteen. I'll give you a name at Fire Island who will fix you up with a job. I stole drugs from the guy, so I'm burned."

"And he will hire me because?" Leif asked.

"He loves good looking boys, that's why."

"Come with me, Rodney. I'm sure the guy will forgive you," Leif spoke with certainty.

"You don't know drug dealers, dude. They forgive through vengeance. I'd be tasting cock and balls—my own." Rodney put on his best act of happiness for Leif, but even this couldn't hide the affection he felt for this boy in his life. "If you can, Leif, stop by some time and we'll go to the library together."

Leif reached out and touched his friend's face. To be so young and vulnerable to the whims of adults and a cruel world. There was a morbid thought that Leif would never see this boy again, so he tried to convince himself that his mind couldn't possibly know the future.

"I'll make us some money and we'll find our own apartment," Leif said with optimism and saw a glimpse of hope in Rodney's eyes.

Leif had been successful in escaping his confines before the director had notified the police that he matched the picture and profile on this list. Finding Fire Island wasn't all that difficult, but getting around a half-dozen men who wanted to do him favors took a considerable effort. He'd learned a lot about what to expect from Rodney's experiences and this served him well. Within the hour he was in a string bikini, dancing on top of a platform. Men swarmed around his area and, though he wasn't all that professional with this jiggling and swaying to the music, men put all sorts of money in his strap. Leif put his mind like he was dancing on top of his bed in front of his web cam and began thrusting his groin to the stares of the customers. Notes were passed and given to waiters to give to Leif. Offers of hundreds of dollars to pieces of jewelry accompanied these requests. Leif waited for the right offer.

Dancing for gay men was a real turn on for a boy who sought attention. His endowment had mouths agape and tongues swirling around lips. On occasions he felt an erection growing and almost caused a riot, until the manager had to call his bouncers to throw a few lecherous men out the front door.

"Check this out, kid, you've got to control that wiener of yours. You're going to cause a riot, got it?"

Leif promised to do better. Because the weather was so nice he slept the first two nights on the beach with another dancer. The boy had AIDS and a bad case of acne, but he also taught Leif new moves and how to dance to the music to entice men's eyes. All over the beach there were couples doing it under moonlight and groups of men who had boisterous orgies.

By the next night Leif noticed a well-dressed, young executive, who kept his eyes glued on Leif's eyes. Leif was soon dancing for just this man's eyes, a thrust of the pelvis, followed by a spin and a pair of butt cheeks inches from the man's face. Leif swept the rear of his bikini brief down to show the crowd his hairless crack. When Leif felt an erection growing, he would grind against the chrome pole on stage, which helped hide the protrusion.

On his break Leif accepted the offer to have a drink with the man.

"How much? The guy came right out with an offer.

"You a cop?" Leif asked.

"Married with three kids," the man replied. "I'm Michael Rush, and I think you're the most beautiful boy I've ever seen. Look, I have a hotel room in the city for the weekend. You name the price. I'm not a jerk or a druggie, nor do I want to share you."

"Five-thousand for the weekend. I'm a virgin."

Leif knew it was ridiculous and expected a counter offer. The young executive paused, then agreed. "I bet you say that to all the guys who hit on you. Okay, but you're all mine and you act like my son."

"I can do son," Leif said with a gorgeous grin. "The truth is I've never been fucked. Close, but my mother came in." Leif laughed with his memory.

Leif soon learned the guy was a stock broker who had created an image for himself by marrying into money. Though the man showed Leif a picture of his family, the man knew he had a passion for guys, but being married had given him prestige and a position in the stock market trade.

After the first night with a few comical efforts of rolling on a

133

condom, Leif was convinced the guy was legit and about as much of a virgin as himself. By the next day they were fucking without taking the extra few moments of sliding on an umbrella, as Leif called it. Going bareback was a lot more fun, especially after Leif did a seductive striptease and had Michael chase him around the suite until they were both in hysterics and hot for each other's arms and bodies.

The weekend was far more than sex; it was dinners out at fine restaurants, a ballet, a Billy Elliot play, and a bar where Leif was paraded in front of drooling mouths and wanton eyes. Michael had taken the boy to a men's clothing store and had him fitted with an Armani suit. When they stepped out on the town they were the handsomest duo, and also the sexiest. The two males were often the talk of the bellhops and waiters at the hotel. These servants even drew straws as to who got to take up the morning breakfast to room 222, knowing that they'd be asked to just come on in and place the trays on the bed, where the males were propped up on pillows without a stitch of clothes on. For a man playing daddy it was the ultimate role to have a platinum blond, blue-eyed boy by his side and totally devoted.

They would arrive back at the hotel at midnight and be greeted with a dozen smiles from everyone from the desk clerk to the elevator man. There were more gays working at this hotel than beach goers on Fire Island. They all wanted to assist or cater to these guests' every wish. New towels or bathrobes were delivered at all hours in hopes to catch a single glimpse of this blond boy in various forms of dress or undress.

On this Saturday night, Michael had stopped at a music store and bought a big band album. No more than they were in their hotel room when Michael requested a CD player and three employees delivered this item to their door: one to carry the player; one to plug it in; and the other to supervise. When the music started up, Michael had delayed in tipping the three aspiring bellhops; instead, he began to dance with Leif; a tango, at first, then a fast waltz, as their audience smiled with the boy's innocence and submissiveness to his partner's control and brilliant dancing.

Leif felt helpless with these new dance steps and his lack of technique. It didn't matter; all Leif had to do was stay in his man's arms and be swept to the rhythm of the music. He was laughing and had the hiccups as he was moved around the carpet to the beat of this dance band. His body was at the mercy of his partner, lifted up, then lowered to where he was sure his head would hit the floor, only to find Michael's lips pressed against his. They quick-stepped from one end of the suite to the other, the rumba beat kept them giddy with their eyes locked on one another with an implied lust that a bed awaited this conclusion. A final sweep and Michael arched the boy backwards and kissed him square on the mouth. The applause was worthy of the dancing skill of this married man, a man who was more in love with this boy he'd met only 36 hours before than his wife of seven years.

A half-hour of dancing had Leif dreamy-eyed and in awe of a man who made him feel so alive and desirable. Their exertion had their hands stripping each other of articles of clothing, until they were down to their underwear, then nothing. Another knock at 2:30 in the morning and all three of their admiring, young employees twisted their heads around the door.

"Is everything satisfactory, sir? Can we get you wine or a late snack?"

"Thank you, but no. Things couldn't be better," Michael said and moved his groin into Leif's, followed by another kiss to tease these voyeurs.

Seeing this nudity in the moon's glow, radiating through the sheer drapes, had these lucky lottery winners from the kitchen area giddy with excitement when they returned to tell all their friends what they had seen. No one believed that this hairless boy had a nine-inch dick, only because it was just common knowledge that everyone exaggerated their masculinity.

When Michael and Leif woke up at noon, they bathed together in a tub filled with bubbles. A reach to the phone by the bathtub and Michael ordered their breakfast. The hotel manager and his young assistant had rigged the lottery that morning and brought up the trays themselves. Placing the breakfast trays across the tub required ingenuity, but they aimed to please.

Michael wanted to keep this boy forever and even paid a bellhop to rent the movie, PRETTY WOMAN, so the two of them could lie in bed and laugh at the similarities. For Leif, this is how he had envisioned his life with Sveinn, dancing and dining between rounds of heated sex. Instead he found himself in the arms of another man, his head resting on a masculine chest, with a strong hand smoothing his hair. This time it wasn't so much the feeling of love as it was satisfaction and feeling totally desired.

He had used Sveinn's education by rimming Michael and making him purr. It was the constant discovery of how another male could love another that fascinated Leif so much. He knew for certain that another mere boy like Gisli would have never sufficed once he had experienced a man's love.

By Sunday evening Leif wasn't so sure he wished to leave this man. Michael had been really good to him and so very kind, even cradling this lost boy from Iceland in his arms as they went to sleep. When handed the $5,000 in cash Leif felt guilty for accepting it and offered all of it back.

"I don't make a very good hustler, Michael. I've never had more fun with anyone."

"You know what, Leif? I believe you now that you were a virgin. Sex with another male has made me feel whole. I've really screwed up my life by getting married, but you might not understand why men do the things they do. I'd take you home if I could because I don't believe you're eighteen and I don't want you out on the streets or dancing in front of licentious men. One of these times a guy is going to hurt you and the thought of that breaks my heart. I'm madly in love with you, but I think you know that. I'm not sure how my wife would accept it if I told her to go sleep in the guest room. By the way, I noticed you bought one of those tabloids last night, and I really enjoyed you telling me about all these grave robberies. I think your idea is sound and any tabloid would be interested in your perceptions and information on this subject. I have a friend in Florida who will make sure you see their editor. I'm not sure what they can do for you, but it gets you out of New York. Use this five-thousand to stay in a nice hotel room, but don't pick up any men. When you run low, I'm going to give you my phone number, so don't be afraid to ask for help and know I'll come wherever you are."

Leif was appreciative and gave Michael a red face when he kissed him on the lips for several seconds in front of two-hundred other passengers waiting to board this flight to Tampa, Florida. Leif wasn't sure if this was the smartest move to fly to Florida, but, thanks to a stock broker, he now had a chance to put all his efforts into action. He would never forget two close friends who had meant so much to a runaway boy discovering life.

Chapter Thirteen

Alex and Travis hustled to the volleyball arena and watched as the Kings' players ran off three straight games to shut out a fine St. Bonaventure squad. Jon Lincoln, at 6'6", had 22 kills and a dozen blocks to lead the Kings Academy. Travis asked about the strange looking shoes the players had on, similar to the ones the gymnasts and wrestlers used.

A composite shoe is molded on, much like our suits, and is computerized to make it feel like you're standing on grass. It's all programmed to the ankle movements and bone structure and won't allow any turning of the ankles. The foot is as natural of a limb as a hand, if you leave it alone. That's why they look like they're barefooted, though they're not. I'll show you later, but let's go take a shower."

The Kings' locker room was even more luxurious than the visitors. There were no locks because theft was unheard of. Travis watched Alex fit on another type of pad over his swimsuit. A whiteness appeared and then the swim suit was literally taken off in frozen chips. A small aluminum-type wrapper was removed to release the penis and with it came a tiny, needle thin probe from the rectum.

"My dad was right. You had something up your ass," Travis spoke while eyeing this undressing.

"Our swimsuit is like a second skin, Travis. A Kyros pad, which means ice, is used to change the chemical balance of the suit; thus, it's why I'm able to take it off and throw it away. This little probe is part of a computer system that analyzes my homeostasis during competition. I can take the results and program in my next workout or what I need to improve my performance. See, it programs all my bodily functions and temperature. Want to try one?"

Travis agreed and slipped off his own bodysuit. Alex squeezed a dab of lubricant on his finger and reached between Travis's legs to spread this film around his anus.   Travis's penis flooded with blood.

"I like what I see, but you have to be flaccid for this to work," Alex noted with laughter.

"You did it, not me," Travis humored.

"Think of something gross or non-sexual," Alex suggested.

"My sister," Travis thought.   It took a few minutes but the erection finally subsided.   Alex found a razor that the only female at the academy used and removed all of Travis's pubic hair.   This was made more humorous by a penis that sprang to attention again immediately. Moving the penis to the left, right, down, and sometimes pressing it forward finally allowed the removal of all hair.   When Alex slid the pre-come over the rod, Travis' knees nearly buckled to the sensation.

Another five-minute wait for the erection to subside and a paper-thin short was slipped on to just above the base of Travis' penis.   Alex made sure the probe slid into the rectum.   All new sensations to Travis, he gave warning that his penis had a mind of its own before it sprang upward again.   The teens laughed over this seemingly impossible task, and Alex knew only one thing would remedy this libido.   Thirty seconds of masturbation and sperm flew in looping shots.   It worked; the throbbing member was soon flaccid.   The short was fitting back on, then Alex crushed a capsule inside the short, which caused a heated sensation that melted the silicone to the skin, leaving a swimsuit exactly like Alex's.

"I don't even feel that thing up my ass," Travis admitted.   "This feels like being nude, but what if I get a hard-on?"

"The suit is very flexible and will expand, but it won't contract back to its normal shape.   Best thing, don't get aroused for at least another twenty minutes."

"You mean like when I saw your ass in your suit.   Your crack is so cute.   Uh-oh!   I feel my blood wanting to take over."

"I love horny boys, but wait until we get in the water," Alex said, grabbed Travis' hand and whisked the teen off toward the pool area.

The pool was vacant when they arrived, so they swam a fast 200m freestyle, pushing each other through the last 50 meters with a dead tie at the pad.

"I feel it," Travis said. "My body felt so smooth and warm, I feel the water is just perfect."

"Exactly! Now let's go look at the results."

The two athletes returned to the locker room where Alex showed Travis the computer results of the swim. On a LCD screen they watched the replay of their swim and how their bodies reacted to the energy demanded.

"You see here, Travis, your teres major is imbalanced on the right side. You're pulling too hard because of the muscular deficiency on the left. You're low in glutamine and arginine. Your ATP/CP is out of whack with your krebs cycle. I'd recommend a whole new nutritional supplement before your next meet."

"This is too amazing, Alex. We don't have anything like this at St. Bonaventure."

"Why not?" Alex asked without any understanding why schools weren't like theirs.

They used Kyros pads to remove their suits and headed for the showers. Travis kept staring at this sleek, hairless body walking next to him, and there was no way he could subdue an erection. His member swung straight out, so Travis made sure to keep his back to his new friend, so as not to make Alex think he had no control over his attraction. The water running over his shaven skin was an invigorating sensation. His boner might not ever go down. A slap sounded from the rear of Alex, as Travis turned to see a naked female in the shower with them.

"Nice going, guys! Loved your swim," Wendy complimented both of them.

"Thanks, Wendy. I want you to meet, Travis."

Travis wasn't sure what to do. He'd never been this close to a naked female who knew he was watching her. Travis' initial instinct was to turn and put out his hand, but to do this meant removing the hand over his groin. She did the favor by kissing him on the cheek and telling him he had a cute rear and not to get a hang up over the nudity or his erection.

Wendy departed to her locker and left a stunned boy who had yet to remove his hand from the obvious arousal.

"They actually allow girls in the boys' locker room?"

"Wendy grew up with all of us, so nudity isn't a big thing here. If we weren't all gay it might present a problem."

Travis felt silly with his hand covering, so he began to relax and waved when Wendy glanced again. "You're teasing me, right, Alex? I mean, you guys aren't all gay, are you?"

"Not only that, but I picked out the cutest boy at your school—you!"

"This is totally weird, a school for gay teens. And here I thought I had fallen for a straight boy I'd never be able to tell that he has the hottest body I've ever seen."

"I think your hard-on didn't need words," Alex said. Alex took Travis to the Jacuzzi where five other boys were enjoying the comfort of the swirling water. Travis felt immediately accepted. At least in the Jacuzzi his penis finally relaxed. After this pleasure they went to a chamber and stood on a pad with their backs against another long vertical tube.

"Now don't move, Travis, and if you feel a prick, don't jerk. Are you allergic to anything?"

"I don't think so, but I trust you on this thing. What's going to happen?"

Instantly Travis felt his body freeze. It wasn't his choice if he was going to move or not anyway. The prick in his butt didn't really hurt. A few seconds later it was all over.

"Okay, what happened?" Travis asked.

"You've been in a cryogenic chamber. It promotes recovery by freezing. The shot you received was all the nutrients the computer figured your body was lacking."

"I feel like I'm on an alien vessel and my body is being used for research," Travis teased.

The boys wore each other's sweatsuits as a sign of friendship. Alex reached in his locker and pulled out an iPOD with a 5" vibrator attached. He went to a computer panel on the wall and hit a few buttons.

"Hi, Coach Rigby. Is Zach competing?" Alex waited a few seconds, then, "Hi, Zach. Can I use your iPOD for a friend?.........Thanks."

Alex reached in and found his tube of KY Jelly, then lubed the vibrator. "If you're game this is the way to enjoy a match. Let's go watch the tennis players."

"What do I do?" Travis discovered quickly the feelings of having a 5" tube up his rectum. To the sounds of the rhythm of the song on the iPOD the vibrator pulsated to the beat inside. Travis couldn't stop laughing at first.

Down the hall and outside to the court Alex was being entertained at the sight of Travis cracking up. "This is too much!" Travis spoke, but was enjoying it immensely.

At the courts, 2 grass, 2 clay, and 2 hard court, Kings Academy had allowed the St. Bonaventure players to pick whatever surface they wished to play on. Jacob Tilden, the academy's best player, sprinted over to say hi to his brother, Alex, and meet his new friend. Travis was cordial but giddy, trying not to smile too much with a vibrator in his body. It wasn't easy having something up your butt and talking at the same time.

"You guys drunk?" Jacob asked with suspicion, but he knew Alex would never consider alcohol.

"It's the iPOD. A new kid on the block," Alex said and the three broke up in laughter.

"I'm super horny," Travis admitted.

"You're feeling the effects of how your prostate appreciates music," Alex joked.

Watching Jacob play tennis was like watching a junior professional. He had no trouble with his match in singles and doubles.

After Alex suggested a tour of the academy grounds they were soon by the lake when Alex casually put his hand inside Travis'. A momentary shock took Travis by the very core of his emotions of being this close to another boy.

"Your parents are okay with this?" Travis asked without releasing.

Alex again chuckled at this absurd question. "Of course. When

142

we were four our fathers told us how they met.   It was so cool when they held hands between their beds and feel asleep.   All of us started holding hands after that and have never stopped.   We've progressed quite a bit in our love-making."

"Your fathers?   I'm confused," Travis admitted and realized they were somewhat secluded behind a few trees.   Travis pinned his new friend against a tree.

"May I kiss you?"

"Don't see why not."   The kiss was soft, at first, then passionate. Travis admitted he was new at all this but wanted to prove a point to himself.   The erection in his outfit proved the secret Travis knew all ready.

Alex obtained his friend's iPod and punched the bass button.   The pulsation went right to Travis's prostate and had him ejaculate in his shorts seconds later.   The kiss took on more passion with the obvious orgasm.

"I think I qualify for the Kings Academy," Travis said with a grin.

****

Though Travis's heart and mind were above the clouds and dizzy with love, his better judgment had him dress in his own sweat clothes for the banquet.   While tossing their shorts in the locker room's washer, Alex teased his new friend that two orgasms were probably better than winning.   Travis had to agree with that one.

The elder Franco brother wouldn't relinquish his desire to sit at the table with Alex and their eyes absorbed each other's words, as if life had begun when the two met.

The banquet was more an informal gathering to relax and unwind for the athletes, coaches, and parents.   For the Kings' students it was an opportunity to socialize, learn skills that were desired in business and being adults.   If the spectators and athletes were disappointed with the concession stand, which did not sell soft drinks or fast foods or sweets that didn't have a positive nutritional value, they were impressed with the tables of avocados, caviar, chocolate, lobster, milkshakes, shrimp, and, for the adults, wine.

"You guys have lobster before meets?" Travis asked.

"Lobster tops off your calcium, zinc, and vitamin B levels. Doesn't your coach tell you these things?" Alex inquired in disbelief that every athlete would be treated like them.

"Our coaches push creatine and weights," Travis said with a sigh. "They also recommend no sex and eat spaghetti."

"Old myths," Alex informed. I highly recommend sex before, during and after competition, though I believe we'll have to take up wrestling to do it during." The boys laughed and created curiosity from the other St. Bonaventure athletes looking on.

There were a few ooohs and ahhhs as a pretty young girl strolled in. Wendy was the only girl at the academy but she'd long thought of herself as a boy. Toy was already to throw up his arms in surrender with the belief that his only daughter desperately wanted to have a penis and be male. According to Oshi, when he was brought on as the school's psychologist, the young lady suffered from an identity crisis. She had lived with and around boys all her life, identifying with their moods, desires, and sexuality. Though a year younger she had reached puberty before any of the boys, only to watch them grow into puberty and desire their own sex for their attraction.

Wendy spent many a night at the boys' dorm under the scrutiny of her fraternal brother, Lance. Lance was just one of the boys and blended in with these students as if he was one and the same. Though he had come to the conclusion he was probably straight, that never stopped him from enjoying sex with his peers and not caring one bit that his sister took part. There was one specific code of honor amongst them all: his sister's virginity was kept in tack. That didn't eliminate Wendy from participating in all other forms of sex.

Now, fourteen, she was finally accepting her blossoming body as a woman's blessing. Her white, silk gown was designed by her and Zach, who had aspirations of being a clothing designer. The sleek, almost see-through garment defined her physique like a shear strand of onion paper. There was a complete absence of jewelry which gave the naked neckline a very sensuous appeal. The soft chiffon hung loosely at her breasts, appearing to be kept up by only her nipples, which it was—

glued by strips around the areole. Her back was completely bare with the V revealing the top inch of her rear. A pair of firm glutes from years of athletic conditioning was easily visible. With no pubic hair the visual was pure skin with the disappearance of sex.

Most of the adults gawked at this nubile charm, a smile of a princess and a body that had the boys from St. Bonaventure fumbling with their crotches underneath the table cloth. Wendy moved to those who commanded her attention; daddy's little girl playing her stage as if she was at an awards show. A year ago she would have wanted to blend in with the boys as quickly as possible; now, she was glad she was different.

The students of Kings Academy barely gave her recognition. They all loved her and knew her from a child. She was fun loving and had wrestled, swam, played games, and had intimacy with most of them. They just took this for granted that she'd always be one of them.

Travis bubbled with his own secret that he wanted to tell all his teammates, but knew he couldn't. He'd seen her naked and up close. Even been kissed! What he was more proud of was what he considered the handsomest boy in the room sitting next to him. He squeezed Alex's hand to reinforce the truth; he was more attracted to a male than a sexy female.

Travis whispered in Alex's ear, "She's really hot, but you're hotter."

****

Senator Franco decided to use this banquet for his own rubbing of shoulders and, what he was best at, charisma. It was a distraction when a beautiful girl paraded by him and had the men lose their concentration from what the senator was conveying about health care. Robert was impressed himself and could smell male hormones, the sexual tension that grew instantly amidst so many males in the room. It was somewhat embarrassing to the adults watching their sons' eyes rape the girl's sleek body. The gown was something one would see on a Hollywood starlet, but even a star with broad hips and a more robust breast couldn't give it justice. There was something about a teen's beauty, her slim hips, perky breasts, a butt like a boy's, perfectly shaped and firm, and then the way she walked without a sway or trying to appear sexy.

"Isn't she glamorous?" Mrs. Franco whispered when she leaned toward Robert.

"Breathtaking, my dear. I can't believe she beat our son in wrestling. She's given every kid in this room a hard-on. No wonder a boy can't beat her, she has them all drooling with her see-through clothes." Robert's humor offered his wife a funny relief and the senator hadn't even meant to be humorous.

Maria was awestruck. She thought the girl in the white dress was the most beautiful creature she'd ever seen, even prettier than their cutest cheerleader. Maria's eyes were on the same parts as her male counterparts, and she noticed that only thin silver threads around the groin kept the vagina somewhat obscure. That is until Wendy began to walk, which made the thin fabric sway and the slit become very visible. Maria knew a girl had to shave to have the smoothness of tender lips visible between her legs. She wanted more than anything to meet this beautiful young lady, and instantly Maria became aware of her own nipples pointing and a warm glowing sensation crept between her legs. More than ever she was really glad that her mother had talked her into coming to this academy to watch her brothers compete. And here she thought it would be so boring.

Wendy had strolled over to her father sitting with other adults. She plunked down on the man's lap and gave him a hug and kiss on his lips. If she meant not to cross her legs it was well appreciated by a dozen boys sitting at the adjacent table.

"My favorite wrestler is beyond words," Toy spoke to the other adults at his table and received a chuckle and tacit agreements with smiles. "A year ago you couldn't have gotten her to wear such an outfit. Possibly a few more layers of silk around the hip area might better preserve a lady's vanity."

"Oh, Daddy. I suppose a penis wouldn't look right in this dress, but I no longer consider myself a boy." Wendy received hearty laughs from the men and women around her.

Dr. Kerho loosened the hair band from his daughter's pony tail and allowed the blond hair to cascade over her bare shoulders. Wendy didn't mind and shook her head to the freedom. She had never minded not having a mother and fully accepted her father and Kami in their marriage. To Wendy, Kami was her second father and she made it a habit to treat him with the same respect and affection.

The informality permitted the athletes and cheerleaders the freedom to explore and traipse through the hallways and the venues. After the main course and desserts were served the students began roaming and discussing amongst themselves the different exhibits and areas of the academy they'd noticed earlier. Kings' students began to blend in and among these guests, attempting to thank them for coming and enjoying the competition, though it had often been one-sided.

Brittany had walked over to the piano and struck up a humorous strum of western music. What the Kings' boys knew, this was an invitation to sing the school's song of allegiance. They all stopped what they were doing and stood in absolute silence, while Wendy walked over to a separate storage area where musical instruments were stored and pulled out a violin.

When they were ready the boys gathered together and sang **Wild West Hero** to their own pride of being a student at Kings Academy. It sure brought the St. Bonaventure students to attention. It was when the chorus was sung that Dr. Kerho and Dr. Marion stood as proud directors of this amazing academy.

> By the rays of a new day
> To the first fading light
> He was my Western Boy
> From the fire oh so bright
> I am the son of a king
> A king not so much unlike me
> Come, let us rise and sing
> By God's strength, we will be

By the second verse Alex had grabbed Travis's hand and the two sang the verse together with the broadest of smiles between them.

Robert Masters was so impressed he yelled, "More! Encore!"

Brittany glanced at her many sons and received a nod of approval.

The boys had done a rendition of a composite of ELO hits when they were nine years old. Their performances at the Dallas Convention Center were something the city looked forward to each year. With this in mind the boys sang a variety of ELO songs that had their visitors awestruck with the boys' talents. The performance received a standing ovation from all the adults and the females; though a few of the St. Bonaventure boys were reluctant to give too much credit to these peers. Many sure eyed their peer with disdain that he had held a hand of an opponent, an opponent who was male!

Wendy was one of these who moved from person to person, often discovering more than a few boys surrounding her. She saw the athlete she had defeated on the mat and made it a special treat for him when she embraced the cute son of Senator Franco. Wendy admitted what an excellent wrestler he was. Jay smiled and blushed, not used to having females pay attention to him. At fourteen he was just now growing into his body and lacked the confidence to approach any girl. He found himself being accompanied by this pretty girl to the hallway, followed closely by other St. Bonaventure athletes who were puzzled that this sprout got all the attention by this dynamo female. Wendy whispered in Jay's ear that he should have had this type of support when he wrestled her. Jay laughed and glanced back with a smile, loving the attention and the envious looks on his peers' faces.

"Maybe the girl will let you try on her dress, Franco," one of the boys spoke with an attitude that was beyond envy. "I wouldn't let a girl beat me."

Jay was nervous in his response, but didn't want to show his fear of the senior wrestler in a much heavier weight class. "The way girls stay away from you it's no wonder, Johnson."

"I have a girlfriend, dude, and she doesn't pretend she's a boy. Bet she even jacked you off while she put you in that cradle hold," Johnson replied and received support for his antics from his sidekicks.

"Jealous?" Wendy asked.

Ralph Johnson liked to use his weight and status to impress and bully other students. He certainly wasn't going to take this from a girl. "Jealous of Franco here? If I'd wrestled you, bitch, your face would've been kissing mat while I humped you till you squealed."

Wendy waited until these delinquents had their jollies. "I can see why you might be in to pigs with a nose like yours, but I doubt if you have enough dick to make anyone squeal. By the time I got done with you, you'd be my bitch."

"Whoa!" several boys yelled in amazement at this challenge. This was something they hadn't expected.

Johnson began to fume, then clinched his fist. "Bitch, if you weren't so fuckin' dressed up I'd make your face look like it was a road kill."

Wendy laughed. "Oh, please! Where'd you learn that from, a cartoon? You are a girly man who can't even slap a girl when he's been insulted." Wendy stood her ground and hoped the ploy worked.

The boy's hand barely swung halfway when Wendy gracefully turned and caught the boy's wrist, allowing the momentum to continue forward, then quickly reversed the flow of action. She could have flipped him but watched the hefty kid swing around in a half-circle, then be jerked back, then forward again right smack in to a wall, which was more fun and entertaining for the boys watching. The smack of the boy's face hitting the tile made all the boys' faces frown with the pain. Almost knocked unconscious Johnson rolled on the floor, holding his face in agony. Wendy grabbed Jay's hand when she heard a security man hustling down the hallway toward them.

All Jay could say as they briskly jogged down the hall was how cool it was, then asked Wendy if she could show him sometime how to do that crazy karate stuff.

"It's Aikido, Jay," Wendy told him. "I'm going to hear from my father and probably be restricted, but the guy had it coming. I don't like bullies."

A few minutes later Lance had found his sister showing Jay some of the exhibits in the main corridor. Lance had made friends with Jay's sister, Maria, and the four of them had their laughter as Jay recalled the incident with a few exaggerations.

"You're in for a lecture. Dad wants to see you, like now," Lance said.

A level down, Ralph Johnson was delighted in playing the victim and getting all the attention by a cute nurse from the academy's medical department. The sympathy gained by the parents with the sight of blood also delighted him. Blaming this pissed off girl for striking him when he wasn't looking had been believed so far.

Coach Venegas recognized the team's bully when a security man brought him in the dining area to put ice on the left side of his face. Art figured the boy had gotten in a fight with another member of the team. Nothing new, but it was embarrassing to be at this occasion and see this from a St. Bonaventure athlete. Like other athletes Art decided to wander about instead of being involved in the chaos. He, for one, didn't believe a word the kid said about a girl pushing him.

Wendy heard her name being broadcast to report to the security office. This was something she had no intention of doing, since they'd simply make her wait for her father, anyway. Instead Wendy went right to the dining hall to face the repercussions.

While the senator didn't involve himself in the disturbance he did continue rubbing elbows and practicing his PR. Mrs. Franco got tired of the political double-talk and strolled over to make friends with a woman she figured must be part of the Kings Academy. Debbie felt an immediate connection with Brittany Makawa, wife of the academy's psychologist. Brittany's sense of humor had the two women talking about anything and everything within moments of introducing themselves.

Debbie Franco was well used to the social circles of the political elite, so the effort required to converse with loyal supporters and interesting women made her feel right at home. She quickly discovered that Brittany had none of the pompousness and arrogance of aristocratic housewives. It's what made the conversation that much more enjoyable.

Debbie had many of her questions answered with the absence of parents for these Kings' boys. Brittany used the word "adopted" as her means to explain, except for a few of the teenagers like Wendy and Lance, and Dr. Marion's sons. The demands and varied curriculum explained the many talents and successes of the students. Mrs. Makawa made it sound so simple and routine, like any academy could develop gifted scholar athletes while maintaining discipline and cohesiveness. The academy's liberal views and freedom of thought had created a wide appreciation and openness concerning important subjects and aspects of growth in the children they nurtured. Without being restricted or treated as unequal all the boys and one girl grew with uninhibited knowledge and personal development.

Mrs. Franco nodded and then played the devil's advocate. "But, Brittany, don't you have to give your children boundaries and have expectations that are influenced by their parents, teachers, and, possibly, a coach?"

"Oh, absolutely, but parents often have unrealistic expectations of how their children should act without giving any consideration for the child as an individual. Too often the parents emphasize caring for their daughters and rules and justice and might for their sons. Boys have feelings, too; they laugh, they cry, they struggle with their identity and their father's façade of masculinity. They recognize how genders are treated different and these conflicting beliefs become a cancer for generating emotional chaos.

"If there really is an evil, Debbie, it is the ignorance of the laws of nature that causes our children to harm others and poorly create their own life. Our boys and one girl here at the academy aren't perfect. Their flaws don't occur because they're bad but because they're hurting, insecure about something, or beaten up by life's challenges. Listening to my husband, Oshi, which I give him permission to speak his mind once in a while,—the ladies giggled—his biggest challenge is getting these adolescents not to entertain their peers, because that's how they think they'll be accepted, but to be themselves. He tells them that personality is what you want the other fellow to think you are, and individuality is what you really are."

"I see that so much in my own sons," Debbie admitted. "They're scared of expressing themselves in fear of rejection. Their father has such high demands for them, I'm afraid they'll never measure up."

"So typical," Brittany agreed. "I dread this perpetual masculine ideal fathers put on their sons. If before puberty they did not know what to make of the masculine codes their fathers suppressed upon them, they certainly get confused when their hormones kick in. Boys at the academy have learned that the touch of another boy and the fantasies derived from this companionship and love, much like other boys find with girls, doesn't have to be kept as secrets of great consequence and private from other children and adolescents, and, of course, their parents."

"Well, yes, but isn't that more for the occasional homosexual? I mean, not that every school doesn't have one, but I can see my own sons and their interest in girls. I suppose you have an exception at Kings, as well."

A few years earlier Brittany would have leaped on such a statement with a vengeance. She'd grown in maturity and knew that to talk about homosexuality being a problem really means as much as to talk about humanity being a problem.

"Actually, Debbie, to be a straight boy at this academy is an exception. My brother is gay so I'm sensitive to the issue. For their choice of sexual partners gays are, on the whole, no different from straight people. The desire for love, sex, companionship, comforts, creative growth, and intellectual stimulation are universal. We teach here tolerance and empathy, rather than have the boys purge themselves of so-called dangerous mannerisms that felt right in their childhood but can be marked by alienation, deception, and assaults, both physical and spiritual."

"Yes, I hear your point, Brittany, but I didn't mean to imply, well, you know, that being gay is wrong. My husband and I were brought up Catholic, so we're rather dogmatic in our viewpoints."

"Religions, sadly, do this to adults. The Roman Catholic opposition to homosexuality was not an original or inherent element of the Church's teachings but was actually a medieval development. Here at the academy we don't recognize cultural norms and spiritual truths as being necessarily the same. We teach that a spiritualized homosexual enjoys God's blessings as fully as anyone else. All our boys are aware that centuries of biblical interpretation by fallible, often hard-hearted men, have turned organized religion against them."

Debbie could only smile without further efforts to debate what Brittany found as a very serious topic. "You certainly defend gay rights exceptionally well, but you make it sound like all your students are gay." To herself she was saying that, to support a brother was certainly understandable, but to ride the band wagon for another boy or two appeared over doing it.

"My apologies, Debbie, I have a tendency of overreacting, given the boys I love dearly."

"No, I should apologize," Debbie replied. "I'm so impressed with your group that my curiosity has gotten the better of me. I do suppose I understand better why the boys switched the pronoun to masculine in many of those songs they sang. They are so professional for their age, and their range and versatility are amazing."

"Thank you, Debbie. The boys have changed the lyrics to fit their own needs and desires. There is such history in many of those songs.......it brings tears to my eyes when I think of them. I was a big Electric Light Orchestra fan as a teenager, so I had the boys sing many of the songs when they were about four or five. They loved Wild West Hero so much, rather than have to play this every day, we agreed to make it their school song to be played only at special occasions. It's why their mascot is Wild West Hero. Their favorite at five was Latitude 88 North, only because they could say hell and get away with it. I'm rambling, aren't I?"

"No, Brittany. Please continue. I'm enjoying hearing about how these boys grew up."

"I do have to tell you something about Wild West Hero. There are two guitar solos where the beat picks up and the boys used to go wild with dance and their karate moves. They don't get into this as much as they used to, but, given the chance, they will go wild with their expressions."

"I loved their rendition of Turn to Stone and Night in the City. The way they combine their talents and echo each other's voices, it was amazing."

"They have practiced those techniques since they were children. When they were twelve they put on a performance in Dallas, a ballet,

actually, to the ELO album, **Out of the Blue**. It went over so
spectacularly that the boys were invited back for five more shows. The
gay community has always been so supportive of the boys."

"Do you think that's healthy, Brittany? I can imagine the boys in
tights and the attention they received."

"Debbie, our children are sexual human beings and they know it.
Yes, I can imagine that many men were aroused by the boys' outfits and
dance, but the boys have amazing talents and voices that go far beyond
their physical presence. Every year tickets are sold out for the
following year immediately after the last performance. I'm sure the
City of Dallas doesn't want our boys to grow up, but they're in for a
surprise. There's another generation coming."

"We must talk more," Debbie said with fascination and cupped
Brittany's elbow.

Across the room Dr. Kerho and his brother, Reece, a history teacher
and the school's baseball coach, were surveying the doctor's cellphone.
Toy's gadget had access to every camera on the premises at a second's
notice. In this case he had the security department relay the incident.
It didn't appear that his daughter had done anything wrong as this St.
Bonaventure boy had proclaimed. In walked his daughter as Kami had
also found amusing the final results.

"Ah, the academy's Bruce Lee," Toy greeted his only female.

Jay spoke up first. "Sir, may I mention that Wendy reacted only to
defend my charm and innocence, for which I'm truly grateful."

Those words had Kam and Toy busting up laughing, as well as five
other men at the table. The boy had certainly added an element of
humor, given the intended discipline.

"Well, young man, I'm pleased your charm is still intact. I
suppose if the reverse had happened, you'd deserve to be knighted for
your gallantry. I want you both to know that I don't condone violence
at the academy. The sex of the aggressor or victim is not of a major
concern, but I have looked at the video and feel my daughter could have
defused the situation or escaped without relying on her martial arts
skills."

"Sorry, Daddy," Wendy spoke with a sad face that always melted her father's disposition.

"Your coquettish pout won't deter your responsibilities to this school, young lady. You will apologize to the boy for your behavior. Do we understand each other?"

"Yes, Daddy." Wendy scanned the room and saw the culprit recovered and talking with a few of his supporters on the far side of the room. Jay accompanied her for support. Her approach had the boy with a dread of panic on his face.

"I'm sorry for my behavior," Wendy spoke. "Perhaps I should have used better discretion and diplomacy."

"Pretty big words for a cunt," the kid spoke. "Some cute tricks you did on me, but I have my own tricks. Don't get caught around my school."

"I'm disappointed you don't get it, but, here's to maturity," Wendy said and walked off with Jay's hand in hers.

There were a few chuckles behind her, yet the words had left food for thought. At best, their school bully was stuck for a comeback.

****

Niqui, pronounced Niki, actually, Enrique, was Robert's middle son, an excellent gymnast at 5'8" tall, four inches shorter than his oldest brother and two inches taller than Jay. He was also the quietest and most handsome with strong Spanish features and thick black hair. Girls at school thought he was the sexiest brother with eyes that made people mesmerized to his matador demeanor. He was surprised to see his brother involved in a dispute, but Niqui was beyond defending even his youngest brother's character. Though he liked girls he had yet found the courage to want to spend time with them. Gymnastics was his life and this day had been a downer with losing to, not one, but three other gymnasts with skills that were far above what he was taught.

Niqui considered his Johnson teammate an embarrassment to the school for his involvement in another fight. He wandered by himself to a different floor and perused the art collection for a while, then admired the painting of Apollo holding his friend Hyacinthus in his arms, blood draining from the boy's forehead where a discus had struck him. The nudity was embarrassing and never would've been allowed at St. Bonaventure. Niqui read the story of the myth and wondered if it had really happened. Moving further down the hall he scanned the various motivational quotes scattered above the displays and doors.

NEVER FOR THE SAKE OF PEACE AND QUIET DENY OUR OWN EXPERIENCE OR CONVICTIONS

Niqui liked this philosophy, though it gave him great thought of what it might entail. He began to read all the signs he could spot:

TEENAGERS DON'T REALLY TALK TO THEIR FRIENDS, THEY PERFORM FOR THEM. UNTIL YOU LOVINGLY EMBRACE YOUR OWN EXISTENCE, YOU HAVE NO CONTEXT IN WHICH TO LOVE ANOTHER PERSON OR THE WORLD OR INDEED LIFE ITSELF.

Niqui thought about this and wondered if he was happy with his life. Lonely and confused about adolescence he was keeping up good grades because he was afraid of his father and the repercussions by not meeting his father's expectations. He smiled, not because he was happy, but because it was expected.

"Pity party, Franco," he told himself right there, as if acting his own psychologist would pull him out of his slump. In the reflecting glass he saw his long face and dark skin, curly and short black hair that the girls loved, but he wasn't too proud of himself right now.

Further down the hallway, next to a scientific display on genetics, Niqui was taken back with this poster of two boys kissing:

BEING GAY IS NOT ABOUT CHOICE. COMING OUT IS ABOUT FIRST BEING A PERSON WHO RATES HIS COMMITMENT TO HONESTY HIGHER THAN FEAR. HE IS CLEAR AND COMMITTED TO THE RIGHTNESS OF BEING WHO HE IS.

At his school those words would have been defaced in no time, along with the poster. He didn't know any gay kids and wouldn't know what to say if he did. He liked girls, but lacked the courage to ever kiss one. Shrugging his shoulders Niqui stopped at a Greek exhibit from the academy's vacation to Greece. "Wow! Their whole academy went to Greece?" He said out loud.

"Hi, Niqui!" Bobby Newton greeted and nearly scared the boy out of his shoes.

"Oh, hi, Bobby. I was just admiring all these displays. Must be like a fair or something, right?"

"We do this year around," Bobby admitted.

"That must have been really cool to go to Greece," Niqui said.

"It was fun. We go to a different location every year. Last year we were in New Guinea and the year before that to Powder Island in Japan. Europe before that and Australia earlier."

"Okay, I get the message. This is a school for rich kids."

"I don't think so," Bobby admitted. "I have about six dollars to my name. I think you're an awesome gymnast, Niqui. With a few more techniques I think you'd be aiming for the Olympics."

"My coach said that once but then I came here. You guys are too good."

"Good coaching and conditioning. Come on, let's talk about how you can improve." The boys walked to the main entrance and Bobby used his eye identification to access the outdoors.

The sky was aglow with stars and Bobby pointed out a few constellations, which Niqui readily agreed. They went around to the other side and meandered back to the all-weather track next to a baseball diamond with all the amenities of a major league part except for the stands. Tiny strobe lights glistened upward from bases of stone pedals to illuminate three-meter high statues.

"Whoa! What is this? Olympia?" Niqui asked as his eyes scanned the assorted figures surrounding the circumference of the track.

"It was our fourteenth birthday present," Bobby said with delight. "There is one here for each of my brothers and sister, representing the sport of our choice."

Niqui reached up and felt the smoothness of marble, its texture and perfection in defining the human body. "They're beautiful, though I'm thinking the sculptor over did your penises."

"Actually we each posed for pictures, so it's quite accurate. They're supposed to represent youth and optimism, a world beyond time."

"Where's yours?" Niqui asked and felt his friend's hand guide him on the shoulder to a statue near the starting line.

Bobby's features were precise and his pose was similar to a gymnast at the end of a routine—perfectly balanced with a satisfaction of completion.

"They sure captured your bubble butt," Niqui said and had them both busting out with laughter.

"It's my best side," Bobby admitted. "Course, you have an adorable one yourself."

"That's what my mother says, too," Niqui admitted and felt a sense of happiness and contentment for the first time that day. "You don't really have a penis like that, do you?"

"I really do, but it's not such a big deal when all my brothers have one just as long."

Niqui scratched his head and wondered how so many boys could be so gifted. "If my sister sees this, she'd freak out. Sometimes I have to wonder if girls have any sexual awareness at all, especially my sister. She's such a prude," Niqui explained.

"I don't know why she would act like that," Bobby replied. "My sister's statue is over there."

"Are you saying that girl in the banquet room posed nude?"

"Why wouldn't she? Wendy is really proud of her body."

This time it was Niqui who laid his hand on Bobby's shoulder. "Where? I just gotta see it."

The two boys sprinted to the far corner of the track and stared up at the imposing statue of a nude female.

"Awesome! She's beautiful, even in marble." They laughed together while Niqui moved his hand up on the inside of her leg. "What I'd give to see her in the raw."

"I have that pleasure almost every night," Bobby said with satisfaction, but not knowing why this was any big deal to see a girl nude.

"Sure you do. And I have Jessica Beale in my closet. May I ask you something, Bobby? How do you get away with having all those nude paintings, statues, and writings about sex? You must have one of those gay-lesbian clubs, but you only have one girl and she's hot. Man, at our school you can't even mention sex without being sent to the parish for confession."

"So you read my report on Apollo and Hyacinthus, huh? It's kind of sad what happened, but Hyacinth's blood formed the flower the Hyacinthus to remember him by. Sex is no big deal around here. I mean, we're all into each other and it's not like we keep secrets from our brothers. If you want to have sex you simply ask."

"Now there's a concept!" Niqui chuckled with this idea that had no realism to him. "That was a cool story about Apollo and Hyacinthus, but I don't know about two boys loving each other. Way too weird for me. My mom has a garden of Hyacinthus, but the stem doesn't always support the large flower. I'll have to tell her that she has a queer flower." Niqui smiled.

"Maybe that was what Hyacinth left in his memory, that all the love he felt for Apollo, no one man could conceive or hold. In answer to your question we're taught that writing is a safe place to explore past memories and new ways of thinking and acting. See, writing sexuality is about power and control, a fantasy created to make me feel a viable teenager and in control of my thoughts and emotions. The wonderful thing about writing is that it affords me the opportunity to reflect. And the paintings of nudes aren't about sex, they are about human form. From the Greeks we learned about aspiring to embody heroic ideals of honor achieved by action and beauty, expressive of power and distinguished solely by their superhuman performance. For the original Greek Olympics, athletic virtuosity and physical beauty were inseparable."

"You guys are all geniuses, aren't you? My teachers are the only ones I know who talk like that." Niqui felt intimidated by this boy's intellect, but the words were also logical and had a deep meaning to him.

"What did I say wrong?"

Niqui laughed. "I've never heard anyone be that positive about sex. My father thinks that walking around in our underwear is a sin. He's got this thing for the Pope and confession once a week. I hate it. You know how difficult it is to tell a priest that you masturbate every day? What faith are you guys?"

"We don't practice a religion. Being a Christian isn't just a spiritual commitment but a total release of our intellect, our bodies, and our souls. We're each God."

"Wait! Are you saying you each think you're a god, like those Greeks had all those gods?" Niqui had hoped Bobby would laugh at his honesty with masturbation, but the boy took it in stride. They began to walk past the various statues and Niqui realized he had a new appreciation for the artistry and masculine details of each boy. There was a rush of blood to his groin at the sight of a male's genitals carved in solid marble.

"No, it's not like that. You're God, too. See, God is everywhere, not nowhere, and if God is nowhere, then He has to be everywhere. We're all one, you and me. To harm you is to harm God, is to harm me. Having a Pope, as like a mediator between you and God, doesn't make sense to me. Why should you worship a man who is no more than yourself? It's like making an image that becomes your God. But if you have for your God that which is within our own individual self, you yourself being a portion of the Creator, you will continue to build upward to it."

"That makes sense, so why am I confused. Okay, I think what you're saying is that our relationship to God, then, becomes our relationship to our brother."

"Exactly. You know you're already saved by Grace, so you should be praying to thank God for all the good things you have. God didn't create your body to have you apologize for feeling good and happy with masturbation. He doesn't want you to apologize to another human for what He gave you to enjoy. Simply the realization that you are that which you are, and part of the eternal scheme of life, that you are everlasting and perpetual and will never die, is tremendous."

"This is a church school then, right?" Niqui asked. He received a laugh from Bobby.

"Did anyone ever tell you that you have amazing eyes?"

Niqui felt a sense of friendship for just that second, but then he saw what appeared to be his sister and younger brother over by the baseball diamond with two of the Kings' students. The more he looked at their giddy behavior and closeness he realized that his sister was wearing the dress that the Kings' girl had on at the banquet.

Wendy was wearing his sister's clothes. Even with the faint lights on the grounds he'd never seen his sister look sexier and appealing. His father would find out about this and chew her butt for good, Niqui was thinking. The thought repulsed him with anger and it wasn't his nature to think ill of his sister or brothers. His father had a tendency to bring out the worse in his patience and tolerance of others.

Bobby and Niqui strolled over to the diamond, where introductions were made and Niqui's eyes adorned the beauty of this girl who had beaten his brother at wrestling. He knew the resemblance of the statue was a perfect match to what was underneath the clothes. A quick hand in his pocket helped suppress an erection. Trouble was, Jay had his hand entwined within her's—his little brother?! This was a hard thing for Niqui to fathom; Jay had never even been to first base with a girl and, only lately, had just learned where the batter's box was and what a bat could be used for, thanks to an older brother's nightly sex education course. Niqui knew what was to follow when he faked sleep; within minutes Jay would lower his sheets and jack off. It was kind of sad that, as brothers, they never talked about their sexuality and kept secrets from each other about such things, yet lived in the same room. In this case, Jay looked like a young boy with his hand being held by a woman.

Niqui wasn't paying attention to the conversation going on around him. His thoughts were on this beautiful young lady holding hands with his kid brother. It wasn't like Niqui had ever been to first base either, but he had experiences with kissing when he was thirteen when he attended a slumber party at his best friend's house. Their truth and dares led to kissing and fondling, and Niqui discovered that his friend, Kirk, a boy he'd known since first grade, had come out to him that night with the revelation that he'd known he was gay since fourth grade. It totally flustered Niqui when the boy had said that he loved him.

As much as Niqui enjoyed the night, it scared him with this newfound pleasure at the beginnings of puberty. He began to avoid Kirk at school and in their leisure time, but guilt soon crept its way into

Niqui's mind as it usually did being a Catholic kid. By the next day he was eating at the same lunch table as Kirk and praying that the other kids wouldn't pick up on Kirk's attraction to his own sex. To make matters worse, in English class, where they were discussing people who were judgmental, Kirk raised his hand and told the class that Niqui was the least judgmental person in the entire school. Though his classmates and teacher were impressed by this, Kirk's comment had a profound effect on Niqui's conscience.

The new school year had brought Kirk a new friend who had moved into the neighborhood, a blond boy from California that thrived on the affection which Kirk readily gave. In some ways Niqui was relieved, but he also had to admit he really missed being around his one and only friend and sharing their hobby of collecting baseball cards and practicing new gymnastic moves together.

The mutual masturbation was something he looked forward to, but always justified to himself that it was Kirk's idea. He didn't want to admit he was jealous of Kirk's new friend, but the emotion was revived in the presence of these Kings' boys. After all, Niqui realized he'd gotten an erection in front of a boy statue and not in front of the only girl; yet, he knew he was also attracted to females and was now envious of his youngest brother. He wanted to believe this matter would be finally settled if he could only see a girl like this naked and be able to kiss her.

His glance to his sister created another animosity. This Kings' boy was handsome and maybe his age, a year older than Maria. He shouldn't have been jealous of a boy holding his sister's hand, but he was. For an opponent of their high school to see his sister's body through this sheer garment wasn't fair. Just like girls to have switched outfits and now they probably even liked each other. It was just hard to believe that his sister would want to wear anything that sexy. He'd had his own glimpses, but not with his sister's permission.

Niqui's eyes went to the boy's groin to see if there was a lump, a sense of being the protectorate for the time being of his sibling, who didn't appear too much like she wanted someone to protect her vanity. What would his sister think if he knew a boy possessed a penis that even

162

awed him? Niqui had come to the conclusion that all boys had the same lust that he did for sex.

This sense of jealousy of a boy he didn't even know made Niqui frown with annoyance at himself and the envy he had for Jay's fortunate luck. It wasn't like he didn't have a good looking penis himself, though it was barely seven inches hard. He knew one thing, he'd have to talk to Jay about girls and how to proceed to first base, maybe doubles and triples, though a girl like this wouldn't even consider a kid like Jay hitting a home run with her. For the time being, at least one of them was getting to hold a girl's hand.

Chapter Fourteen

Leif had left a trail that a blind man could have followed. If it was snowing it wouldn't have given Sveinn any better clues than he had already to guide him directly to New York City. Mrs. Bjornsson had called the man that following morning and expressed her concern for the absence of her son. She all but threatened the poor teacher if her boy was hiding out at his house. Sveinn guaranteed her that this wasn't the case.

A few hours of calling friends and the local places where teens hang out proved that Leif had taken more drastic action. Mrs. Bjornsson was in a panic and promised she wouldn't pursue any reporting if Sveinn found her son, and to apologize to the boy for any words she had spoken. She knew she'd hurt him deeply and was quite willing to accept Leif as the gay son he thought he was.

Not expecting any results of locating Leif around the neighborhood, Sveinn called the airport and received confirmation that the boy had bought a ticket for New York City and had left that morning. Sveinn was on his way to New York the next day. He walked around the many city blocks, not knowing where to look, but asking a lot of questions.

A picture of Leif Bjornsson was all that Sveinn had to give to the police. This led him to a boarding house for runaway teens when the director reported he'd housed a boy of this description for several days. By the time Sveinn had actually arrived Leif had long taken off. A conversation with a boy from Ohio didn't produce immediate results. Two days later this same boy saw Sveinn again wandering the streets in desperation.

"He's at Fire Island," Rodney admitted. "He's too good of a kid to get into this profession. Take him home, but you better take care of him."

"I love him, too," Sveinn admitted.

Once again he was too late. Getting a tip from the bar's owner,

Sveinn went around to different hotels, pretending he was an investigator until he found the right hotel. Thanks to a bellhop, in room 222 was a left behind copy of the latest tabloid Leif liked so much. A Florida address was circled on the inside and Sveinn had a hunch that the boy was heading south.

****

Leif arrived in Tampa and was surprised when a man came up to him, cupped his elbow and asked, "You wouldn't be Erik Eldjarn, would you?"

Leif was stoic for the first moment but saw a harmless expression from this man dressed in very casual clothes.

"Yes. Are you the person that Michael sent to meet me?"

"Not really," Joe Rush admitted and directed Leif to the baggage claim. "My brother and I have this same friend from college, but Sam was busy and called me. I'm Joe, Michael's brother."

"Thanks for coming. Your brother is real cool," Leif said, then remembered he had all his clothes and belongings in the backpack. "I'm ready."

"Great! You travel light. Where do you know my brother from?"

"We're old friends."

"You strike me as way too young to have old friends. I have socks older than you. Do you guys work with each other?"

"Yeah. In the stock market."

"Really? Where exactly is this located?"

"In Greenwich Village," Leif replied with the only location he knew.

"That's Wall Street, I presume?"

"Yeah, Wall Street."

Joe grinned but didn't embarrass the boy. They went directly to Joe's house where his wife made them dinner and Leif was glad to climb in bed early so he wasn't interrogated anymore about what his relationship was with Joe's brother. Mrs. Rush had been just like her husband—way too many questions. Leif had evaded every curiosity by talking about his venture to the ballet and to the play they'd seen

165

together.   He'd made a mild faux pas by admitting how cute and talented the boy who played Billy Elliot was.   This alone was enough to convince them that he definitely liked boys.

By acknowledging the arts, they knew he was friends with Mike, because Mike loved New York plays and the arts.   Only other trouble was, Joe's wife had trouble believing Leif was eighteen because she couldn't find even a tint of peach fuzz on his face.

By morning Joe took the boy to the offices of this tabloid magazine and introduced him to one of the editors of the tabloid, another friend from college days.   Leif quickly brought forward his ideas about all the recent articles of the tampering of tombs in Europe.

"I think it's possible that only gay men are being stolen," Leif said. "I've done my research and it appears all these leaders and great artists and poets—everyone—they're all gay."

"That's quite an assumption, young man," the editor said with a disbelieving smirk.

The executive went right for the jugular.   "The question is, who's stealing them?"

Leif leaned forward with this secret he'd been just waiting to reveal.   His enthusiasm was obvious. "The way I see it I think someone in America is doing it. There were reports of several boys accompanied by teachers from America, in Europe a few years ago.   They wore the same jackets as this kid did when he went to visit St. Petersburg.   If that's the case I think there might be American heroes who are gay and have been dug up as well."

This mere fact had the editor far more intrigued, being this was potentially becoming an American story.

"Impressive viewpoint for a young boy.   And how do plan on proving this?"

"I'm not a young boy, I'm eighteen.   I had hoped you would want to hire me and I'll do the investigation."

A subdued chuckle from Joe and the editor didn't sit well with Leif. "We usually don't hire people on assumptions or because they're cute. What I can do is run this by my boss and see what he says."   The editor excused himself while Joe and Leif eyed the ambience of the office.

"What makes you so interested in gay people?" Joe asked.

The question caught Leif unaware. "My brother's gay and I respect this sexual attraction; plus, I need the work."

Joe nodded but didn't buy the lie. Five minutes turned into fifteen and finally the young executive came back with a smile on his face.

"You're lucky, youngster, my boss is in a good mood and believes this has potential. He's offering you a thousand dollars for proof that any important person in American history--assuming he's gay—has had his grave broken into."

"I want five-thousand dollars for everyone I discover, plus travel and expenses," Leif responded back which surprised Joe and the editor.

"You're more street wise than I considered, Mr. Eldjarn. Tell you what, my boss gave me some leeway on this. How about $2500 a story and I'll pay your hotel bills, bus tickets, and meals for a month."

"All expenses and I travel by plane," Leif demanded.

"Okay, but you better be on the level with this. I want a daily report on where you are, who you're looking into and what you find."

"I expected as much," Leif said and stood up. "How do we do this?"

"I'll give you a company credit card, but if I see even one purchase of a video game or any other toy, I'll cancel the card immediately and send the police out after you. That includes Internet sex sights."

"Agreed," Leif said and stuck out his hand.

Joe was quite impressed with the boy's tenacity. As they were leaving the building he noticed a shocked look on Leif's face. "What's wrong, Erik?"

"That man over there, I think he's after me." Leif darted to the right and sprinted for the parking lot.

Joe wasn't sure what was going on and walked to where the man was prepared to pursue after Leif. "Excuse me, buddy. What do you want with the boy? You some kind of pervert?"

Sveinn halted his decision. "Do you know that kid? He's a runaway. I was sent here to bring Leif back home."

"Wrong kid, buddy. The boy's name is Erik, Erik Eldjarn."

Sveinn reached in his wallet and showed the man his Iceland

driver's license. I'm Sveinn Eldjarn. The boy is using my last name. He's Leif Bjornsson, fifteen-years old."

"Okay, I'll handle it," Joe said and departed for his car. Parked six rows back, Joe found Leif hiding in the back seat.

"All is clear," Joe said and had Leif join him again in the front seat.

"I think the guy likes boys," Leif said and glanced around.

"Good thinking," Joe replied. He began to drive out of the lot, then saw his target. Pulling up alongside Sveinn, he stopped. "Out! I'm going to call my friend and cancel that card."

"Please don't, Joe. I swear I was telling the truth. Okay, I'm not Erik, but I can do this job."

"Leif, you're a runaway which means you're not eighteen. How'd you fool my brother? He's usually more alert."

"I like Michael. He was really good to me. Okay, let's make a deal. I'll go with Sveinn because he's my teacher, but give me a chance. I'll spend a month working my ass off just to show you."

Joe had a soft heart and was in deep thought while Sveinn waited outside. "One month and I want a call daily. Additionally, I'll write my number down and talk to Sveinn before you guys take off. I have only one question and it better be answered truthfully. Is my brother gay?"

Leif hesitated then nodded. "I really like your brother, Joe. Please don't tell his wife about me, okay? Mike doesn't go around and pick up boys; I really like him and he likes me."

"Leif, thanks for being honest with me and, don't worry, if that's how he wishes to ruin a great family that's up to him. I've always been suspicious, that's all. I'm just glad he found the likes of you."

"Thanks, Joe."

Leif didn't look at Sveinn, still too pissed off to give his teacher the satisfaction of forgiveness. They rode in silence for many minutes.

Sveinn didn't want this silence to continue. "Be mad at me, I deserve it. My talk with your friend there convinced me how dedicated you are to this crazy idea you have. I went to college in Miami, so what do you say we go down there for a few days and get reacquainted. I can stall your mother off for a while. She wants you to see a psychologist when you return."

"I'm not going back," Leif said with determination.

"Yes you are if I have to drag you back kicking and screaming."

"How'd you find me, anyway?"

"You left a map.   Don't ever apply for the CIA."

"I had sex with a man who really loved me.   Stick that where the light doesn't shine."

"I bet his mother didn't interrupt."

Leif pouted that he hadn't hurt Sveinn as much as he had been.   He still couldn't subdue a smile.   Without a response Leif was quite willing to allow Sveinn his travel plans until he could find the opportunity to break away.   His research in the library had given him several names and burial plots he needed to investigate.

Chapter Fifteen

The evening was winding down and the banquet was, for the most part, a success. Coach Venegas had parents and other coaches rounding up the students for the trip home. Three of the St. Bonaventure athletes and two of the cheerleaders had had their fill of food, and there was no question that certain exhibitions of science perked their hormones and humor. It was one of the cheerleaders that dared the boys to show them this stupendous locker room they all talked about. Once inside the visitor's locker area one of the boys dared this same girl to take a Jacuzzi with them. This escalated to a double dare when the girl whipped off her cheerleading outfit, bra and underwear and jumped into the churning water to the amazement of her peer and the three boys.

"Okay, chickens, put up or shut up!" she challenged, her boobs floating in the sudsy soak of hot water.

There was a quick game of rock, paper and scissors to determine who raised his sweat top, lowered his pants, and dropped his supporter. The first loser turned his back, whipped off his clothes, spun and nearly jumped on top of the shocked female. The second cheerleader giggled and stripped as the boys did. When the last boy removed his remaining article his boner flipped up for all to see. His cannon ball splash wasn't welcome.

Relaxed and pretending to be the coolest teenagers in Texas the conversation turned to who was dating who, the latest gossip and breakups—all while the boys' eyes were glued on two sets of tits. There wasn't a boy in the tub who wasn't hard as rock.

"Hey! What in hell is going on here?!"

All faces swung to the imposing figure not more than 6 feet from the Jacuzzi. Coach Venegas stood defiantly with his hands on his hips. The girls scampered out first and were quick in placing their skirts around their waist then slipping on their panties. Dressed and red face they hustled right out of the locker room to leave these boys with the wrath of a coach.

"Out!" Coach Venegas barked.

Embarrassment was the mildest word to describe the three athletes who did their best to hide their erections and fumble to put on their clothes. Not a further word was said until the coach ordered them all to the tram. He followed his three horny students right out of the locker room, badgering their every step.

"That's all I need, a couple of our females getting banged up on my time! Your parents would have my neck!"

****

Senator Franco was disappointed with the change of itinerary. He'd just introduced himself to Dr. Kerho and found the man intriguing. Robert was shocked several minutes earlier to see his daughter in the sexy gown that showed more than he'd allow around his house. Maria's pubic hair stood out under the sheer garment. At first he was adamant that she change into something more presentable, then Dr. Kerho had said that his daughter had the figure of a model and the status of a Spanish princess. Robert's pride swallowed those words with an agreeable smile. He was actually thinking that his daughter's boobs were going to be bigger than his wife's and twice as perky.

There was quite an endeavor to find all the athletes and tell them there was a change in plans. Senator Franco decided to return with his wife and daughter. A return trip to San Antonio with a bunch of rowdy boys didn't excite him after such a defeat. Coach Venegas was nowhere to be found, so it was relayed that the senator wouldn't be on the bus. As word would be passed around and quickly forgotten it would cause eventual confusion.

Given the surprises of one of the boys being blasted by a girl, naked teens of both sexes in the Jacuzzi, and parents whispering and trying to keep their teenagers from certain floors and exhibits, Coach Venegas was just glad to see both buses jammed with athletes and cheerleaders, no matter what they were laughing about, pushing and shoving, or asking each other if they'd like a dose of creatine. The coach didn't quite figure that joke out.

Chaperones were tired and had lost count, but then someone had mentioned that the Franco family was traveling together so the count

171

was screwed up again.   At the exit the ankle bracelets automatically fell off by a computer signal and tossed in a receptacle.

As the buses pulled away there was a humble realization that their sport teams had lost to superior athletes, a trumping this school had never before witnessed.   For many of the athletes this aspect had been forgotten for the moment with each and every boy and girl more interested in what happened in the Jacuzzi.

A short lecture from Coach Venegas on one bus let those before him sit for a minute in reflection on this devastating loss. A minute later a wadded up missile smacked a girl's head and chaos resumed. Laughter and chatter rose to decibels of teenage energy and Coach Venegas was just glad he'd said his peace, for whatever it was worth. He reached in his pocket and took out an envelope that the Kings' trainer had given him before their departure.   It was an appreciation note for the competition and good sportsmanship, but, at the bottom, there was a list of athletes with medical findings: three boys with marijuana in their system, two wrestlers with anabolic steroids, one with cocaine, and a cheerleader was pregnant.

Coach Venegas sat back with the weight of the world seemingly on his shoulders.   He hurriedly found the contract both schools had signed and, sure enough, St. Bonaventure had given tacit permission for drug analysis of all individuals using the facilities of the Kings Academy.   It probably wouldn't stand up in court as an invasion of privacy, but the coach didn't wish to create animosity between the two schools.

There was a list of half his team, it seemed, one boy on Dalmane, another with type II diabetes—one of his heavier wrestlers.   An asterisk was beside the name of an adult—a suggestion for Mrs. Debbra Franco to see an oncologist for abnormal white cell count.   This alone appeared more important than confronting a boy for his marijuana use.   He just hoped this report didn't start a string of lawsuits because one school decided to medically diagnose the participants.

****

"Do you really have to learn five languages?   Spanish and English are enough for me," Niqui asked as he and Bobby began to stroll back to the main complex.

"Most of my classmates speak more than five, but five is required. I like Italian, German, French, Spanish, Dutch and Japanese, but I know nine all together," Bobby admitted without appearing to brag.

The other four kids spotted these two and hustled over. Jay was extremely happy and holding Wendy's hand. He made sure that his older brother noticed this. Almost inside the main pyramid, here came Alex and Travis from the lake.

"We're right on time," Travis said with a grin that he'd enjoyed his tour of the grounds. "Coach said eight o'clock."

Inside the building silence prevailed and Alex walked over to the first security officer he spotted. "Hi, Mr. Faulkner. Where is everybody?"

"Are these kids from St. Bonaventure?"

"Yes, sir, but we're on time," Alex assured.

"Their coach decided to leave earlier. I can't believe they didn't count their own athletes. Let me check with your father," the man said and punched a button on his handheld computer. "Dr. Kerho, Walter here. We have a problem."

The situation was brief and to the point. Though the boys and two girls couldn't hear the response through the headset a solution was in the works.

"Alex, take the students to your quarters and we'll drive them home tomorrow. Dr. Kerho will notify their parents immediately. All privileges are to be extended to our visitors, except your father wishes to see you, Lance, and Bobby at nine o'clock. He'll be coming to the dorm."

Alex and his peers gulped, but there was also a hyperactive euphoria on having their first ever guests that night stay at the academy. Dr. Kerho hadn't fully understood that Maria was one of the students left behind. As the tram took the kids to the student quarters Alex discussed with Bobby and Lance how they should approach this. They choose Alex to be the spokesperson.

"Okay, there's a lot you don't know about us. We have a very

173

open atmosphere in our quarters, an independence granted to us by our parents. Anyone who wishes to stay with my parents is welcome to do so because you might not feel comfortable in our setting."

"I'm sure my brothers and I will like it," Maria stated.

"What could we not like?" Travis asked, his heart already on full throttle.

"Here it is. We don't wear clothes. You don't have to worry about anyone pawing you or making you feel uneasy. To us, this was the greatest means for us to create an environment that we all felt comfortable in. Each of you will have a room, so you're invited to participate in all our activities, or you can stay in your room and watch television or play on the computer."

Niqui's face was white. "Huh, we don't exactly go around nude at our house. I'm not sure Maria is okay with all this."

Maria acted insulted. "Trust me, Niqui, I've seen naked boys before."

"Five-year olds you baby-sit don't count," Jay said to humor the group.

"Very funny, my little fraternal twin. Don't forget we bathed together until we were eight."

Jay had to tolerate his sister's belittlement. "It's because she was born two minutes ahead of me. Big deal."

"We're here," Bobby stated. "Last chance." No one spoke.

Above the entrance was a sign written in Latin:

HIC HABITAT FELICITAS

"What does that mean?" Niqui asked.

"Here lives happiness," Alex said with pride. "We hope you agree."

They moved into a wooded chalet which had a multitude of wooden lockers and carpeted floor. Each locker had a picture of a boy but no name. The Franco children were guided to their own locker by Lance. There were no adults around but music made them feel relaxed. While the Kings' students removed their clothes without hesitation, the Franco kids took their time and turned their backs when the last few articles were removed.

Eight naked teens moved toward a set of sliding doors into an

174

interior hallway of shag carpet and tiled floors. Heating coils kept the floor at a perfect temperature and the soft lights were low and relaxing to the senses. Light-colored woods guided them to a long hallway of curved windowed rooms. The Franco kids were trying their best to be cool and not look at each other, as if this new dimension of being naked in front of each other wasn't all that traumatic.

Alex touched a wall panel and the curved glass swept back, revealing a room with a sunken floor with warm and soft colors. Against one wall lay a futon and several electronic appliances. The eyes of the visitors began to wander downward to examine the nudity of their peers. Jay was the first to take a glimpse of Wendy, but had to distract his mind because he felt blood flowing in all the wrong places. He threatened his mind with torture if he got a hard-on. Maria tried to be nonchalant but was as nervous as a girl on her first date. When she took a peek at Alex's body her eyes glued on a hairless torso with a penis that shocked her knowledge of a boy's anatomy.

"Maria. Maria. Maria!" Alex said and had the kids laughing that it had taken three times to bring Maria back to reality.

"I'm sorry," she apologized. "I....I..was thinking."

"It's not too late if you want me to call my parents," Alex informed.

"Oh, no, I'm okay."

All the kids stepped into this room, a full 180 degree view through one-way glass scanned the grounds below. The room was very aesthetic and functional. A computer terminal next to the futon had more functions than the kids had ever seen.

"Bobby, do you want to explain while I take Jay, Travis, and Niqui to their rooms?"

Bobby nodded, but then thought Lance should do this since he had nudged him in the side of the ribs to get the message across. Lance proceeded to give assistance to this very attractive girl he'd met only hours before. His sister had been the only girl their age to be in the dormitory at night.

Maria was glad when everyone else left and was glad that Lance was the one who stayed to show her all the amenities. She pretended not to look, but it was too tempting. Lance's body was so athletic, smooth, and with a penis that kept her eyes darting back and forth. Every time Lance would point and look away, Maria glanced at the unique appendage. She dared to ask this boy a question that was on her mind.

"Don't any of you have body hair?"

"No. It's a long story, no pun, but we're happy not to have something very archaic to human evolution."

"We don't believe in evolution," Maria stated.

"Neither do we, to a point. I mean we were created by God, but evolution still exists in the development of mankind and improvements. Anyway, there are many choices you might wish to consider, but most of them are meaningless for you now. I'd rather show you around."

"I'd like that," Maria said and hoped by lifting her chin she might receive a kiss. It wasn't to be.

Niqui was the only one who had second thoughts about staying in his room. It took Jacob Tilden to stop by and introduce himself.

"Hi! I'm Jacob. I've heard about the mix-up. This is a fun place unless you wish to hibernate all night. We can stay up till one o'clock. Care for a swim or we can play video games? You name it and I'm your guide."

Niqui ruminated and was again mad at himself for acting this way. He had too many questions on why everybody had a super long penis and were hairless. He'd do a quick tour and return, he decided. Jacob guided Niqui on a complete tour of this massive dormitory and recreation center. Past several fish aquariums and a music room, filled with every instrument available and private playing areas, the boys went into a museum that astounded even Niqui. He was already forgetting he was naked, his mind so focused on these rooms and facilities that were totally not expected.

"This is our museum, a history of our many vacations and many of our handicrafts we've done since childhood. It means something to us and every boy can add or delete what he puts in here."

Niqui was already examining a group of naked black men and several of the Kings' boys. "Where is this place?" Niqui asked.

"Just last year. We visited New Guinea, the second largest island on earth. This is south of Asmat, where headhunters used to be the norm. Yes, they're naked, so we felt right at home. There they greet you by rubbing chins and grabbing your testicles. They're very affectionate."

"You let them grab your balls?" Niqui asked in disbelief.

"Why not? They love intimacy and thrive on male bonding. It was a great trip."

"I bet," Niqui said sarcastically. He glanced further at all the vacation pictures and saw a whole lot of nudity. "Where is this place?"

"Oh, that was in Germany when we were twelve. The letters stand for Freikorperkultur, or "free body culture"—nudism, in other words. The location is Sylt, a twenty-five-mile-long sand spit in the North Sea. Kind of a getaway for the rich and famous, we swam there, then St. Tropez. The low red cliffs were amazing."

"The only time I've gone skinny-dipping is when my parents and brothers and sister weren't home," Niqui admitted and scanned the bodies of boys around twelve years of age, with penises longer soft than he had when he was hard. He spotted another vacation setting. "Cool, skiing."

"That was in Niseko, Japan. We travel to Japan quite often. There are a lot of hot springs there."

Niqui was both amazed and apprehensive about this academy and what it represented. The museum gave a full account of their upbringing. Jacob explained how their physical talents had blossomed through soccer, swimming, volleyball, track and field, baseball, basketball, and all sorts of minor sports were played on a daily basis. Martial arts had become part of their curriculum by the age of six, and most of the boys had their black belt by age nine.

"So, you like tennis. I see your own display, Jacob."

"I believe God made me for a purpose, but he also made me good at tennis. When I play I feel God's pleasure."

Niqui smiled for the first time that day. "I feel that way when I do gymnastics."

"I know, I watched you do gymnastics, before I went out to my match. You need to learn mental control such as maintaining an energy level that is neither too excited nor too relaxed. In other words you need to train your body, especially your nervous system so that you can automatically do your best on every technique. Then you get your mind and its anxious chatter out of the way, go on "autopilot" and let your body "fly itself," Jacob lectured.

"You can teach me that?"

"I can help."

Niqui had another bewildered expression, but he had to ask. "My father said your mascot is Wild West Heroes. Is that true?"

"We voted on it when we were five. It comes from an ELO song, but we all loved what it signifies, and maybe someday one of us will be a hero."

"I think it's cool. We're the Golden Knights at St. Bonaventure. I'd rather be a Wild West Hero than a Golden Knight." Niqui laughed at his own conclusion. He eyed the many group photos and realized that these boys and one girl had grown up together.

In another location Alex and Travis were enjoying a massage. Alex said this was expected by all the coaches after workouts or competitions. Two Japanese had bowed to the boys when they entered and Travis made sure to bow like Alex did. Alex had adjusted immediately to the nudity, though he was hoping not to run into his sister. The full body massage was very invigorating and his first ever rubdown.

Alex took his new friend straight to the pool. The swimming pool resembled more a pond with white sand beaches surrounding the huge rock enclosure. On one side was an old wooden pirate ship with a slide from its deck to the pool below. On deck was a Jacuzzi, and there were all sorts of ropes and climbing apparatuses for fun amidst assorted sails. Other boys were diving off boulders and enjoying this massive water enclave. A pair of frisky teens were wrestling on a beach towel when Alex and Travis approached.

"Tim, Dow, meet Travis," Alex introduced. Both boys stopped and Travis saw that neither boy was mad at the other. They were simply having fun. Dow reached up and shook Travis's hand and Tim followed.

178

"Good to have you here, Travis. We heard what happened," Dow said.

"Really cool being here, guys. Didn't mean to disturb you," Travis teased.

"Didn't faze us," Dow replied and picked up Tim's legs to continue to pin the boy on the white sand with a hand cradling the other's balls. "To the victor go the spoils!"

Travis laughed at the feisty play and walked away with an awed expression. "Is that usual? I mean, is it okay?"

"Why wouldn't it be?" Alex asked.

"Hope my sister and brothers don't see it. They'd freak. Jay is so naïve and Niqui is very uptight. He even writes poetry. What kid writes poetry, huh? Maria has no idea that boys do that."

"Don't you guys have sex education?"

Travis was humored by this question. "I think in the twelve years I've been at St. Bonaventure we're up to a blurry picture of why birds like bees. If I may ask, why do all you guys have what every male dreams of? And why no hair?"

"I suppose it had to do with our doctor. He had a great sense of humor," Alex said.

Travis appeared puzzled but didn't pursue this line of questioning. "How did you guys all end up here? You're not all brothers."

"I can't tell you that, Travis, as much as I would like you. Let me give you a riddle. It's ten miles long and can be rolled up inside a fist. The truth is, what I was, I am. What I am, I never was. I define a moral Rubicon."

Travis was in deep concentration and was only distracted when Maria and Lance came into the pool area. They waved and smiled at each other, before Maria got a great lesson in two boys making love on the beach. She appeared in a trance in this new lesson of males in full copulation. Travis waited for some hysterical reaction, but, instead, Maria simply held Lance's hand as they walked down a stone ramp into the pool.

Alex grabbed Travis around the shoulders and they were quickly swimming to meet other Kings' students. There wasn't a minute that went by that Travis wasn't trying to figure out the riddle Alex had given him.

Wendy also wanted to show Jay all the sights and areas of enjoyment. They visited the arts and crafts room where a few boys were making pieces of sculpture or drawing. Wendy explained that for many of the available things to do the boys could only work or play in such areas for an hour.

"Why's that, Wendy?" Jay had asked.

"The academy wants everyone to be well-versed in many areas. This is so someone doesn't get all wrapped up in music or the arts, or even all the video games."

"You have video games?"

"You won't believe it," Wendy said and guided Jay by the hand down the corridor, past a dance floor and theatre, and into a room filled with virtual reality games and about ten teenagers.

"Totally cool!" Jay spoke as he eyed the vast array of games and equipment. His eyes glued on the three miniature aircraft in the center of this large room, the size of a gym. A Stealth bomber was nose to nose with an F-22 fighter. Beside these two jets was a Blackhawk and all three aircraft had dual cockpits of regular size. He recognized that the cockpits were equipped with simulators.

Against one side wall were three climbing stations with foam pits below each one. Two bare-ass boys were halfway up one of these moving walls, freestyling with their legs dangling and using just their arms for support. Each boy had a headset on, which had Jay ask Wendy what that was all about.

"That's the advance wall. Most all of these games and sports are done with virtual reality. To those climbers they're climbing El Capitan, and though the wall looks to us to be rather simple with all those grips and indentations, to them it's a highly difficult, vertical ledge that every hand hold represents life or death. No one much uses the beginner wall anymore, but if you fall it's like experiencing a thousand foot drop in your mind, yet, thankfully, you land in the pit."

"Wow! They're so good at it. Is that a chalk bag hanging from their hips?"

"Yes. It helps keep their hands dry. If you'd like to learn how to fly I can sign us up. We're allowed twenty minutes every hour; one time being a pilot, the other being a navigator. All my brothers have a pilot's license and can fly jets."

"You're kidding me," Jay said with a smile.

"No, really. See, you start by learning in a Cessna and then gradually develop your skills to other aircraft. The simulators in each jet have various degrees of difficulty and can represent various types of planes. Don't feel too bad, my brothers have been flying since they were five. Dad took us overseas to a place where we each got to fly a Mig-29. It wasn't that difficult after playing with these babies every night. Over there beyond the planes are the sport virtual reality games. Each requires a headset and you start by learning mechanics of the sport you've selected. One station has all the racquet sports from ping pong to tennis and squash and racquetball and badminton. The pad is convex and moves to simulate running and quick moves. You can play the world's best or just a beginner. It's very realistic."

"Who's that? He's so good!" Jay pointed to where another boy bounced on a six by six foot pad with a katana in his hands.

"That's Rom, a black belt in three different arts, though everyone at the academy is a black belt. He's practicing kendo, probably against a samurai, depending on who he choose to work with. You can choose any weapon or free sparring, such as fencing or assorted blades. The apparatus allows you to fence like Peter Pan against Captain Hook, or fence saber at the Olympics against the world's finest."

"No wonder you guys are so good. That boy over there has such an awesome swing," Jay noticed. "I bet he's good."

"Spirit Owens. He won the Texas Amateur last year at fourteen. Spirit also holds our course record at sixty-four, though he shot a sixty-three at the Amateur tournament and won by seven strokes. He's our only boy right now who wants to turn pro. My fathers say he has to wait until he's eighteen because education always comes first. The virtual reality device allows him to play any golf course in the world. Every boy here is a scratch golfer."

"This is too cool! I've never seen training aids like these: baseball,

181

surfing, mountain biking, track and field, basketball, even ice skating!" Jay pointed around the room as he scanned two boys using the sports training devices."

Jay had long past the nervousness of being nude, but seeing so many boys with long penises had him feeling a little under-developed. Wendy certainly didn't give it consideration and the two jumped into a two-seater jet fighter and had an air fight against another aircraft. Jay was too busy laughing to remember what he'd thought a minute ago. To this fourteen-year old, this was a boy's Disneyland, and he could spend days at a time and never require food or drink.

A few minutes of other cyberspace devices and the two ran over and watched a boy jog on a pad with a pair of virtual reality glasses on, and then leap up into a Fosbury Flop over an imaginary high jump bar into a soft pit of foam below the bounding sensor. The bar stayed balanced in the display on the screen. Jay and Wendy clapped.

"That is so cool!" Jay said and was invited to try.

Jay wasn't going to refuse and began to jog in place on the pad, only to feel when the time came the pad accelerated his body upward to where he practically did a flip in the air. There was laughter and fun from the other boy who explained the mechanics of a flip, helping Jay spread his body out into a backward position.

The youngest Franco also wasn't used to having other boys put their hands on his nudity, but in no time he felt a whole lot more comfortable and was able to clear 5' 6" on the screen's display. The boy played a replay of his jump and Jay realized that Wendy had pretty much got a great view of everything he possessed, from flapping balls to an exposed anus.

Another machine had a mechanical swimmer motion. Jay said he'd have to tell Travis about this. "So this is how you learn so many skills," Jay said with a smile and was glad he'd missed the bus.

"My parents know these type of games produce distinctive psychological features while helping us learn physical techniques. They reduce or alter sensory experience, the opportunity for identity, flexibility and anonymity, the equalization of social status, the transcending of spatial boundaries, the stretching and condensation of time, the ability to access numerous relationships, the capacity to record permanent records of one's experience, and the "disinhibition effect", which makes kids feel free to engage in conduct that they might be afraid to try in real life. I know this because I did an assignment on it last year. Pick your sport and there's a machine for you somewhere in here."

"I want to try golf. My father plays it and has taken us to the country club a few times," Jay said as a confidence booster.

Spirit was quite willing to help Jay fit the headset on, and Jay was pleasantly surprised to realize this course was on the first tee at St. Andrews golf course. A ball was quite visible below him on a tee, as he used a wood and swung at this object. The ball immediately dove left and caromed more sideways than it did straight down the fairway.

"What happened?" Jay asked in awareness that his first drive was in the deep rough.

"You sliced it, Jay," Spirit told him. "Here, I'm putting on the Coach for you. Pay attention and do exactly as you're told.

An instructor began to give commands and Jay followed the advice with exceptional obedience. Slowly he swung back and corrected his left arm, balance, knee and hip positions and kept his head still. As he swung through, the driver hit the ball about 150 yards straight down the fairway.

"Much better," Spirit complimented and then had Jay remove his headset to watch himself on video replay. Jay saw a marked improvement in how his swing had improved from the previous one. "I could have you beating your father in no time."

"Boy would I like that!" Jay replied with new inspiration of beating his father in something.

The youngster enjoyed going from one game to the other: hitting a baseball pitched by a major leaguer, ice skating against a NHL player, mountain biking down steep paths, and using a light saber against Darth Vader. Jay was sure he'd lost his arm until he swept off his headset and was glad he still had his limbs. A figure of an adult appeared right beside him and Wendy.

"Your time is up, Jay and Wendy." The image disappeared

immediately.

"What was that?! I thought it was real," Jay said.

"It's a telepresence, the illusion of another person being physically present."

"Then we don't have to go if it's only make believe."

"Not quite, my cute boy. There are repercussions for even spending one more minute in this room," Wendy informed without explanation.

They stopped to check what was playing in the theatre. Jay saw a few boys kissing and another pair head to head. Other boys were sprawled out on top of stuffed animals that appeared to be real comfortable. One boy was lying on top of another, his head propped upon the other's head and a slow grinding action at the top boy's rear proved the penetration was pleasurable to both.

"That doesn't look like a G-rated movie to me," Jay said seeing nudity on screen.

"The Dreamers," Wendy said. "I've seen it but we can stay if you want."

Jay's eyes strayed from the screen. "Whoa! These guys are having their own fun. Your parents let them do that?"

"Of course. You don't have to worry about getting your shoelaces tied together while watching a movie. My father and Dr. Marion don't allow us to watch violence or horror, but sex is okay. You've never seen boys having sex before?"

He broke up with a giggle. "I…uh….they sure know how to do it, don't they?"

"When you live around boys who have the same sexual interest it's the norm, trust me. I feel like odd man out, but they allow me to intrude at times."

"Odd woman out, don't you mean?"

In an hour Jay was beginning to expect to see another two boys in some heated excitement as he turned the corner. Even Maria was now accustomed to not being shocked with this expression of male sexuality.

Watching two boys give fellatio beside the pool had Niqui whispering, "Mama told me not to come." Niqui laughed at his own remark.

Jacob overheard and replied, "Mama just didn't want you to experience pleasure any earlier than she learned. It's a parent thing; envious, I believe is the correct word."

Jacob elaborated as the boys continued to tour. "This is our way of life here, Niqui. Loving others is a profound way to add love to our lives. Just as important, it makes anger, hostility, resentments and stress disappear. Loving not only brightens each day and makes us feel good about ourselves, it also makes others naturally want to return that love to each other."

"Aren't you self-conscious about what other boys think?" Niqui asked.

"Niqui, we've never known anything else but being ourselves without being judged by anyone. There isn't a boy here I haven't slept or had sex with. Dad taught us when we were nine the importance of hygiene and about lubrication. If you use our bathroom you'll see a cabinet with enemas and tubes of lubrication. We each had to give someone an enema and, in return, get one from them. The lesson was in caring for our brothers and that it's more fun when someone assists us than when we do it ourselves. When you grow up feeling comfortable in your skin, what's there to fear?"

"I guess those who aren't gay," Niqui answered.

"I don't understand," Jacob said honestly.

"I guess you wouldn't, if this is what you're used to," Niqui replied. "But, aren't women better?"

A grin creased Jacob's lips. "A choice of a partner in a sexual relation becomes more significant only because society demands that there be a particular choice in this matter. When you make love to yourself, do you not love yourself in your thoughts and feelings?"

"I guess so, though I think about sexy things, sometimes girls."

"We are attracted to people who make us laugh and who are interesting, no matter what sex they are. Sex is mostly in the mind, so that which you think you have to love only restricts your possibilities. We can choose your responses to everything in life. Make a conscious effort to feel love toward yourself and everyone you meet. But think of

185

yourself as an actor or athlete who is rehearsing or practicing his skill. When you feel love, you allow joy and a sense of fulfillment into your life—and into others' lives as well."

"That makes sense. Accepting and loving ourselves allows us to reach out and love others. I'll have to think on the boy part, though." Niqui smiled again at his statement and felt good that he was able to have a lot more fun than he did during the day.

He was completely comfortable with his nudity now, even after seeing his sister a few times. Travis puzzled him by holding hands with that other boy, and watching his brother kiss a male had him even more curious as to what his older brother thought about all this.

"How do you want to change the world, Niqui?" Jacob asked.

"Wow!" No one had ever asked him that, not even a teacher. Niqui thought about his answer. "I really dislike politics, so don't get me started. I'm a confirmed peacenik who would probably find fault with any war. I write poetry and cry at the end of sad movies, so I guess I would like everyone to do the same. I'd have my own room so I could get naked more often and jump on my bed. I want to understand if I'm really part of God, like Bobby says I am, then I'd promote peace and harmony."

Jacob was impressed with the boy's honesty. "We don't believe in war here either. My fathers said we'd never be in the military because they give medals to men who kill men, but if you love one they don't want you. Why would I want to serve a country who didn't respect me for who I am?" He got a nod from Niqui.

They put their feet in the side of the pool and touched elbows.

"How about you, Jacob? What would you do if you had power?"

"I want to destroy religion as we know it and have people really know God. I want people to love each other and think independently, to share and experience to help God know abundant life. I want to experience sex as a way of not proving something, a way of being non-competitive, a means to eradicate fears, and a source of love in not-controlling, but sharing myself with others. I want to write poetry that can make people laugh and cry. Maybe make films about boys in love and how this love can bring happiness to others."

"That's beautiful," Niqui said.   "Do you guys really always have this much fun here?"

"See Alex, there?   He's our comical jester, a practical joker who you have to watch out for.   He was the first boy to reach puberty and was more than happy to exhibit his new talent of ejaculation for everyone to envy.   Alex would lay spread eagled on top of the water slide and yell, 'Omnia vincit amor!' Those words mean, Love conquers all!   It was one thing to masturbate at the top of our hundred-foot slide, quite another to slide down on his knees, backward.   When his twelve-year-old legs began to plane on the water it flipped him over and he fell thirty-feet down into six feet of water.   The fall was so spectacular even our lifeguard thought sure Alex would have nothing less than a broken neck.

"When we ran with tears in our eyes toward Alex and pulled him from the water, all the boy said was, 'Was that the coolest exit, or what?!'   His smile didn't last when our lifeguard escorted him to Dad's office."

Niqui was laughing but wanted to know the repercussions.

"It's not that Alex didn't know that it was against the rules to slide down on his knees, but Alex said there was nothing in the rules about going backwards.   Dad's hand print was on Alex's rear end for three days.   So, yes, we have a lot of fun with each other and there's never a dull moment."

The boys found this amusing together and were telling each other stories of their past and the fun times of being caught.   Their heads arched back and hiccups started to develop.

"Do your fathers allow you to watch television?   There are a lot of shows that are, sort of, homophobic, if you get my meaning."

"Sometimes people who act like that are just uncomfortable and afraid that they're one of us.   When we were kids we watched a whole lot of gay shows: Bert 'n' Ernie from Sesame Street, Teletubby, Tinky-Winky, and SpongeBob.   To us they were all gay," Jacob said with a smile.

"Yeah, I got that same impression watching those shows.   I was always suspicious of those guys, but when you're a kid you don't know any better," Niqui said with a smile. "I wonder why they never put any lesbians as characters?"

"Oh, but they have," Jacob assured. "It's interesting you say that. You're right, you didn't know any different because you accepted people as they were--no judgments, no prejudice, no decisions based on hatred or biases. You, as a child, were as God wanted you to be, free of mind and open for love. I don't mean to preach, but in class we are taught to defend our positions. Sometimes our teachers make us take the opposing viewpoint to gain a different perspective. We can't do it with humor or sarcasm. You can imagine having to debate boys like Bobby or Toni. Oops, I guess you haven't met them yet. Anyway, about lesbians," Jacob laughed with Niqui.

"Did you know that Peppermint Patty from the Peanut comic strip, was clearly in love with Lucy? Marcie was hot for Peppermint Patty, always calling her "Sir". And then there's Velma from Scooby-Doo. Her hairstyle is really butch, and she always wears sensible clothes and shoes. Our favorite was Spin and Marty from the old Mickey Mouse Club. Dad told us that Marty got caught making out with another boy when he was fourteen, and then Walt Disney fired him when his parents went to Walt for some advice. Anyway, we all loved Marty after that."

"I have some really great material next time I do my poetry," Niqui said a grin.

Jacob ran to get pen and paper in hopes of doing a poem with Niqui right there and then. His love for tennis initiated the poem, as the two boys sat by the side of the pool and Jacob wrote the first lines:

Flushing Meadows
I parked and walked:
Ticket, paid in cash
I saw a unique statue of Arthur Ashe
Interesting pose
I had no idea he didn't wear clothes

Niqui helped with the last line and roared with Jacob at the outcome. Their faces a few inches apart, Jacob kissed Niqui on the lips. Niqui paused, thought for a second, and decided it was okay. "Let's do another verse," Niqui said excitedly.

188

Ladies and Gentlemen!
The announcer shouted
It's quite a show, you know
The entering of foes—
Majestic, faces of truth
Inspired physiques sculptured in youth
Tennis bags, towels, sport lutes
Gratis autographs and photographs
Omnipotent salutes

Sunken chests, lean arms, firm glutes
Colored headbands, white shirts and shorts
Skintight bodysuits
Cheeks of brown, legs tanned, faces mellow
In hand, two fuzzies—yellow
I'm scopin'
Here at the U.S. Open

The boys drew back and smiled at each other with their fine work.
"You are going to teach me tennis for helping you with this, right?"
"I promise you'll be the best player at St. Bonaventure," Jacob said.
"One more verse, okay?" Niqui pleaded.
To Stadium Court I moved in haste
I drank a Corona, watching Sharipova
Playing Serena from Covina
Bodysuit in black
Every male longs for her striptease act
Suddenly, a bare body of pink, a running freak
Why is it always the geek who wants to streak?
Where's the physique?!
Serena, rather pale,
Not much into males
"Listen pal! You better bail!"
Serena yells—a trifle perturbed
Maria's sayin', "This is far more fun

than Wimbledon!"
With her critique of this hasty streak
Maria's leer assesses a cute rear
When this streaker finally fled,
Serena hit this overhead,
Topspin, for a winner
Maria's foamin',
Here at the U.S. Open

Another uproarious crack up had the boys give a high five to each other.   Jacob laughed so hard he was on his back, then felt the lips of Niqui pressed against his for a mini-second.

"You're wonderful.   Thanks for your help," Jacob said and stared upward into Niqui's eyes.   Sitting up he noticed that Niqui's penis was enlarging quickly.

Niqui hadn't noticed this until he saw Jacob's eyes.   He slipped into the water just as several boys asked the two if they'd like to play water basketball.   The game was far more fun than Niqui had expected, the slipping of bodies and laughter made the sport not about scoring but who had the most fun.   Even in the midst of play Niqui was examining the maleness around him.   Physically he measured up to every boy there, but he also realized what was missing in his heart—the love of life.   These boys expressed a joy, a contentment with just being alive and around each other.   The boy guarding him, Dane, was exuberant with every expression, every move evoked caring.   A shove or an accidental elbow received an apology from the boy, as if empathy was the framework for every action.   What Niqui came to realize he had begun to idolize this gift from God for this accidental evening of being able to stay at the Kings Academy.   He'd never kissed anyone before and here he kissed a boy.   It didn't make sense, yet he was happy.

Alex and Travis were paged to the dorm office.   Alex gulped and knew what this was all about.   When they entered, Dr. Kerho was laid back on a couch without a stitch of clothes on.   It sure had Travis smile.

"My father should take a lesson on this type of lecture," Travis humored.

"Isn't it great that we see adults as vulnerable as we are with our clothes off?   It's a dorm rule that no adult enters without equality," Alex said.

"I still wouldn't trade my wisdom for that thing between your legs," Dr. Kerho kidded.

"My father puts things in perspective.   Don't sweat the small stuff," Alex assured his friend.

Dr. Kerho was smiling with the arrival of two handsome youth and teenage wit.   "I hope you're enjoying your evening with us, Travis."

"Yes, sir.   Do you think I could be considered for your academy?"

There was an understanding smile from the director.   "Would you believe there have been times a few of my own boys would have loved to trade places with you?"

"That would be difficult to believe, sir.   I get almost all A's and I don't eat much," Travis humored.

"We could use a pet, Dad," Alex said.   The three of them had their chuckle.

"I'm honored you think so highly of our academy, Travis.   Our curriculum here is in contrast to many parents' expectations, I'm afraid. If you were ours to raise perhaps the adjustment could be made. Knowledge does not necessarily help the boy to discover who he really is.   Sadly, schools don't teach what it means to be honest or responsible. They don't teach what it means to be aware of other people's feelings and respectful of others' paths.   Your parents teach the same stuff their parents taught them.   Mistakes compounded into more mistakes.   Each person is responsible for detaching themselves from parents' prejudices, preconceived notions and, possibly values."

"Did I do something wrong, sir?"   Travis asked.

"No, you didn't.   I was just rambling."

"Part of being a parent," Alex said softly.

"I heard that young man."

The boys laughed with Dr. Marion's smile.   "I'm sure you are wondering why I allow my boys such freedom to create their own lives. Here we teach our boys to embrace their own sexual feelings, curiosities and urges, which, we hope, might cause them to connect this new and expanding experience of themselves with an inner sense of joy and

191

happiness, not guilt or shame.   There's a Bible verse that says, what you condemn, you will become.   Be careful not to mock those you feel are different, only to become one of these in your next life."

"Dr. Kerho, I am one of those.   I mean, gay teens.   I really like Alex and want to take him home if I can't stay here."

Dr. Kerho chuckled.   "That does present a conundrum, of sorts. The purpose of all your relationships with every other person, place, or thing is not to figure out what they want or need, but what you require or desire now in order to grow, in order to be who you want to be.   This is the act of God being God.   It is Him being Him—through you. Through you, He experiences being Who and What He is.   Pretty cool, huh?   So, am I refusing God here?"

"I was just kidding about taking Alex home, sir, but I do love him. If I can come back to visit, I'd really like that."

"That's a given, Travis.   I only wanted to make sure everyone is adjusting and there was no panic to get away from this madhouse."

"My sister and brothers like this place as much as I do.   Thanks for allowing us stay."

"Get your butts out there and have fun.   I only came over to enjoy a Jacuzzi.   You guys were just a good excuse."

****

Eventually Lance and Maria made it to the arcade.   They'd kissed on the beach for several minutes and Maria loved her moment of exhibitionism and feeling an erect penis from Lance poking into her hip. She wasn't sure how far she wanted this to go, only that she was as heated as she'd ever felt as a young woman with budding hormones. Though she had promised herself to wait until marriage to lose her virginity, her body had other ideas and swiveled until she felt this hardened member press next to her vagina.   Seconds away from whispering, 'Fuck me,' words she'd never used in her life, it was Lance who simply released this bond and stood up.

"Let's go to the arcade," Lance suggested, his erection stood proudly pointed toward the ceiling and then swayed back and forth as he helped Maria to her feet.   It was the most beautiful thing Maria had ever witnessed. She expected the other boys to make fun of this, but they just all took it stride without a second glance.

Scattered around each room were various lofts for the boys to read or relax with food or beverage from the juice bar. The unique and seemingly real stuffed animals of various sizes and shapes had Maria laughing. They sure put her miniature dolls and Winnie the Pooh collection to shame.

Maria's eyes caught one of the boys sitting back on this huge grizzly bear, a cub, half its mother's size, which was snuggled up next to his mother. If one didn't know better it was easy to think that the bear would growl any second. The boy was sitting back on the bear's rump, with his left leg rested on the cub and his right knee bent upward to reveal most everything male. Lance noticed his friend in this state of relaxation and took his first girlfriend in his life to introduce her to Tomas.

"Hi! Pleasure to meet you," Tomas greeted while peeking over the top of this paperback. He lowered the book to his waist.

Maria was all engulfed with this naked youth before her. She pretended to read the cover of the book, but her vision was on a rather extended penis hung over the boy's hip like a pink floppy ear and the exposed anus below his testicles. The testicles hung loosely and Maria noticed they weren't really balls at all, like she'd heard girls describe them at school. They appeared more like eggs hung loosely in the sack.

"**Man Without a Face**," Tomas said in case the girl was having trouble reading the cover.

"Oh. What's it about?" Maria asked and switched back and forth from looking at Tomas, to the book, back to Tomas, though her eyes couldn't wait to lower and inspect the limp penis.

While the boy was explaining the plot of this novel, Maria was totally intrigued with the underside of a penis, its meaty urethra and assorted veins crisscrossing this lengthy tube. For some reason she remembered her giddiness a few days earlier when Travis was working on their father's car and she had reclined on her back to slide in under the frame beside her brother. She'd never seen the underbelly of a car and Maria laughed with all the mechanics Travis tried to explain that was part of this undercarriage. This nudity was far more blatant and far more revealing than anything Maria could imagine.

She had seen her brothers naked, at least two of them, and held this secret to reveal only to her closest friends at school. That secret didn't seem so important now. Babysitting a set of brothers, seven and nine, and bathing them, had given her an education to boys' erections and their bodies, but this was the adult version. She found that the head of the penis was like a turtle's head, an art form she found really cute. It was when the book moved that she broke the spell of concentration.

"…..and this teacher saw in the boy a projection of his own desolate longings for ideal beauty and friendship, but was too afraid to act on these feelings because of the pain and suffering he'd endured before," Tomas finished.

"That's so beautiful," Maria replied, but was really praising the boy's sex. She tried to remember what the boy said about the book. "So this teacher……what did he end up doing?"

Tomas answered unaware that Maria hadn't paid attention. "It was bound to happen. I mean the boy wanted this guy to make love to him, explore his body and validate what he had felt in himself. They finally made out in bed, but neither could deal with what it all meant to both of them by the next morning."

"You okay, Maria?" Lance asked as he glanced at her facial unease.

Maria had an overwhelming sense of hotness, a radiating glow in her body. When she glanced at Lance, he had a worried expression and was examining her face, which were felt flush and very red. "I don't know it's just so hot in here."

She subconsciously put her fingers to her clitoris, but that's when this radiance shot a jolt of pleasure through her loins that made her knees shake, her limbs tremble, and her whole body near collapse, until both Lance and Tomas leaped to catch her.

Slumped to her knees, Maria lifted a pair of droopy eyes and barely spoke, "I'm okay. I'm okay."

As if inebriated she put her hand around Lance's head to kiss him longingly. Embarrassed and exasperated, Maria stood up with the assistance of the boys and collected her equilibrium. Lance and Tomas had the most curious expressions on their faces.

"I'm so sorry, guys, but I think I just had my first orgasm," Maria said honestly.

They laughed together and quietly wondered why a girl would just have discovered what an orgasm was. Tomas suggested to Maria that she sit on this grizzly bear, and Maria complied with the same sitting position that Tomas had prior, picked up the book and began to read to the boys. The wetness between her legs was quite evident, but this was her way to show she was just one of the guys now. The humor relaxed an awkward moment and Maria felt a quick relief and acceptance after her loss of control.

Another boy had raced to bring Maria a cup of water. She felt rather embarrassed over this whole scene.

A few seconds later another boy called Tomas to compete against him in a video game. Tomas agreed and glanced at Lance. "We on for fencing tomorrow?"

"Probably be tomorrow night, Tomas. If that's okay?"

"Last time you stood me up, remember?"

Lance reached for Tomas's penis and held it. "I promise."

Tomas returned the gesture and darted away.

"What was that all about?" Maria asked.

"Just something we've done for years as a way to show we mean what we say. See, Rev. Harris, he's our minister here, gave a sermon several years ago about what they did in biblical times. Genesis states that Abraham ordered his servant Eliezer, 'Put your hand under my thigh and…swear, by the Lord.' In those times the word thigh was used to mean penis, so men grabbed each other's penis as a means to give oath. Its where our word testify comes from and the same reason we swear on a Bible before we give oath in court. So we all use it here because we're usually naked and it's always a promise to uphold, though we are very honest here anyway."

"That's so funny," Maria admitted and loved it when Lance sat on the fluffy bear to where they were practically sex to sex.

Through an archway from the arcade, Jay and Wendy were at the juice bar when Jacob and Niqui came in, their broad grins and arms around each other's shoulders was a pleasing sight for Jay.

"I want to see you at confession tomorrow," Niqui said with a straight face and had Jay worried. "Just kidding, bro. We can skip our appointment with our divinely anointed male leader who rules his flock of obedient and submissive sheep. I think I'm not being brainwashed anymore."

"Are you my brother, Niqui?" Jay asked and got a punch on the shoulder.

"What's next?" Niqui asked Jacob and the two of them ran to the dance floor when they heard the music.

It was pleasing to Niqui's eyes to see Dane on the floor; his talents, litheness, posture and pose were beyond anything the boy had witnessed at plays or dances at St. Bonaventure. The other boys encouraged and praised the performance, and then Dane, sweaty and breathing hard, came up to Niqui and asked him if he danced.

Niqui chuckled. "I danced a Spanish number in an elementary school play once. Without castanets I'm worthless."

"Someone go get the maracas and castanets!" Dane yelled, and within minutes these musical instruments were available.

A mere press on the wall's panel and a Spanish number commenced; the dance floor was vacated for a guest with amazing Spanish countenance. Niqui caught the rhythm of the Spanish piece, and all the practice he'd done in sixth grade came flowing back like a speech he'd learned and put aside until the right time. The pair of small, hollowed pieces of hard wood clicked together in his hands in synch with the music. His legs had a memory of their own, as did his arms.

With the smiles and clapping of a dozen naked boys Niqui flowed with delirium to the Spanish beat, the rhythm flowed through his limbs and his eyes fell on a boy in the front row, encouraging him the most-- Dane. The expression was more than delight or intrigue; it was one of longing, like one soul finding the other. From that moment on Niqui danced for Dane alone. When the number was finished the boys did more than clap, they rushed and held Niqui aloft to praise this guest for his courageousness in allowing his body to express his heritage.

196

"Initiation into the Company of Kings!" Dane shouted and they all sounded like a squadron of young marine recruits in their unity of brotherhood.

As one of the boys ran out of the dance room, the other boys humored Niqui by spreading his body out on the floor. Too dizzy to contemplate this fun Niqui laughed and enjoyed this moment of peer acceptance. He never thought his Spanish heritage and culture would be so appreciated. A feeling of coldness swept over his groin, with the sound of carbonation erupting from a can of whipped cream. A minute later, and the wetness of a towel, Niqui stood up and scanned his hairless groin.

"Now you are a sexy Spanish conquistador," Dane told him and the boys patted Niqui on the back.

Niqui admired his lack of pubescence, then wiggled his hips to the beat of maracas. "I am in the company of kings," Niqui said in appreciation. He had no idea how close to the truth he was.

Dane bent forward and kissed Niqui, before he told this boy that they'd never had a Spanish dancer before. Niqui's eyes melted to the words, his lack of control dictated his instant attraction.

Dane gave his peer a gift of pleasure by dropping to his knees and engulfing this sign of arousal. Niqui almost panicked and swept his eyes around to see if other boys were laughing or making fun at this blatant display of gay sex. He was relieved when no one seemed to really pay attention or care. His worry about Jacob being jealous was forgotten when he saw Jacob on the dance floor with another boy doing the jitterbug to the laughter of other boys. A high Niqui had never felt was only increased by an orgasm, which was drained by the mouth of Dane.

Niqui assisted Dane to a standing position, with a smile from this boy he'd only known for a few minutes. His brain was fuzzy in what he should feel: Repulsed? Invigorated? Honored? Niqui tossed these thoughts with the realization that he was having the best time of his life; yet, he didn't know why.

A final kiss and Dane said, "I'd like to show you how you can improve your ring routine tomorrow morning, if you like."

Niqui nodded as his body was only then recovering from the jolts of release.

"I see you two have met," Jacob announced when he leaped off the elevated dance floor. "The dance floor is where we usually end up after video games or other activities."

Niqui gained around and verified the sheer number had multiplied.

"Come on, I want to show you our library," Jacob suggested. "If it's okay with you?"

Niqui was in a trance and hoped Dane would come with them. When he left the dance room, his erection had yet to recover, the world still spun and his senses had a need to come back to earth.

The library was the perfect setting, with so many books and computer terminals to find any subject or category available. Niqui paraded down the shelves of poetry books: Keating, Hawthorne, Whitman, Lord Byron, and Owens. He read a few passages from each book and wondered if he'd discovered heaven in the last three hours. Another shelf had Oscar Wilde, Herman Melville and any other author Niqui had heard about and some he didn't. On the wall were paintings by Thomas Eakins; The Swimming Hole defined everything right about the night. A picture book by David Hamilton had Niqui's perusal, a book he would have given his right arm for a day earlier. Now he looked at young nubile girls with subjective tastes and admiration and wondered if he would feel as horny with them as he did with these boys here.

The two young men spent an hour in their research, and then Jacob announced that it was getting close to one and they had to return to their rooms. A shower preceded their trip to the bathroom where Niqui found he didn't need a toothbrush. He placed his mouth over a disposable rubber piece called an Oral-cleanser and discovered that his teeth were being cleaned automatically. A reading that he had no cavities was a welcome relief. His brother Jay wouldn't be so lucky. Niqui saw the cabinet filled with hygienic items and toys that were definitely for sex. He reached for a small clear tube with a rubber nozzle. A quick step to a stall and he gave himself his first enema to clear his rectum. To Niqui's rationale, he just wanted to see what it felt like.

"Waaahooo!" Niqui shouted when he pressed on a computer pad by the commode.

Jacob came around the curved glass divider to where Niqui was sitting on the toilet with the biggest smile on his face. "Experimenting, are we?"

"Yeah. But I wasn't prepared for that spray of water up my ass," Niqui said.

"That's a bidet. The bowl not only cleans you, but sanitizes with a disinfectant, and then dries you with a ultra-violet light and heat. With everyone naked around here it's important we have the best hygiene."

As the two boys passed the cabinet again on the way to their rooms, Niqui reached for a tube of KY Jelly. It sounded better than saliva.

Jacob walked his new friend to the guest bedroom, where Niqui asked Jacob if he wanted to come in. They lay on the futon talking about the competition and the evening. Jacob touched the pad by the bed and the wall turned into a real life scene of being on a South Pacific beach. The smell of salt and sand, with the aroma of actually being by the waves, had Niqui believing they really were by the ocean.

"Thanks for everything," Niqui said in a moment of tranquility, then kissed Jacob on the lips with his new confidence. He swung his hand to his front and showed his new friend the tube of KY Jelly. Niqui really thought Jacob was kind and handsome, though Dane had struck a chord in his heart.

What started was a half-hour of sexual play with Niqui discovering how his body responded to the touch of another boy. Jacob's tongue explored the most intimate parts of Niqui, before he sat over the boy's waist and lowered his own passage onto the dripping manhood of a more than receptive Spaniard. Two orgasms each and they separated. In their relaxed state, Jacob reached out to hold Niqui's hand, but received a rejection.

"Don't!" Niqui rebuked. Guilt swept over his state of sexual release, a dread of shame that his religion had so indoctrinated within him that he should have never experienced such feelings of love and appreciation.

"Tell me what you're feeling, Niqui. Is it guilt from something you did, or shame for how it made you feel?" Jacob asked.

"How did you know that?" Niqui asked.

"You have been exploited and coerced to believe in a certain morality for years. It can't be easy to change."

Niqui quoted a touch of Latin that he remembered. "Credo quia absurdum—I believe because it is absurd." They both laughed. "Dumb Catholic upbringing!"

Jacob thought for a second and reached over to press on music. "It's interesting how man feels most secure when others do their thinking and tell them how to behave. Is it that man needs lies? Moral life is reduced in their ideology to personal, individual piety. Their conditioning of children to fear nonconformity and blindly obey ensures continued obedience as adults. The difficult task of learning how to make moral choices, how to accept personal responsibility, how to deal with the chaos of human life is handed over to God-like authority figures. Our minister here preaches that to us every Sunday."

"Why haven't I seen this all before?" Niqui asked.

"Because you're a kid, and kids just do what they're told. Our parents here at the academy give us free will and the knowledge of how our decisions affect others. Dad has told us for years about other children, but we've never had a chance to experience other boys until we started to compete with other schools. Other boys and girls we've met seem so unsure, so tentative about allowing themselves to feel and explore their needs. It must be cruel not to be who you really are."

Niqui thought about this and something Bobby had told him. "If we're really God, per se, what would He feel about this?"

"God's body is full of pleasure and gifts for you to enjoy. I dare say he'd be embarrassed if you made him feel shameful and guilty for something that He created. Is it not in your nature to express your desires and feelings?"

Niqui agreed but remained taciturn. He listened to the music Jacob had picked: Foreigner. The number was **I Want to Know What Love Is**, which gave food to thought for a boy searching his mind. "I think you know," Niqui said and confused Jacob.

Jacob smiled in the darkness when he figured what those words meant. "I have my favorite story in mind. It came from our minister here; I know you'd like him if you ever want to speak with him. It's about Huck Finn, the time he was in trouble and didn't know what to do with an escaped slave, his friend Jim. He knew what was expected and wrote a letter to Miss Watson about where to locate Jim. After he had written the letter he'd felt good and all washed clean of sin and had honored the expectations of those who made the rules and detested slaves. He thought about how he'd almost lost his soul and gone to hell for wanting to protect a black man. Then Huck thought how he and Jim had spent many a night, floating down the river, their talks, their laughter, their singing. How Jim had always been there as a friend and shared what they each felt and thought. How Jim had caressed and taken care of him, and they'd escaped trouble several times by working together. Jim was the one person Huck could count on and was always there. So Huck looked around and saw the paper he'd written, held it in his hand and knew a decision of right and wrong had to be made. Short of holding his breath, he said to himself: "All right, then, I'll go to hell"—and tore it up."

Jacob didn't see Niqui's tears, but he felt the boy's head land on his chest, a right arm swept over his body and the wetness sliding down his rib cage. In seconds they were both asleep.

Travis had never felt more comfortable with a person in his life. He had personally invited Alex to sleep with him, and they explored every aspect to increase their knowing of the other's body. Travis's first gay experience far exceeded everything that he'd fantasized and dreamed about with other boys. He had no idea that his body had so many reactions and pleasurable spots. The reversal was as exciting as the passive role. Travis made sure to return every bit of caring and touch that Alex had brought him to such heights.

Alex had switched on the telescope relay from the observatory. Above the boys was a world they could only imagine in the heavens. Occasionally a shooting star could be seen which made their love even more real and blessed. After their fourth orgasm and wondering if they should go for a fifth, they fell asleep in each other's arms.

Maria was not going to allow this night to escape her memory or this opportunity. She'd met, according to her, the man of her dreams. All the fun and excitement of the evening was trivial to her aims, the moment of her union with this gorgeous male climaxed her goal. She felt like a woman and delighted in her second orgasm of her young life. Lance had been as gentle and caring as Maria dreamed of her first man in being. He knew how to caress and bring a female to the height of desire, of almost pleading to be entered and loved.

Maria had no clue that a male's mouth could bring her sex to such heights, and she had returned this love technique by discovering how to pleasure a penis. With a little instruction of how to position her teeth and use her tongue, she was an expert in fellatio by the time she was done and had Lance begging to unite them in intercourse.

What made this so special to her, this was Lance's first intercourse with a woman. To discover that Maria was a virgin gave special meaning to this first love for both; at least it was the first time they had both experienced the opposite sex with sexual intercourse. Lance was not sure that a girl could take his full length, but Maria coaxed the boy with every stroke to penetrate her fully and her words were music to this boy's ears.

Jay and Wendy had stayed together because they wanted this night to last forever. Jay had been passive, which was his nature, not sure of his next move or how to act around a girl. He had only felt this comfortable around other boys, yet envied their talk about girlfriends or things they did—most of which he knew were lies and exaggerations. When Wendy began to read Shakespeare he just knew he'd be bored. Instead he was engulfed in the mood and humored with the words. He lay on the bed naked next to her, and even Wendy's nudity had long been absorbed in his brain.

"This is a scene where Romeo kisses Juliet. I'll finish the words, then we'll kiss," Wendy said as a choreographer to assist a boy who wasn't very assertive. She touched the wall panel and they sprawled out on a meadow of grass on a mountain slope.

"Who's going to play Romeo?" Jay asked very innocently.

Wendy giggled. "You are, silly. Okay, here goes, good-night, good-night! Parting is such sweet sorrow, that I say good-night till it be morrow. Sleep dwell upon thine eyes, peace in thy breast! Would I were sleep and peace, so sweet to rest!"

Wendy turned on her side and closed her eyes. Jay's kiss was warm and sweet and short, then Wendy's arms wrapped around this young Spaniard's body, which was as virgin as her own. She pressed her lips against his and taught him all she knew about the art of kissing. Her compliment on his kissing was all it took. His hardness was pressed against her thigh and her experience with boys took over.

A mere shift of the hips and their union had erased all this. Jay lasted less than six seconds. A few minutes of Jay relishing a woman's body with his hands and he lasted over two minutes on his second effort. Their words expressed all they felt and ever thought of feeling for just enough time to recover. Forty minutes later and his third orgasm at the same time as Wendy's, Jay now considered himself a connoisseur of making love to a woman. They spent the next thirty minutes removing a small patch of pubic hair so Jay would feel initiated into kingdom like his brothers.

Jay giggled the entire half-hour, then made love like a porn star.

****

At nine o'clock the next morning a robot came to each room and, with a soft voice of an affectionate grandmother, told the boys and two girls that it was time to rise and shine. The robot only stood 4 feet tall but its potential was mind boggling. Besides being able to vacuum, it possessed facial-recognition and announced each and every name.

Alex took direction and knew the protocol. "Travis, if you want to wait, no problem, but we have a routine every morning. You're welcome to join us, but it might not be something you're used to. We run about two miles, swim three-quarters of a mile across the lake, then run the last mile back to our dorm."

"Fine with me," Travis replied, not to be left out of anything.

The other Franco children felt the same way and followed these Kings' students down two chutes that became an exciting slide for those not used to it. Landing on soft mats, the group of thirty-seven naked teenagers were soon gathered together on this bright and sunny morning. The air was still brisk with a morning dew, a chill that would soon be forgotten with a sweat.

Wendy turned to all the Franco kids. "Don't try to keep pace with anyone. We've been doing this for eleven years." Though the guests nodded, they had no idea.

Running on a carpet of soft grass the woods surrounding the academy was soon filled with the pounding of bare feet and the sounds of lungs taking in air and expelling it forcefully. Thirty-three bare buns moved ahead of the Franco children who had tried to keep pace, only to find that their own conditioning was not even close to these teens.

Through the emerald-green path, lush with grass, flowers, and pine trees, running naked was an escape from the rigors of academy life where platoons of leprechauns would not have looked out of place. The Franco children ran as a family, their nudity breathed in fresh sweet air and the aroma of the great outdoors.

They had first settled in as a group, though Travis wanted to be a step ahead of his siblings. Coming to the lake all four of them were equally out of breath, their lungs gasped for air, their eyes stung with sweat, as their eyes caught sight on the school of naked bodies already halfway across the lake.

"Let's go swimming," Travis announced and the four of them dove in to test their swimming skills.

The last charge up the grass knoll was nearly a walk. Alex, Jacob, Lance, and Wendy were waiting at the top of the knoll, their recovery was as if they hadn't yet started. They greeted their guests with kisses and hugs, smiles of accomplishment accompanied this satisfaction of completion.

"You do this every day?" Niqui asked with his hands on his knees.

"Haven't missed a day in eleven years," Jacob said.

Everyone took a hot shower and adjourned to the cafeteria. For Maria this was way too weird. Here she was showering with dozens of naked boys, yet none of them outside of Jay gave her a stare. She stood in front of a mirror, like any morning, but to her left were five boys in front of urinals, another next to her with an Oral-cleanser, and everyone was naked. This was no time for vanity, that's for sure. One thing definitely stood out, and she wasn't thinking it was the boys' penises. She was the only one there with pubic hair. A simple request to Wendy, and Maria stepped into a stall and shaved herself perfectly smooth. Jay gave her a high-five when she stepped out.

Wendy and Lance invited the Franco children to a side room, filled with hundreds of kimono and yukata, the smaller version. Little did they know that several of these were worth thousands of dollars, yet they had been all gifts through the years from Miyato and Mr. Kamito. Many of these were glass encased, way too small for these teenagers because the boys had been wearing them since they were two. Patterns of flowers and grasses, insects, fish and shellfish, and Western landscapes, in addition to daring colors and delicate embroidery, made these garments to the level of art.

"You must each pick one for breakfast," Wendy instructed. "No one has their own, as we each wear something different each day for our meals."

Maria went first and found a print that had a delicate sense of beauty. Jay picked one with a playful spirit and creativity. Each fabric featured a variety of woven motifs not produced in years, including fine, delicate crepe, woven from threads of astonishing strength.

"Beautiful selection, Maria," Wendy complimented. She was an expert on these kimonos, having designed several herself.

"Miscanthus and butterfly patterns are embroidered on this Ojiya chijimi. This fabric was from the Momoyama period."

Wendy picked out an obi, made from jofu, which looked perfect around Maria's waist. It actually felt strange again not to be naked.

Jacob waited patiently and watched Niqui picked out a kimono with goldfish patterns. He helped tie the obi around Niqui's waist, showing the boy the proper method. A kiss was a thank you for his help.

"This sur-yuzen technique, a kind of kata-yuzen, was made during the Meiji period, when artificial dyes were introduced to Japan. This kimono is worth over ten thousand dollars," Jacob said.

"I can't eat in this," Niqui replied all flustered.

"Ah, but beauty within beauty deserves only the best. One should never insult their host." Jacob kissed the boy right back, took his hand and led him to the cafeteria that was more like an expensive restaurant with exquisite woodwork, tatami, and soft lights.

The other kids had politely waited and allowed their visitors to move to the front of the serving line. The serving options were so varied it had the Franco kids wide-eyed with what to select. Various fruits, some exotic, gave a multiple selection with whole grain cereals with nuts and dried fruit. Yogurt and English muffins had packs of honey alongside. Eggs and choices of protein were options. Quarts of orange juice replaced the small glasses the Franco kids were used to at home. By the time they sat down Travis wasn't sure he could eat everything he selected. Two cantaloupes and kiwi were appetizing enough, but his stomach spoke hunger after that morning wake up run and swim.

There was just something calm about eating with three dozen teenagers in various kimonos. Alex teased Travis with the selection of a beautiful landscape on his fabric.

"Your kimono was reserved strictly for the emperor's costume, worn at small ceremonies or on his formal visits. The brilliant feathers of the wild duck are an iridescent bluish green referred to as "tail green.""

"Is it okay to wear this?" Travis asked very concerned.

"At Kings Academy, emperors come in assorted sizes and handsome faces. I would have picked this one for you, as well." Alex had his penis squeezed for his compliment.

Jay was excited to find out about his kimono. Wendy held the boy's cheek in her hands. "You, my love, have chosen a pattern combination of willows and swallows which symbolize spring and a pattern of waves and swallows denotes autumn. In Japan, the swallow was often depicted in betrothal gifts as an auspicious bird, since a pair of swallows work together to raise their little ones. The swallow is considered love's messenger and a symbol of the raising of children. As much sperm as you pumped into me last night we may have many children."

Jay blushed, but his self-esteem went sky high when he knew his brothers had overheard this. "I'm a stud," he whispered to Niqui. Niqui thumped his younger sibling on the head.

The meal was interrupted by a page that Travis had a phone call. Everyone at the table listened as Travis took the cell phone from a Kings' student.

"Oh, hi Dad. What's up?.......Of course we're all behaving ourselves. Sorry about last night.............Yes, we had a great night. We went swimming and..........Where'd we get swimsuits? Ah, it's a boys' dorm, Dad. Not like we needed any...........Oh, yeah, well, Maria was in a different area...........Jay? Oh, he was hanging around Wendy...........Yeah, you know Jay and Niqui, all right. They're way too modest to go swimming like that..........Yes, I've not become friends with the boy who beat me and I barely see him. (Travis winked at Alex)........I don't know why Wendy likes my brother, but you're right, Dad, Jay is so naïve I'm surprised he's comfortable around a girl...........I'll make sure to keep an eye on Niqui, I know how shy he is and don't worry about Maria, she's doing her best to stay out of her brothers' way.......Yes, I'm keeping an eye on my sister and protecting her well-being. We'll be home in a few hours. Bye, Dad."

The Franco children laughed when their brother hung up and agreed to keep this overnighter a secret. The Kings' boys thought this father was really out of touch with kids their age.

After breakfast Jay found his presence being required to a room with a dentist chair. In minutes his cavity was repaired. Jay found humor about visiting a dentist in the nude. Wendy had taken the time to find Maria, and, without inquiry or judgment, she handed her friend a pill.

"Morning after, unless you're on the pill," Wendy said.

"No, I'm not. I couldn't bring up such a subject with my mother. I know I should be more careful, but we don't think of those things at the moment until it's too late," Maria apologized.

"That's why they call it the Morning After pill. My brother told me, so don't think I'm getting personal. He thinks he's in love with you. Don't all boys think through their penis?"

Maria laughed and was glad for the thoughtfulness. "My youngest brother is kind of shy, but you've done wonders for his confidence. I hope he wasn't a nuisance last night. He's really very innocent when it comes to sex."

"Nuisance? He's a little bunny when he gets going. Jay is so sweet I want to marry him right now, but I think my father wouldn't approve of such a commitment. He went from instant ejaculation to making love to me for fifteen minutes before he couldn't hold out any longer. Those were his words. He's so cute and honest."

"Jay had sex?" Maria had trouble believing her sweet, innocent brother had it in him.

"Four times in an hour. He's an animal. We woke up and he wanted one more time to prove that it wasn't a dream. He loves giving me pleasure orally." The girls laughed and Maria just had to tell Wendy what a great lover Lance was.

"He brought me to a height I've never experienced. Having a boy's cock in me was something I could only imagine what it would feel like. I want it every night," Maria chuckled. "I still can't believe my twin brother knew what to do."

"God made that one of the easiest things to do. For some reason a boy gets the hang of it real quick. Something I've long learned from my brothers, Jay loved having my finger up his butt when I gave him head. He's a born lover."

Wendy covered her mouth in her hilarity. "I must do that to Lance. He won't mind, will he?"

"My dear friend, my brother hangs out with twenty-nine gay boys every night. One of his favorite positions involves something far bigger than a finger."

"Three fingers, then," Maria said and the girls laughed.

Wendy and Alex soon escorted three of their guests to the dressing area and began wandering the grounds. Dane had found Niqui earlier and taken him to a trainer's room in the dorm where they also had had their massages. Dane asked the computer for the gymnastics' film from the competition. On a large LCD screen the boys watched the previous day's events while the automated system gave feedback. Niqui learned a great deal from the analysis and where he could improve.

"I love your suits," Niqui said and saw Dane move to a counter and pull out a pair of heat-sensitive shorts.

"Here try this on. It's very, very thin silicon—100 nanometers, to be exact, or 1/1000 the thickness of a human hair. It contains a transistor, but the key component is in the silk pad that covers your groin and the same hair-like rod that goes up your rectum. You're basically naked except for this covering. The silicon has just enough color to give it an impression you have something on."

After Niqui allowed Dane to coat his anus, Niqui stepped into the short and then removed the outer cover when the heating component was turned on. His groin and rear were now coated in Kings Academy red with no feeling that a sensor was internally placed.

"Cool! I feel naked, but so mobile," Niqui admitted admiring this suit.

"All I have to do is check the computer for your readout." They watched the screen and an assortment of physiological assessments began to come up. "Because you're not in a workout or competition your vitals and body composition remain calm. Your body hasn't recovered from yesterday. See? You need creatine monohydrate, L-glutamine, white willow bark, adaptogens, herbal insulinominmetic formulas, bifidobacteria, and mumie—it's an immune booster."

"Can I have a printout of all that?" Niqui asked.

"Increase your protein to 108 grams per day. You're a growing kid and you're having way too much sex with one of my brothers."

Niqui laughed. "Jacob is cool. He was really sensitive and gentle, considering I'm not gay and carry around a bunch of hang-ups. It's not like you couldn't have come over last night."

"I thought about it, but Jacob wanted you to himself and we respect each other around here. Tonight he'd love a three-some with you."

"I don't think we're staying another night, though I'd love to. I never thought I'd enjoy making out with a boy before."

"Straight or gay, you're a human being who deserves to feel pleasure. Here, let's take off the suit. This is a Kyros pad; it will get ice cold for a second."

Niqui watched his red suit turn white and then detach from his skin.

Dane peeled back the cracked layers and removed the probe. He was delighted to see the penis before him move straight up. A minute of attention took care of this need, and Niqui didn't panic with worry that someone would come in the room and catch them. A prolonged kiss had this Spaniard taste his own sperm. Niqui wanted to know if they could call each other after the Franco kids returned to San Antonio.

"Sure, I'll ask my father, but he's really good about that. My brother Trevor has a friend in St. Petersburg he calls all the time."

"You aren't going to tell him about us having sex, are you?"

"Niqui, our fathers know and condone all sexual behavior as long as it's consensual. We don't lie around here because there's nothing we have to keep a secret."

"Wish my family was like that," Niqui said.

****

Brittany had walked in on Niqui and Dane as they were in one of their passionate kisses, their groins pressed together. "Sorry to distract this training method but the kids' transportation home will be departing in fifteen minutes."

Niqui jerked back in shock, his erection very evident. What amazed him most was that this woman was totally naked herself.

"Thanks, Mom," Dane said and grabbed Niqui around his butt cheeks and pulled him back together again.

"Aren't you worried what she saw?!" Niqui was still panicky and unsure.

"Why would I be? See, any adult who comes into our dorm has to be unclothed. It's a rule. Mrs. Makawa is very accepting of all of us and, as far as we're concerned, she's our mother. Trust me she has her own kids and knows what a penis looks like; she even has a young boy who is gay. Her husband was boyhood friends with my father."

"This would really take a lot of getting used to," Niqui admitted and devoured his friend's mouth once again.

Niqui dressed hurriedly and was just in time for the tram that would take everyone to the front gate. Brittany was the chaperone for the day and Alex persuaded his adopted mother and father to allow him to ride down to San Antonio. As the tram passed the lake there were already a dozen boys surfing in the wave pool or windsurfing.

"I wish we could stay here forever," Jay said with a frown.

Using the academy's helicopter Brittany sat in back with the kids, though they were quiet and somewhat worn out by the last twenty-four hours. Mrs. Makawa asked how the kids had enjoyed their stay and competition as they lifted off from the ground.

"It was the most perfect night of my life. Even our bully got beat up," Maria joked.

"Wendy overreacted," Brittany admitted. "There is a point where self-preservation allows justification of force. But to damage another is to damage the Self, and the kids are all taught this at the academy. That is why children don't raise children or we would have a Lord of the Flies mentality. I ask you to remember that nothing can happen, and I repeat, nothing can occur in your life which is not a precisely perfect opportunity for you to heal something, create something, or experience something that you wish to heal, create, or experience in order to be who you really are. Wendy could have used a comical reaction, logic, even removed herself and the target of sarcasm from the focus. Instead she chose anger to save her the trouble of thought and wisdom."

Brittany waited for some response, then glanced backward. Every boy and girl were asleep.

Chapter Sixteen

Leif was offish with Sveinn when they arrived at the Elysium Hotel in Ft. Lauderdale.   Surrounded by gay men everywhere Sveinn glowed again with having such a beautiful male beside him.   Leif knew what was happening by now, didn't smile nor give credence to the fact that he even liked this guy.

"I'm sorry, sir," the desk clerk said.   "You have to be eighteen," as he pointed toward Leif.

"He is. Show him your passport, Leif."

Leif brought out the forged passport and the clerk eyed him closely.

"The boy doesn't even have one whisker.   If he goes swimming here they're going to think he's twelve-years old."

"Sir, trust me. The boy will bring in more customers than he will chase away," Sveinn assured the clerk. "Anyway, the boy had a father who was albino.   That explains the hairlessness."

Leif almost cracked up but for his precarious mood.   Like a scorned child Leif followed three steps behind and had no intention of giving in to Sveinn's romantic efforts.

Sveinn was at his wits end after a few hours of trying to make sense to a stubborn boy.   He finally said he was going to the pool and if Leif wanted to join him, so be it.   Leif pouted and didn't budge.

The teenager had only one focus, get out of there and find a ride to the nearest airport.   Now that he had a credit card he was Mr. Independent.   With the room on the second level he slid up the window and jumped the ten feet to the soft grass.   He landed in a crouched position, then checked both directions. Leif picked up his backpack and moved toward the front of the hotel. Barely on the sidewalk, a heavy hand fell on his shoulder.

Sveinn had an exasperating expression on his face.   "Okay, maybe I'm approaching this all wrong.   Leif, you're fifteen-years old and you have no idea of the dangers of traveling alone.   Tell me what you want."

Leif wanted to tell his teacher to just leave him alone but there was a hint of common sense that Sveinn was correct in this warning. "Travel with me and promise you won't interfere until I'm done."

Sveinn agreed, only because he saw no way out. Settled again, Sveinn still had no success in getting Leif to go swimming with him. Two hours later he crawled in bed and wrapped his arms around this young love that had turned his life topsy-turvy in the last few months. He was correct in his assumption, Leif couldn't resist cuddling, which led to a kiss, then to a complete rapture of a body Sveinn desired every second.

The next morning began a trifle more cheerful than the previous day. Leif knew he'd gotten his way and made Sveinn grovel for the sex. The two of them went to breakfast; afterwards, they traipsed down the beach, a mere football field away. It took a long walk to clear their heads and talk about the previous evening before Sveinn decided to take his protégé into a clothing store. Arms full of clothes, from terrycloth polo shirts to bathing suits, Leif was busy trying on all these new clothes.

Behind a curtain the two of them chuckled with each new exchange of attire. The salesclerk didn't mind running back and forth for different sizes to appease this bright-eyed, platinum blond boy in his store. His own hair was dyed blond which helped his fifty-five years appear more in this thirties, so the salesclerk figured. A small digital camera in the palm of his hand certainly assisted this fun of pulling back the curtain and visually capturing this Adonis in his all-together.

During lunch Leif couldn't wait to unveil his plan. From Camden, New Jersey, to New York, Pennsylvania, and Illinois, Leif planned their itinerary to the tee.

"And how to we get these people to just tell us? Do you think they are just going to say, 'Yeah, we remember that grave being dug up? What do you want to know?'" Sveinn asked with curiosity.

"I've thought about that," Leif started. "I have a plan that is sure to work. You are like my father, see, and you tell them this is a school project and I was given this certain year in the community to do a project on. I read in this magazine while I was in New York that they'd gotten this former security guard at the Westminster Abbey to confess he was paid $50,000 sixteen years ago to take samples from the tomb of Isaac Newton. That is all the information we need. Now we know that a doctor offered all these different people all this money to get what he needed."

"I don't get it. What does that have to do with finding out if these guys' graves were broken into?" Sveinn asked.

"See, when the current owners of the cemetery give us who worked there sixteen years ago we find the guy and let him know we know everything. He'll talk if we promise not to go to the police."

"Interesting. I better do the talking because they'll never believe a kid."

"Oh, thanks. Do you want me to hang onto your leg and stroke your dick while you're talking?"

"Very funny."

The afternoon was spent parading on the sand and watching men drool when Leif pretended he was the boy from Death in Venice. The two males had spent the late night watching this DVD in their hotel room after two rounds of sex. Leif loved the movie and equated it with how he seduced his share of men while dancing on stage. One difference from the movie to this beach, Leif had on a thong, which wasn't fair to the gay world of Ft. Lauderdale with the hanging of this boy's pouch.

From the beach to the clothing optional pool at the hotel every man there attempted to catch Leif's eye to offer him money for just one hour. Leif pretended no one else existed, swimming to and fro, on occasion stepping out of the pool, and then diving back in. And the true teaser—sitting on the pool deck with his feet in the water and arching back to rest on his hands, his own face downward and watching drops of water roll down his chest, over the ripples of his stomach, to the pointed protrusion in his thong. Men gawked at the looping penis which was no less than 7" soft. Hairless and pretending innocence Leif was now an expert at making men drool and act like fools.

Chapter Seventeen

The helicopter landed on the front lawn of the Franco estate on the outskirts of San Antonio. The circular driveway with a half-acre grass lot within this horseshoe made a perfect zone for the academy's pilot to land his craft. Mrs. Franco was pleasantly surprised to see her children with Alex and Brittany. She offered her guests dinner, but Brittany said they should be getting back in short order.

While Travis showed Alex his room and shared a moment of privacy to grope what was continuously on their mind, Mrs. Franco had a moment to express her thanks to Brittany for taking care of her children through the evening. She also admitted that Coach Venegas had given her the medical diagnosis of a possible condition. Her doctor's appointment had been that morning and she was given a breast scan. Fortunately the problem was caught early and the doctor saw no complications.

There was time for a piece of cherry pie, then everyone gathered to wish Alex and Mrs. Makawa goodbye. When Travis was fairly sure no one was looking he and Alex had one last kiss. It wasn't quick enough to escape his mother's peripheral vision.

The chopper lifted and the Franco family ventured inside.

"Where's Dad?" Jay asked.

"Your father flew back to Austin this morning. He'll call later on," Mrs. Franco replied. "Did I see what I saw, young man?" She trained her eyes on her eldest.

"What, Mom?"

"You know what, young man! Since when did you start kissing boys?"

Travis didn't want to appear in panic. "If Alex was a girl you'd be giggling and teasing me how pretty she was. I'd like you to challenge your bias and judgment of me."

Mrs. Franco was more humored than corrected. "My son is using psychology on his mother. For your information I'm not bias; I just thought I knew my children well enough to know that none of you are that way."

"What's, 'that way', Mom?" Niqui asked.

"I think you know what 'that way' is. Boys just don't go around

215

kissing each other. I do think it's time I have a talk with all of you. After our visit to the Kings Academy I realized your father and I have been remiss in discussing the birds and bees with our children. We certainly can't expect the church to be responsible in doing this."

"Sure, Mom, what do you need to know?" Jay asked and had his siblings laughing.

"Sit!" Mrs. Franco ordered and they all gathered around the dining room table--the silence was no less than a morgue. Mrs. Franco finally broke the ice. "I figure you all saw some of those exhibits at the academy and have questions. You're getting old enough to discover masturbation and, though it's a private affair, I don't want you to feel guilty about it."

"Niqui does it," Jay spoke up. "He goes into the shower stall every night and makes these grunting sounds." He had his siblings cracking up and his brother red in the face.

Niqui didn't think it was all that humorous. "As if you don't. You might think I'm asleep, but I see you over there jerking it. Your moans are enough to wake up the dead."

"Okay, you two! It appears you are all aware of each other's sex habits. I suppose my children are at the age where you're curious to know about the other sex. Maybe Maria would feel better if I talked to the boys first, then her."

"Oh, please, Mother. I'm not a baby," Maria scolded.

"Far be it for me to treat you as a baby. Let's start with a simple explanation of a female. Does anyone know what a clitoris is? I'm not asking my daughter here."

"It's a pea shaped thing at the top of the vulva," Jay answered. "Very sensitive to the tongue," he added with a smile.

"And you know this how, young man?" His mother inquired.

"Huh, I'm not dumb, Mom," Jay covered up his faux pas.

"You've seen a clitoris?"

"Sure, hasn't everybody?" Jay smiled with the confidence of an accomplished lover. His brothers and sisters weren't about ready to tell or challenge him.

"Okay, let's move on. If I find out that you're using your

216

computers for pornographic purposes you'll all lose them. I suppose you understand about the penis inserting in the vagina. I hope you realize that the penis you saw in the picture at Kings Academy probably was a dildo—that's an artificial penis. I doubt if a penis can grow that long."

"Yes it can," Niqui said very assertively.

"I've seen a few in my day, including changing your diapers, and it's highly unlikely, son. Anyway, as I was saying, it's never safe to assume you can have intercourse and pull out. I realize our church doesn't believe in contraception, so it's best just to wait until marriage. The movement of the hips during intercourse causes a friction and the feeling is so powerful, don't think you can prevent a release."

"When was your first time, Mom?" Jay asked.

"I'm not sure that's relevant here. Sex is a very private affair," Mrs. Franco replied.

"Why does it have to be, Mother? Why can't it be open and fun instead of this big secret and taboo?" Maria asked.

"When you have sex, Maria, I'm sure you don't want the world to know what happened and all the intricate details," Debbie lectured.

"I have had sex, and I don't mind if anybody knows," Maria confessed.

"You have? But why haven't you discussed this with me? I thought we had a better mother-daughter relationship than keeping secrets from each other?"

"It's interesting you say that, and in the next breath you won't tell us about your first experience," Maria corrected her mother and had her brothers nod their heads.

"Okay, if you want an open family discussion, I can do that. It might prepare you for when you meet that special person. You might find this hard to believe, but your mother was an egghead in high school. What you kids refer to as a geek or nerd, I'm afraid. It might not have got me a lot of friends or peer acceptance, but it did get me a scholarship to the University of Texas, where I met your father."

"The sex part, Mom.....when did that start?" Jay redirected this trivia back to the topic he was most interested in.

"I studied with this boy named Barney," Debbie began.

"Like the dinosaur?" Niqui asked and they all broke up.

Debbie gave her son an annoyed, but humorous stare. "Much cuter. Anyway, the two of us were the brains of our class, and Barney was our Bill Gates, a computer wizard and the most picked on boy in school. We'd study at each other's house and our mothers thought nothing of us two being up in our bedrooms together. We were juniors and my hormones were running pretty wild like my peers, except they had all these stories about making out and these exotic dates with varsity athletes. All I had was Barney. I figured the boy liked me because we'd call each other and share our secrets. It was your mother who initiated the romance; I seduced that poor boy with kisses when I faked not knowing how to do a certain problem. In this case I pretended to be puzzled by a work of literature we were required to read. Are you sure you want to hear this?"

"Yes!" All four Franco kids responded at once.

"The first time we kissed, I think it was really the first time for both of us. Soon we were experimenting with our tongues, and that's when I noticed the bulge in his shorts. I asked him to stand up, before he, okay, I unsnapped his buckle and let his shorts drop to the floor. His underwear poked up and we both giggled. When I slipped those down I was fascinated with an erection that sprang up like a spring. This is so funny!"

"Go on, Mom. This is getting really good," Jay encouraged.

"Well, Barney let me examine his penis for a second and then I cupped it in my hand. The thing exploded and the first stream caught me straight in the eye. Boy did that burn! When Barney relaxed and saw my discomfort he ran to the bathroom for water, totally naked and with my mother downstairs. Thank God she didn't see him. We spent the next fifteen minutes washing out my eye and getting all that goo off my clothes. Sure, it might be funny to you all now, but it wasn't the way I visualized my first experience to be."

Niqui was laughing so hard he barely could get out his question. "But.....did you two finally do it?"

"Well, the next day we studied in his room. Barney kissed me first

218

and asked if he could see my breast. I suggested we both get undressed and climb under the sheets. I let him examine my body in return for my examination of a boy's. I didn't dare touch him this time because I figured that a boy's penis had this quick trigger for release. We were kissing and pressing our bodies together when it just happened. His penis slipped inside and I expected all these bells and whistles, but he humped me for a second and froze. I just assumed sex was like watching two birds doing it. A few seconds and on with life."

"That's so cool, Mom," Maria said as if it was as romantic as her first time.

"Sometimes you don't have to move your hips," Jay said.

"And you know this because…."

Jay knew he'd trapped himself this time. "I've tried it."

"Oh, you have. And how many times, may I ask?"

Jay gulped. "Four, maybe, five. I mean, not with three people, but four times in a row."

Mrs. Franco was completely in shock. "Are you saying my baby boy has tried sexual intercourse with a girl?"

"I'm old enough," Jay proclaimed.

Mrs. Franco sighed in deep thought. "I wish your father had such stamina. Forget I said that. Well, at least my older children have more self-discipline. Maria, may I assume you have resisted these feelings? Maria. Maria? You two haven't….."

"Mother! That's disgusting! We might have taken baths together but we didn't do that. I'm a woman, Mother, and I enjoyed it very much."

"I can't believe I'm hearing that my two youngest have had sex."

"Mother, as you said, you've been remiss. God! We never talk about anything important but grades and how to please our father. That is so lame!"

"Yes, I've obviously neglected my role here. Assure me you're not pregnant."

"I used the Morning After pill, Mother. But I did swallow, so I might have gotten pregnant that way."

Mrs. Franco slumped in her chair and had to laugh with her

children. "And here I thought I knew my babies. I must confess I've looked under your mattresses but all I found was a Spider Man comic book under my youngest son's bed, a book, DRAFT POLITICIANS, NOT OUR NATION'S FUTURE, under Niqui's, and a novel under my eldest boy's bed. By the way, what's Bishonen mean, Travis?"

"It's Japanese, Mom."

"Yes, I could figure that out, but what's it mean?"

Travis swallowed, blood drained from his face and he whispered, "Boy love."

"Like boy loving girl, or boy loving boy?"

"The second, Mom, I'm gay. I know it for sure so don't try to change my mind."

Mrs. Franco took a deep breath. "Okay, I can appreciate your honesty. Are you sure about this? I mean, have you given girls a chance?"

"Mom! I know what turns me on. Don't forget, I swim with twenty other boys, barely clothed. Why do you think I spend so much time in the pool? If I was on deck they'd see my reaction to most of them."

"Okay. I can visualize that," Debbie replied with a smile. "So, Alex and you are…..?

"In love, Mom? I find him irresistible. He's the best thing that has ever happened to me."

"What happened to Barney, Mom?" Maria diverted the subject.

"Oh, he's married with three children. They live in Provincetown, Massachusetts. He calls on occasion to see how I'm doing, and he knows all about you kids. I think he'll be especially happy now."

"Why, Mom?" Jay asked.

"See, Barney is gay and lives with his significant other with their three adopted kids. We both needed each other then; he gave me an education about how males masturbate and think about sex, and I showed him how a girl masturbates and what we talk about behind boys' backs. We agreed to study for an hour, then study each other for another hour. By the time we were seniors, Barney had built enough confidence and experience to discover another boy at school when a varsity baseball player caught his eye. I played hard to get for the rest of high school and never did have sex until college. Barney was enough for me until I met your father. Guess we're all due for confession after this."

Niqui adjusted his posture more astute. "Mom, confession is just a way to control us, pure and simple, using fear. Religion does that, creates a doctrine of an angry, retributive God in order to keep its members in line. Confession was created to give the churchgoer what reincarnation promised. That is, give us another chance. Not bad if absolution could come directly from God, but churches made priests to give "penance" which had to be performed. The church found this to be such a good way of controlling us; they soon declared it a sin not to go confession. If we don't, God will have another reason to be angry. It's even a sin if we go to another church. I'm not going, period."

The other kids nodded and left Mrs. Franco feeling overwhelmed by her son's lecture. "Wow! What has happened to my happy, carefree babies?"

"We've grown up, Mother. I'm in full support of my brothers. Being gay is not an issue. Who cares who Travis loves and kisses?"

"I agree. Love is more important than who we love and why," Niqui said. "I love boys, too, and I'm attracted to girls, as well. I don't care if Jay has sex as long as he uses common sense. Dane told me that there is no objectivity when it comes to sex. We all bring our biases, prejudices, and experiences to judge another human being. The boys at the academy are so comfortable being gay, their teachers make them think out of the box to realize that others will judge them different than their peers. I thought about what he said and I agreed that I had preconceived notions about boys loving boys. I'm okay with it now."

"Where did you learn all this, the Internet?" Mrs. Franco asked.

"Experience, Mom," Niqui answered. I've discovered that all this guilt and shame my school has made me feel is just plain wrong. I feel liberated."

Jay pointed to his roommate. "I'm with him."

"Our church never taught you about reincarnation, Niqui," Mrs. Franco tried to persuade.

221

"'I do not demean any faith but by the truth. Without reincarnation—without the ability to return to a physical form—the soul would have to accomplish everything it seeks to accomplish within one lifetime, which is one billion times shorter than the blink of an eye on the cosmic clock. Reincarnation is a fact. It's real, it's purposeful, and it's perfect.' Bobby said so, and he's smarter than Barney."

Debbie laughed. "So, it's those boys at the academy, isn't it?" Debbie challenged.

"So what if it is, Mom," Travis came to his brother's defense. "One night and I feel a whole lot better about myself than when dad ridicules us for not reaching his expectations. I'm not on this planet to please dad. Actually, I want to be a Wild West Hero."

"God! Don't let your father hear that. Your father lives and dies through St. Bonaventure. He has plans for you to go to Stanford and major in Political Science."

"I might go to Stanford, Mom, but I'll be part of their gay rights movement and love every cute boy there."

"For some reason I thought I'd be the only one talking today," Mrs. Franco said. "Whose idea was this anyway? Okay, my children know more and have done more than I give them credit. Is there anything you are curious about?"

Jay blurted out, "Why do girls like the cowgirl position so much?"

"The same reason boys like it, Jay. We get to watch the other person's expression and can control the tempo; plus, it's easier to reach the other person's genitals," Niqui answered and had his siblings and mother eye him with curiosity. "I can have fun, too, guys!"

"There's one thing you missed, Niqui," Maria said. "I love the penetration from this position and the pressure it exerts on my clitoris."

"Maybe I should take notes," Mrs. Franco humored. She turned to her daughter. "And you swallowed?"

"I'd do anything for Lance, Mom. He's such a dream. It was a bit salty, but when we read that project on masturbation and sperm at the academy, it said sperm had all those nutrients." Maria chuckled at her own rationale. "Why do boys squirm after they come like that?"

"You can't just keep sucking, Maria. You could torture the poor guy," Niqui said and even had his mother in hysterics.

Debbie sat back and let her four teenagers discuss sex with openness and humor. She could never tell Robert that his children weren't these innocent little kids anymore that they had always thought they'd possess and nurture.

****

That Saturday afternoon the Franco kids stood together and were prepared to honor their mother's wishes to attend confession. What their mother directed was not questioned because they knew of the repercussions when their father got home.

They were in the car and ready to leave when Mrs. Franco twisted around and eyed all her teenage children. "You're not exactly behind this, are you?"

Niqui was often the most outspoken of this clan. "Mom, you might as well know a head of time what I'm going to say. I'm tired of telling the priest that I masturbate, like I'm supposed to feel guilty and stop it. I'm also going to say, 'Father, don't forgive me because I haven't sinned. I received a blow job from one boy, before I had intercourse with another. That same boy licked me all over, including my ass and I came in his mouth, too. The truth is, I felt a tinge of guilt and remorse and then realized I wasn't struck by lightning and my body really felt great about the whole thing, so I let him screw me and I licked his body all over and swallowed his cum."

There wasn't a breath taken in the whole car. Debbie had no judgment and decided what was best for the family. "Before we give these priests a heart attack, maybe we better reconsider confession today," Mrs. Franco suggested.

Instead they all unloaded out of the vehicle and Niqui received three pats on his back from two pleased brothers and one proud sister. They went into the kitchen and Debbie served everyone a hot fudge sundae.

Marie was the first to speak up after ten minutes of silence. "Did you confess your little fling with Barney, Mom?"

"Honey, I wasn't brought up Catholic. I agreed to bring you

children up Catholic after your father's wishes."

"I'm no longer Catholic, Mom," Travis conveyed.

"Neither am I," Jay said.

"Got my vote," Niqui added.

"Guess I'm with my brothers," Maria sided. "I think we ought to travel up to the Kings Academy every Sunday so I can hear Lance sing."

"Dane plays in the orchestra," Niqui announced and added, "And he also sings in their choir. I'd like to be in their choir."

"So would I," Jay agreed.

"You do realize we're having a mutiny here," Mrs. Franco said.

"Then you're with us, Mom?" Maria asked.

"I've never felt more liberated."

As a family they toasted their new freedom and loss of virginity, except Niqui said only a partial of his male virginity was taken. To celebrate, Jay and Travis went swimming without their suits on, and Niqui went to his sister's room to ask if she wanted to join her brothers.

"What do you say, sis? No more secrets," Niqui proposed.

"I guess we can make a bond to that." She reached under his shorts and grabbed his penis. "I swear I'll be open and honest with all my brothers."

Niqui laughed. "What was that all about?"

"Ask your boyfriend at Kings," she said and kissed him on the nose.

Niqui was soon by the side of the pool in the buff and dove in to join his brothers. Maria came down next, took off her robe and challenged the boys in a two against two basketball game. Mrs. Franco stood by the sliding glass doors leading to the patio and absorbed what had happened and how their father would react. This new family of openness she hadn't prepared herself for, a stage that was almost forgotten but for her memory of Barney and an adolescent introduction to sex. It was like, in three days, her children had become liberated from their parents and their Catholic school's stranglehold on their lives.

She took out a glass of ice tea and sat by the pool, an occasional laughter to the kids' antics in the pool, but nonetheless speechless with the realization that her children wouldn't have the hang-ups or inhibitions of their peers.   There would be no more accusing each other of spying or panicking because they were caught in their underwear. She was even amazed that not one of them had pubic hair.   Jay said it was about being liberated.   Who was their mother to question that?

On Monday afternoon, two days later, Niqui breezed in to the kitchen, slid his school books on the center island and grabbed a glass of cold milk.   His mother kissed his forehead and received a note from school in return.   She read it immediately:

To the parents of Niqui Franco:

It is of my opinion that your son feels it necessary to correct my teaching methods instead of focusing on the facts I present.   Possibly Niqui's contributions would be best left for his college years where insight and creative hypothesis are expected amongst his peers.

Signed, Sister Anne

Niqui stood with a white mustache around his lips, watching his mother's reaction.

"And how, may I ask, have you insulted your history teacher, young man?"

"She gave us some false information on the Civil War, so I corrected her.   She said Abe Lincoln was never in the military and that Bull Run was the first battle.   I was reading the other night that this wasn't the case."

"Let me guess, Kings Academy.   Why have you chosen this time of your life to challenge your teachers?"

"Because at Kings they expect their students to contribute and give opinions and strategies and personal viewpoints.   I think that makes more sense than believing everything we're told.   Teachers aren't always right, Mom."

"No, but they are your teachers, and they can make life more difficult for you if you don't cooperate.   Such is life.   Don't forget your mother went to college to become a teacher."

"You would've been a great one, Mom," Niqui said and kissed his mother.

"Don't try to butter me up. Your father won't appreciate this. He expects you children to have blind obedience and perfect manners."

"Try a robot! We're not children."

"So I've noticed. My level-headed son has become very liberal in his thinking. Are you sure you're okay with your brothers' and sister's new interest in sex?"

"Sure, why not? Wendy is really gorgeous and they're cute together. What more can you ask?"

"But Jay is so….innocent. I'm surprised he knows what to do," Mrs. Franco said.

"He was innocent. Wendy taught him a whole lot, and I hear it every night." Niqui laughed. "I hit my home run with boys, and Jay hit his with a girl. We both lucked out. Maybe we can switch. I asked, but he's not real receptive at this time."

"You're still my little virgin, right?" Mrs. Franco hugged her middle son.

"Girls don't interest me that much, Mom. Jacob was really nice and funny, but Dane was soooooo cute. Does it really matter who we have fun with?"

"No, not really. If your heart is in the right place, your mind will follow? I had sex once with my roommate in college. You know what? Girls make better lovers than males because they know how to treat a lady."

"Thanks for telling me that, Mom. Though I haven't had sex with a girl, I can't fathom a girl knowing what to do or how to please a boy better than Dane or Jacob."

Debbie twisted her son's ear. "Does a penis that long really exist?"

"I saw 32 of them, and Dr. Kerho has one pretty long himself."

"Dr. Kerho was in your room?"

"No, not exactly. See, adults have to be naked, too, if they come in the dorm. It certainly presented a different view of an authority figure."

"A bit odd. I doubt if one that long can even get up."

"It definitely rises, Mom." Niqui held his hands about 10 inches apart.

"Don't tease your mother, young man. I might ask your father to get one of those operations. How did you know Dane was gay?"

"They're all gay, Mom, except for Lance and Wendy, but they don't mind getting it on either with their own sex."

"So it's a school for gay kids?"

"Don't ask me. I'd just love to go there. I won't have to date my palm every night."

Mrs. Franco laughed and messed up her son's hair. "I noticed in the pool that you kids have shaved your pubic hair. What gives?"

"It's the in-thing, Mom. Feels great!"

"I see."

Niqui set his glass down and ran all the way to his room. There was a satisfaction that he felt he could tell his mother almost anything. His brother was on his bed, enjoying the delights of masturbation. Jay's smile didn't nullify what he'd been thinking about, though that aspect was obvious. Both boys had agreed that it wasn't necessary to pretend this didn't happen and began to share their fantasies and goals for their next visit to the academy.

As days progressed St. Bonaventure talked very little about the results of their competition against Kings Academy. There were rumors about the athletes' uniforms, or lack of, and the escapade in the Jacuzzi escalated several boys' reputations of being daring and romantic. Coaches talked about not scheduling Kings Academy for anymore events, only because it was bad for their reputation as a powerhouse in Texas athletics to be beaten by a school with only 33 students.

Ralph Johnson had his own souvenir from the trip up north. No one dared make fun of the school bully for what many already knew was the work of a female.

After school Travis was in his brothers' room, talking about how he was now ranked number one in the nation by Scholastic Magazine for the 100 meter freestyle. What he was more pissed off about was, Alex Kerho wasn't listed at all.

"I wonder if their coaches just don't know who to contact. It's not right I'm number one when he beat me with a better time."

"Call this magazine and let them know," Niqui suggested.

"Hey, guys!" Jay spoke as he entered and slumped face down on the bed. If it wasn't for the boy's grass stained shorts his brothers would never have questioned his behavior.

Niqui moved over and rolled his brother's body to where he could see Jay's face. "Hey, bro, you been crying?"

Jay fought back the tears and told his brothers how Ralph Johnson had waited for him after school, down by the park. Travis dashed downstairs to the kitchen and grabbed ice from the freezer without telling his mother what it was for. Back upstairs Travis placed the ice bag on Jay's eye and quizzed his brother for answers.

"There were five of them," Jay said. "Two girls were with them but they didn't help. They stripped me of my clothes and said he was going to fuck me because my girlfriend wasn't available. While his friends held me down, that scum bag split my legs and pushed my knees forward. One of those girls said, 'He's going to tell,' and Johnson said, 'No he's not or I'll fuck him every time I see him.' That's when I shot my leg out and managed with all my strength to kick him in the balls. The other boys got scared when I started whaling with my feet and Johnson was rolling on the ground. I broke free and looked for my clothes, but they'd thrown them in the pond with my gym bag. When I saw Johnson rise he said he was going to kill me, so I ran and dove in the pond and started swimming away. Those girls convinced them all to leave me alone."

"Where's your gym bag?" Niqui asked.

"Still in the pond. I managed to grab my shorts as I swam by."

"You just bought that new shirt, and Mom's going to be ticked off because those gym shoes cost a lot of money. Those assholes! I'm not going to let them get away with this," Travis said, then stomped from the room but poked his head back in to get the name of the two girls who were there.

Travis took his bike and rode over to one of the girl's home, fortunately finding both her and her friend there with scared looks on their faces. Travis barely controlled his anger but warned his classmates that if they didn't write what happened and sign their names he'd make sure they were both suspended.   In minutes he had the written confession.

A convincing story was contrived by Niqui and Travis on why their brother wasn't at dinner.   He'd taken a knee to the face and wasn't feeling all that well.   The boys took various parts of their dinner up to Jay's room while Mrs. Franco was preparing for another social affair at a lady's function.

The three Franco boys and their sister met downstairs after their mother left so they could discuss what strategy they should use. "Wendy and Lance's birthday is coming up.   We could invite them down for a party," Maria suggested.

"And that means what?"   Niqui asked.

"Invite your boyfriends, too, and we can maybe ask our friends to scare Johnson."

"I don't know," Travis spoke.   "I'm not sure it's fair to get them involved, but I'll ask Alex.   How about Dane, Niqui?"

"I have girlfriends, you know," Niqui said with a smile.

"Yeah, but they don't make you smile like Dane does.   Jay says he heard you call out Dane's name while you were whipping it the other night."

"I didn't!   Maybe once.   Okay, I'll ask him."

The calls to the Kings Academy were welcomed and everyone agreed to the birthday bash after they'd received permission from the boys' fathers.   A week went by and the Kings' students were to arrive that Friday.

On Friday morning Travis left a note for Johnson in the boy's school locker.   It read that him, and his four henchmen, are to be at the park that evening at eight o'clock.   To Ralph Johnson, it was an invitation for him to beat up all three of the senator's boys at the same time.

There was an anticipation of a family reunion when their guests arrived.   Maria and Wendy bonded in female conversation, but only after kisses to Lance and Jay.   Mrs. Franco pleasantly smiled and was keeping an eye on this foursome.   While Brittany and Debbie adjourned to the kitchen all six boys went directly to the backyard pool. The game of water polo gave Mrs. Franco a glimpse of Niqui's affirmation of the male anatomy.

The seven teens gathered at the shallow end of the pool at the request of Lance. "My brothers and sister apologize if we took advantage of you when you stayed over at the academy the other night," Lance began.

There were quizzical expressions from the Franco kids. "Did you get in trouble, Lance?" Maria asked.

"I guess you can say that. Dr. Makawa lectured all of us on treating others as sexual objects without knowing the person for who they really are and recognizing their needs, as well," Lance explained.

"I didn't feel objectified!" Travis spoke up. He reached for Alex and kissed the boy right on the lips. This is the best thing that has ever happened to me. I'm out and proud, thanks to Alex."

"Yeah! I agree with Travis," Niqui said. "I might not be a gay kid, but I had lots of fun with Jacob and Dane. I'm the one who wanted to experiment that night. They never forced me in to anything."

"My brother misses you guys. He sticks all sorts of things up his butt at night," Jay humored.

Niqui smiled. "I'll admit it feels good, though I hope Mom never finds out I use her deodorant bottle."

The teens laughed and didn't put a lot of emphasis on this apology. Maria stepped toward Lance and grabbed his balls. "I hope you're not apologizing to me, big boy."

"You don't know Dr. Makawa," Lance replied.

"I'll kick his butt for making you guys feel guilty," Jay said.

"Jay, my little rabbit, Dr. Makawa is our martial arts teacher and our psychologist. He eats boys like you for breakfast," Wendy said and hugged her favorite male.

"We'll just have to call Dr. Makawa and explain that none of us felt pushed into sex," Maria said and had her brothers nod.

"You can talk to Mrs. Makawa first; she's in the kitchen with your mom, but don't tell her we suggested it. We all have to do a major essay on empathy and where our priorities are. That's a big thing with our sensei," Lance said.

The birthday dinner and lighted cake was a glorious occasion for these fraternal twins turning fifteen. Laughter and frivolity involved everyone, though several sets of eyes stayed on the grandfather clock in the corner of the dining room. It was quarter-to-eight when the boys dismissed themselves and paraded to Travis's room. Maria stayed behind with Wendy, as the two girls talked about their pool conversation and the opinions of the Franco children.

In the boys' bedroom upstairs, last minute instructions were given as a rope ladder was secured to Travis's bedroom window to the backyard. Nightfall had arrived, the skies covered a half-moon, which presented very little light, if any.

Lance, Alex, and Dane sprinted through foliage and around trees till they came to the front gate of this closed community. Travis had assured them that the guard at the gate was very busy this time of the evening, welcoming men back from work or letting other families out for evening shopping. The guard lacked the time to view security cameras and totally was oblivious to the three boys sneaking through the open gate.

Up and over another wall the Kings' boys ran the two miles to the park in just over ten minutes. They scanned the park which had lighting every fifty meters. Their prey had gathered near the pond, so the three boys approached from the rear. Wearing only ninja outfits with booties and black masks they were soon noticed by the five other teenagers.

"Well, what do we have here?" Johnson mocked. "The Franco' boys are all dressed up like little karate kids, as if this is supposed to scare us."

Alex was grateful that darkness was the norm around this part of the pond. The boys' heights were fairly close to their friends: Lance was as tall as Travis; Alex, the same height as Niqui; and Dane was short like Jay.

"You owe my brother an apology," Lance yelled out, trying his best to imitate Travis's voice.

"You've bumped your fuckin' head, Franco!" Johnson yelled back, then motioned his peers to march toward their victims.

The attack was amateurish and foolish. In rapid movement the first three attacks were instantly repelled with the Kings' boys able to grab limbs and snap two elbows and one wrist. A round house kick, two punches, and a judo throw expelled air from two of the remaining thugs. It took a total of thirty seconds to have all five boys lying on the ground.

Alex took things more personal than his peers. He suggested they strip the boys and toss their clothes into the darkened waters on the pond. The collection of wallets offered the amount of $86, which would help Jay replenish what had been ruined. This took longer than the destruction.

Dane also had a grudge against someone who had messed with a younger boy. He picked out the one the boys had called Johnson, then took the boy's other wrist which wasn't broken and used it as leverage. There was only one more, "Fuck you!" before Johnson felt his nostrils being stretched by Dane's fingers. Dane pulled a paper and pen out of his pocket, already prepared to sign, and made sure Ralph signed his name. Only after this gratuitous endeavor was accomplished did Dane make sure the other wrist matched the broken one.

Four of these boys began to crawl away and then run when they saw they weren't going to be pursued. Johnson was dragged crying and whimpering across the park to a waiting vehicle, which Travis had arranged through a friend at school for a few dollars. They tossed the bully on a sheet in the back seat and tied the envelope with Johnson's confession to the boy's dick with dental floss. It might have been a little tight. With two broken wrists Johnson tried his best to get the string off but the effort only increased the pain on a part he didn't want to lose.

The boys found the St. Bonaventure headmaster's home exactly as the map was drawn. Dragging their rubbish to the front door Alex rang the doorbell, then ran. They drove the car to a street in the back of where the senator and his family lived, but outside the gate, left the keys under the mat, and leaped over the wall to scale the ladder back into Travis's room. All under forty minutes.

"How'd it go?" Niqui was the first to help the boys in and ask.

"A bunch of weenies!" Lance said with a smile. They high-fived each other and proceeded to crack up with each recount of the incident.

****

Mrs. Masters answered the front door and saw a naked boy crouched over, attempting to take something off his groin. She thought this was some kind of joke or teenage prank before she cleared her voice and the boy glanced up.

"You attend St. Bonaventure, don't you, young man?"

Johnson nodded in pain and was helped up with some discomfort. "They broke my wrists," he cried out. "I really need to get this off my dick."

Mrs. Masters examined the swollen penis without touching it. She called her husband down from his upstairs office and their twelve-year old daughter walked into the room, wondering what all the commotion was about.

"Young lady, get me some scissors," Mrs. Masters said.

"Are you sure you need scissors?" Johnson asked with a pleading voice.

"You're right, maybe nail clippers," she had second thoughts.

Phil Masters had taken his time coming down in his bathrobe to notice his daughter and wife working on something around the boy's crotch.

"Stretch it," Mrs. Masters told her daughter while Johnson subdued his pain.

"By God, what happened to you?" Mr. Masters asked but was only handed an envelope by his smiling daughter.

Phil read the letter with three signatures. He ran the options through his mind and decided not to call the police. With Johnson now holding up his penis with twisted, trembling hands Phil made his first decision to get the boy to the emergency room before calling the kid's father.

Mrs. Masters wrapped a towel around the boy's waist and the whole family assisted the teenager to the car. On their way Phil called Mr. Johnson on his cell phone.

"I believe we have a problem, Arch," he told one of the city's

233

noteworthy attorneys, known to handle the affluent—mostly drug peddlers and women seeking settlements.

"Is it my son again?" Arch asked and received confirmation.

Phil Masters received another jolt to an impending headache when he noticed several other St. Bonaventure students in the emergency room with fractured elbows, dislocated shoulders and broken wrists. All the boys had agreed that this was the work of the Franco boys. After Mr. Johnson showed up at the hospital Phil drove over to the senator's home, walking right in on a birthday party with an abundance of teenagers.

Debbie Franco was delighted to have the headmaster pay her family a visit. "Thank you for stopping by, Phil. Robert is in Washington but I know he'll appreciate your kindness. Is there a problem?"

"I'm not sure, Debbie. Have your boys been home all this time?"

"Well, yes. They have guests from the Kings Academy. You remember them, they're such well-mannered boys and young lady."

"Yes, yes, of course. I wonder if I may have a few words with Jay, if you don't mind," Phil asked.

Jay was called over by his mother and the headmaster; his wounds were barely noticeable from the previous week. It did appear to Phil that the girls' statement was correct. "Jay, do you have any idea who might have sought revenge against these boys who assaulted you?"

"Why? What happened?" Jay asked, not falling for the first question. He would have given the impression he knew something.

"Oh, you haven't heard?" Phil asked and again saw a negative head shake. "It's not important for the moment. You get in there and have a good time at the birthday. Jay, these boys will be punished, though it appears someone has beaten me to it."

As Mr. Masters was departing, he noticed a light blemish on Lance's lip. Phil went over to the boy. "Do you mind if I ask what happened to your lip?"

"No sir. We were playing polo tonight in the pool. I got nailed by Niqui with an elbow." Lance looked over at Niqui as he nodded.

Phil was convinced that everything appeared on the level. He had no reason to doubt their mother that these boys never left the house.

Over the next half-hour there was a serious tone by the boys in covering up their knowledge of the evening's event. They played rhythm as a group, then it was time to sort out where everyone was sleeping.

Debbie gave the guest room to Mrs. Makawa, before her scan of the faces of her children. Travis gave this pleading stare and Debbie gave in.

"You'll be in college in a few months where I can't keep an eye on you, so I'm granting you this evening with Alex. For the rest of you, Niqui can sleep with Jay, and give up your bed to Lance and Dane. Marie and Wendy can share. Any questions?"

An hour later this room and bed arrangement had been more like musical chairs. Mrs. Franco had her suspicions when she heard laughter, but she decided she'd rather not know.

Chapter Eighteen

Leif and Sveinn landed in Trenton, New Jersey a day earlier than they'd planned. Leif had made sure to call and tell his supervisor that he was in flight and would report his progress soon. Their first target was America's favorite poet, Walt Whitman. Walt Whitman had inspired many a man and boy with his writing, **LEAVES OF GRASS**. In the middle of the nineteenth century, people weren't ready for such blatant poetry or insinuations about homosexuality, though the word didn't come into usage until the late 1860s.

Their appearance at a cemetery in Camden began their role as father and son. It worked. The director of the chapel gave Sveinn the name and number of the man who had responsibility as the funeral director sixteen years earlier. It didn't mean this was necessarily the man that had the answer, if there were answers at all, but it was a hunch that was worth the effort.

When an elderly woman answered the door it didn't appear they were in time. She was 88 years of age. "Is Mr. Hodgekiss available?" Sveinn asked.

"Why, who are you? I can't recall seeing you around these parts before."

Sveinn did his best to sweet talk the lady because his son needed the information only her husband could provide. It got them an invitation inside of a musty house that hadn't seen daylight in years.

There was an uncomfortable aspect to talking to this retiree with his wife present. There was idle chitchat, then, when Mrs. Hodgekiss went to the kitchen to bring some snacks, Sveinn took the lead.

"Sir, we have information that is very private and won't go any further than right here. It's important and we aren't here to threaten or accuse you. Did you receive $50,000 for a few artifacts from Walt Whitman's grave?"

Mr. Hodgekiss laughed. "You boys have done your research. What on earth are two young studs like yourself interested in old Walt?"

"Let's say a doctor helped us. The proof is all we need and it will be forgotten forever," Sveinn promised.

"Is this the doctor's grandson?" The old man asked.

"He is," Sveinn lied. "Possibly someday this boy would like to take over his grandfather's work."

"How fascinating. I remember the man so well. He was such a gentleman and was quite generous with his money. It wasn't all that difficult."

Leif kept tugging on Sveinn's sleeve to leave but the elderly couple wasn't used to company and kept chatting for the next hour. Stuffed with cookies and milk the two finally were ready to say their goodbye when this senior citizen surprised them both.

"Did your grandfather tell you about old Bill, as well? I sure liked the money your grandpa was offering, so I couldn't resist."

Leif's ears perked up. "May I have another glass of milk, ma'am?"

Never having grandparents to relate to, Leif was patient in listening to this elderly gentleman reflect back on his little escapades with delight and connivance. Such grave robbery sounded almost romantic the way Mr. Hodgekiss described it. At first Leif had trouble thinking that this Bill Tilden would be worth anyone's time to clone, but his mind was soon changed.

"What a player he was, Big Bill," this senior citizen recanted his memory of the towering tennis player who was the last of three players to win a record seven U.S. Championships, including six in a row in the 1920s. Counting doubles and mixed, he won 16 U.S. Championships, the most of any man. "The poor fellow died almost penniless and in disgrace in 1953, after serving two separate jail sentences for molesting teenage boys. They treated him without respect after that, I'm afraid, yet to boys like myself back then the guy was as great as Babe Ruth or Bobby Jones. None of us could fathom this sex thing out, but, God knows, the guy would have had a field day with today's ball boys." Mr. Hodgekiss blurted a coughing laugh that even had his wife tell him to stifle himself.

Leif and Sveinn found humor in this old man. "You had access to Mr. Tilden's grave site, as well?" Sveinn asked.

"Oh my, no. I had a friend in Philadelphia, now with his maker,

God bless his soul. His ashes were buried in Ivy Hill Cemetery, a sprawling 87-acre site near Mount Airy, the northwestern-most section of Philadelphia. I split my reward with him. Dare say you'd have a might rough time finding this back then. See, the finest tennis player who ever lived was buried in disgrace. People just didn't understand or much forgive the old boy's transgressions. Rather sad to think about it, if you ask me."

Slowly lifting himself from his rocker that appeared to fit the senior citizen like a grove, his body took seconds to straighten upward. He had spent many hours conformed to the pillows stacked inside the chair. A turtle might have reached this archaic desk before Mr. Hotchkiss did, but he located after a few minutes of tedious search an old notebook. From his notes he read: "Section D at Ivy Hill. Under an enormous cedar tree you'll find the Tilden family plot. Bill's marker is a flat, 2-by-1-foot stone. It sits at the foot of his mother's grave."

When Sveinn glanced toward Leif, he was surprised to see a tear drop run down the boy's face. For whatever reason, the old man's sadness at this great player's demise had affected a fifteen-year old quite deeply.

"I'd like to visit his grave site," Leif spoke up with little embarrassment over his tears.

As desperate as they were to leave this home at first, their visit left a remarkable honor on both these males from Iceland. Thoughts ran through Leif's mind that maybe bringing back the dead might be a good thing. He kept his opinions to himself, as not to freak out his teacher.

Leif couldn't wait to call Florida the next morning. He refused to give up Mr. Hotchkiss's name, though the editor said Leif had no other proof to verify that Walt Whitman's or Bill Tilden's graves were robbed.

"We knew you'd say that so we had it on tape. You can quote without names—that's our agreement. I don't wish to get people in trouble for something they did sixteen years ago."

"All right, send us the tape and you'll get your money."

Leif was beside himself with his first victories, he called it. His next two weren't real successful, as one cemetery had lost its files in a fire; another head caretaker had passed away. Leif's education of so many noteworthy figures in American history had totally captivated him so much any loss was devastating to his ego. The grave site of Horatio Alger had been a story of controversial incident when it was discovered that, sixteen years ago, it had been broken into with a great deal of damage that wasn't repaired. This was all Leif and Sveinn needed to prove a point.

This writer of the nineteenth century had written very popular books about boys; an early example of future novels, like the Hardy boys, so popular in the twentieth century. The man had been driven out of one town on accusations of having sex with boys, only to become part of an orphanage in New York City and befriending newspaper boys and any other street hustler. These boys often became the models and examples of youth struggling for survival amidst poverty and poor living conditions. Alger happened to be one of the first of many writers to put his love in stories of happy-go-lucky boys who end up having happy endings with success and money.

Leif kept his responsibility and stayed in contact with the tabloid editor, revamping his itinerary as he went from site to site. He had time to stop and find his friend at the boarding house, only to discover that Rodney was using drugs again and hustling for money. He gave Rodney a hundred dollars out of his own money if the boy promised to stop using drugs. Rodney lied to get the money.

An eighteenth century figure in the American Revolutionary War and the founder of The Society of Cincinnati was a choice Leif had studied at the New York library when he lived there at the youth hostel. Sveinn told the boy he was wasting his time with this one. The finding of the retired caretaker was much like the first one they'd found. Old and grumpy this senior citizen wasn't about to give up his secret to Sveinn.

"Sir, I'm a distant relative of Baron von Steuben," Leif began with his fabrication. "I know a lot about my great, great, great, grandfather. He was a military lieutenant from Prussia who had been kicked out of the military for enjoying the company of boys. Benjamin Franklin traveled to France in search of military leaders to assist George Washington in training and recruiting an army, so he met my grandfather and gave him the title of Baron to impress George Washington. He even had fictitious papers to travel to America under, and Washington even made him a major general responsible for designing and maintaining discipline for this new colonial army defending their rights against a demonic King George."

"What's that have to do with grave robbing, boy?" the man asked.

"Sir, I've been given the samples you took, as I wanted to verify that this was my grandfather. It's important to me. I'm writing a book about the Baron and how he arrived in America with his new image and title. The public was unaware of his past, but, even then, my grandfather had at his side two 14-year olds—his aides de camp. The boys served by his side throughout the war, traveled to New York at its end, and my grandfather adopted one of the boys to leave his entire will to. In the end a country welcomed the man's professionalism, his contribution to make America free, and several states even have cities named after my grandfather."

"Well there, laddy, if the man never married, how were you his grandson?"

Sveinn gulped. Leif was as smooth as silk. "Well, sir, that adopted boy took Von Steuben's name, which is mine." Leif was proud of himself for thinking up that one.

"If that fourteen-year old was as handsome as you are, I don't blame the Baron for takin' a-likin' to him. Tell you what, laddy, whatever pieces you possess came from his grave. I'll give you that much. Dare say, you don't want the fifty-thousand back 'cause I've done spent it all."

"No sir," Leif said and hugged the startled man.

This shocked senior put his abrasive nature aside. "Youngster, I think you have a gift of gab, but you're also a sight for tired eyes. Check on Cole Porter while you're playing snoop."

Leif kissed the man a quick one.

****

As the travels continued there were far more strikeouts than successes. With every victory the two males had a special dinner, savoring the discovery while forgetting their troubles and enjoying this special friendship. With new motivation came more research on the man's history, details of his life, accomplishments, who he knew, and who his lovers might have been.

Sveinn began to check cemetery records on who might have checked a registry or signed in or out well over a decade earlier, just to see if this doctor ever gave up his name. They struck gold when a sixteen-year old registry showed a Dr. Jolson with an appointment with the director who proved quite useful.

Thomas Eakins was probably the most controversial man in Philadelphia by the turn of the twentieth century. Married to one of his students, he started an art college and had his students pose nude for each other. Not unheard of with the same sex; it was a shock to find that boys and girls posed for each other. Mr. Eakins took a multitude of photos of his male students, often taking them to swimming holes or had them wrestling in the buff.

What intrigued Leif, the man had been great friends with Walt Whitman. Leif was beginning to put himself in Dr. Jolson's shoes, a man in search of gay men who made a mark on history and deserved far better respect and recognition than what they received at the time of their contribution to mankind.

The puzzle challenging Leif was deciphering Dr. Jolson's logic: athletes, military leaders, poets, musicians, writers—Leif was so excited, it was no longer 'Did someone do this?' but, 'Where are they?'

Fifteen thousand dollars richer these Icelandic adventurers used a rental car to travel across Ohio and into Pennsylvania. They stopped on occasion to enjoy the freshness of spring, the smells of nature in America's heartland.

Into Indiana they stopped in Indianapolis, saw a velodrome in the distance and pulled off alongside the road to begin their hike inward to this outdoor bicycle track. Kids and adults of all ages were training their skills on bikes, and Leif had only seen this type of cycling when he watched the Olympics.

Mission accomplished the two of them began to hike back alongside a dirt road. They paused for a minute and Sveinn felt a need to show how much love he had discovered in his own heart for this young man. Their kiss was long and passionate and, unfortunately, witnessed by a group that got its jollies by victimizing people.

In a split second Sveinn and Leif heard the rumble of motorcycles quickly approaching their position. Bikes broke to a stop and nine men began to surround both Sveinn and Leif. A large thug with a beard, that had rarely seen soap and water, stepped off his cycle and stepped in front of Sveinn.

"I see you're into boys, faggot! From a distance we thought we had a pretty girl here, but Blondie turns out to be a boy, a young one at that."

"What's it to you?" Leif asked with naïve bravado.

"Oh, boy wonder speaks." When their leader laughed they all roared like wolves teasing their prey.

"Make 'im squeal, Bull," one of his gang encouraged him.

Bull moved in front of Leif, his chest bumping the boy's 125 pounds. A cigarette hung from the man's mouth, and when he removed it from his lips he blew the smoke into Leif's face.

"I don't want to smell your lousy smoke!" Leif said and slapped the man's hand with the back of his own. The cigarette flew several feet away near the thug's bike. Leif saw the hand raise and he ducked just as the guy's fist swung over his head. Leif backed up. The previous laughter turned to seriousness.

"Bull! Your bike!" A man yelled.

Bull was on a mission to destroy this blond-haired kid in front of him. No one had ever been that disrespectful to him. His quick glance saw a small fire begin underneath his cycle, which quickly grew bigger in seconds with all the dry grass.

"Get my bike out of there!" Bull screamed and he was the first to run and attempt to roll his cycle away. The fire climbed up into the engine, finding oil and gas to its liking. With the blaze becoming hotter the Harley Davidson caught on fire which caused Bull and his gang to step back and attempt to throw dirt on the flames.

Leif began to run to the rear of where all these bikes had pulled to a stop. He hopped on the end bike which was still running. He'd had experience with his neighbor's motor bike in Iceland, though it was half the size of this machine. Nevertheless, Leif shifted it in gear, then kicked the bike next to him over, causing an avalanche of cycles crashing into each other and crumbling to the earth. A surprise jolt almost had Leif tumble straight over backwards, but he recaptured his balance and roared to where Sveinn was standing in total shock that his life would be short-lived.

"Jump on!" Leif yelled and Sveinn wasn't going to question Leif's motives.

Leif spun the gas just as four of this gang swooped down in an attempt to grab these two. A race through the grass and Leif was back out on the road, heading for their car. Leif knew he had probably a minute to make his escape as he slid to a stop where their car was parked.

Sveinn jumped off and ran to the vehicle while Leif unsaddled himself, gave the cycle a little gas and let it roll across the road and into a creek. The bike rested in two feet of water, smoking and its back wheel spinning. With no sound of impending bikes Sveinn slammed his foot to the pedal and they were soon on the main highway. Only then did they begin to laugh, though Sveinn kept checking the rearview mirror.

Chapter Nineteen

Senator Franco arrived back in San Antonio that Sunday evening a few hours after the guests from Kings Academy had departed. A filibuster on an amendment to the Constitution to define marriage had stopped the momentum to defeat gay marriage. Robert wasn't happy because he'd supported family values and the president's self-righteous viewpoints. His testy attitude was ready for the typical family quarrels and complaints that usually arose during his departure. Though he didn't want to admit it to himself the family seemed to get along quite well in his absence and there was more disappointment on his children's faces when he was a part of their everyday life.

Robert knew what was coming; his wife would try to comfort his mood by dressing in a sexy nightie and expecting sex. He wasn't in the mood for such nonsense, only because one of his new aides in his office, a recent college graduate student from Yale, had discovered her political ambitions increased between the knees of the senator. The girl desired a father figure, so Robert had contemplated or justified and, anyway, the young lady gave a better blow job than his wife did. Robert wasn't disappointed, just surprised when his wife played the cordial part but was hardly acting the neglected spouse role.

He checked his e-mails and saw that the St. Bonaventure Board of Directors was requesting his attention, ASAP. Being on the board was a pain in the ass to Robert, but he was quite willing to do anything for votes. His children all but avoided his presence; an occasional, "Hi, Dad," was all he received. After last week's debacle at this Kings Academy Robert didn't feel like rewarding defeat with the happy father image. He'd show them that, to get his attention, results had to be from a winning position.

Before their father woke up the next morning the Franco children went as a group to school, not sure what to expect from their peers. The best thing about the day, there wasn't any sign of Ralph Johnson.

Ralph's father had made several immediate decisions to forgo complications and long-term inquiries concerning his son's responsibilities over this incident. His boy was facing expulsion and, to avoid this, Arch's son was on his way to a military academy in South Carolina by the morning. One of the city's prestigious lawyers was glad to get rid of his worst headache and the school's perpetual menace.

Rumors had spread like a virus through St. Bonaventure. Boys sidestepped the Franco boys, patted them on the shoulder and gave them a respect that usually only their star athletes garnished from the student body. Jay's mind was off in another world anyway. He'd received two apologies from boys in casts, and he humbly accepted their excuses for being involved with Johnson.

There were other things on Jay's mind. He was in love and the role of lover had given this young man a new sense of himself. For someone to actually like and respect him was something totally new to his identity. His wrestling practice wasn't extremely focused, and he heard about it from his coach.

"Where's your head today, Franco? No wonder you lost to a girl, you've lost your fight."

Jay pondered this but knew the truth. Wendy was an excellent wrestler and a great lover. Problem was he missed her already and wondered if she'd discover that other guys could be far more fun than he was. What did he have to offer? He'd been accepted into a group that he considered far more hip than he was. He'd made love to two girls and a guy—something he would have thought as a miracle and never believed a week before. He had trouble admitting it to himself that, with all the bodies switching positions on Maria's bed Saturday night, it wasn't always Wendy who he was either laying on or who was laying on him. With mouths finding every orifice Jay wanted to believe it was real, but it was better to think of it as a dream.

There were teachers and coaches up in arms with so many varsity athletes having broken bones. The athletic program had taken a hit that it couldn't recover from. Travis had heard the rumor that the headmaster, Mr. Masters, was handing out five-day suspensions to the other four boys with Johnson. One of Johnson's punks was in the dumps because he'd heard his friend was on his way to a military academy. St. Bonaventure had an abundance of smiles with this rumor floating through the hallways.

Maria enjoyed her blogging with Wendy, even during school hours. Her relationship with her brothers had taken an impressive and surprising turn. With everyone seeing each other naked there were no more secrets that now appeared silly and immature to begin with. Her brothers had great bodies, though they lacked the length of the boys at Kings. This she couldn't figure out.

Saturday night had been the strangest of her life. Four of them had been naked in bed in total darkness. It started out in pairs, then the humor of twisted limps and a love-out, Wendy had called it. Though not always easy to figure out who was kissing who, she forgot that one of these boys was her brother and enjoyed herself immensely. It was kind of cool knowing that two sets of fraternal twins were sharing the same bed. She had always felt close to Jay, just not this type of bonding. But, in the end, she felt no remorse for sharing herself with people she loved. Whatever had happened she felt closer to Wendy and Jay. There was something about sharing an intimacy and a private moment that made the relationship even more special.

Niqui withstood the school day by his usual way, he was in his thinking zone, and a world he lived in when he wanted to believe school was just for him with no distractions. Dane kept creeping in his mind. His body came alive with each thought, but yet he knew he liked girls. With renewed confidence Niqui was ready to experiment and show how experienced he was now.

Jacob had shown him how to kiss and make love to a penis, but Dane had won his heart. It was now all a matter of taking the time and sharing his gift of love with the right girl. He decided right then and there to present a new image, one of availability and suave. Dane said he was beautiful, so, obviously girls must think the same thing. Their walk on Sunday morning to the nearest pharmacy had been an education in itself. He'd never known that this place contained so many things to add pleasure to his life.

Travis accepted the unwarranted praise with modesty. He knew if

246

his classmates knew he was gay, this type of friendship would be dismissed, if not turned to animosity and ridicule. Travis wasn't ashamed of his new identity, only the fact that he couldn't share it with people he wanted to share it with—his father, for one.

Instead, during literature class, the opportunity arose when his teacher asked them to write about their perception of love in the classics and how it affected their own lives. Travis thought about the story of Apollo and Hyacinth, the vision of Icarus, in surrendering his heart to Apollo for love. He wrote his paper with passion and a new sense of standing up for something he believed in. His first few words flowed from his heart: *My love embodies emotion, sexuality, the erotic, passion, and romance, making me feel and express my heartfelt emotions. It is my creative aspect. My Eros, my love of energy, my libido is not just about my sexual appetite but a general appetite for life. My spirituality resides in the realm of my lover, with whom I access my Higher Power.*

While her children were in school, Mrs. Franco felt overwhelmed with her awareness of her children's sexuality. Debbie wasn't sure if she had failed or if she was being one of these modern day parents who permitted their kids to experiment and find who they were in life. It wasn't that her children were delinquents or irresponsible, only that there was an envy that kids this age should know so much when, at their age, Debbie was clueless about boys and love.

There were also new emotions, feelings of seeing not only her boys, but the Kings' boys naked. Her sex life was almost nil, and she wondered if she wasn't the problem and not Robert's. Maybe she'd let her figure become a little sloppy, but her own daughter complimented how she was still looking good for a mother of four children.

Debbie knew other women in her social groups had teased about making it with their daughter's boyfriend or the kid next door, even hinted of such affairs. Being a senator's wife she was forced to be discreet. Though she wouldn't think about doing anything with her own sons, the Kings' boys were handsome and so well disciplined. She'd felt the heat of sexiness that she hadn't experienced since college. To have one of these boys in bed would truly satisfy her for months.

Their pride in their bodies left no imagination and what woman would not glance at Lance or Dane in all their glory? Alex was her son's, this was something she had to accept, but to fantasize wasn't causing anyone harm.

"Not feeling well?" Robert asked.

Debbie shook her thoughts and saw her husband come into the kitchen in his bathrobe. She realized breakfast was hardly prepared, which meant a grumpy husband. She hurriedly started frying the eggs.

"Long weekend and I'm tired," Debbie said.

"I've never noticed you to be remiss in your household responsibilities. It's not like I haven't given you every amenity available to make your life easier."

"Except yourself, lately," Debbie snapped back and felt good for saying it.

"I've been busy with a lot on my mind. Then you have your own children fail the way they did last week. I have to wonder if we're spoiling them too much."

"My children are just fine. Can't you give credit to a group of boys who were just better athletes?"

"Don't forget the girl. What's puzzling, those athletes were mostly younger than our kids, yet they acted like miniature professionals. There's something strange with that school and I'll find out what it is."

Robert ate without further conversation from his wife. He felt he was losing his whole family. The previous night he'd used his secret surveillance system to spy on his children, something he'd rarely had time to use. The phone conversations, especially, were interesting. His kids talked about orgies, a girl in the shower room, playing video games while naked, and boys having sex while others swam. What kind of academy would allow that?

Phil Masters had given the senator information on all the turmoil happening within St. Bonaventure. Though Robert enjoyed his sense of power, except when his own sons were involved, he didn't want to step on toes with the uppity-ups of the community. Mr. Johnson was certainly one of those. Robert's youngest son wasn't a street fighter but a little fight might put some man in him.

248

Niqui's feistiness with his teachers was the type of rebellion Robert encouraged, only he didn't want his boy getting too big for his breeches. Travis was another problem, a boy who was getting way too independent for his own good and not accepting his father's advice. The senator decided he'd have to take control of this one before the boy went off on his own. His eldest son was soft, as Robert saw it, and he wasn't going to have a patsy for a son in his house.

He pulled out a memo from the headmaster and spread it out by his plate. It was infuriating to read about teenagers from this school his children attended: reports of steroid usage, the pregnancy of one of the cheerleaders, drug and alcohol abuse. Maybe it would have been better not to know, and who asked Kings to report all these? The Johnson boy had accused this Kings' girl of provoking a fight and punching him first. Robert didn't want to be bothered by such nonsense; anyway, the kid's father was a lawyer and he'd take care of any litigation.

Robert decided he was going to do the State of Texas a big favor by investigating, if not closing this Kings Academy. A school shouldn't be run by two gay men. What kind of example is that? One trip and his own family was abandoning the family values and common sense the senator had spent years in developing, and worse, he sensed a few of those Kings' boys had hit on his boys for sexual pleasure. Just a guess, but those phone calls were way too suggestive. Robert wasn't having any faggot-thinking under his roof. Unfortunately he was due back in Washington by Wednesday, still enough time to start the paperwork for an investigation and to speak to his eldest son about his future and a father's expectations.

The day didn't improve for the senator; actually, he received a call from Travis' literature teacher on an essay the boy had written. She had concerns that his analogy showed an inclination for homosexual behavior and wondered if this was a home issue that Travis' parents were aware of. Senator Franco said he'd handle the problem.

Another call came in shortly after that which was both a positive and a negative, considering the timing and the mood of the senator. A Stanford recruiter was in town and would like to visit the Franco household that evening, since he was there only to see Travis Franco.

The senator played his role perfectly and welcomed the visit.

After dinner Robert decided to meet with his son in the library. Travis stepped in wearing his swimming sweats, more than curious of what his father wanted to discuss. His father pointed to the chair the boys knew only too well when it came to a lecture.

"It's been awhile since we've talked, son," the senator started. There was small talk about swim practice and what Travis was working on in the pool and in the classroom.

"What did you think of Father Flannigan's sermon a few weeks ago when he made the point that two males playing house didn't make up a family?"

Travis felt blindsided and sucker punched. To him it was a set-up question. "I don't understand how the Father can preach hatred one second and love thy brother the next, Dad."

"It's like this, Travis. Gays set a bad example for our children when it comes to family values and the roles of a mother and father. They undermine the integrity of the American family."

"I don't buy it. You can't show me a reputable study that shows any harm whatsoever to children living in same-sex households. This is about moral code, nothing else. You always say we may have our own opinion, but only if it's agreeable to you, it appears. Personally, if heterosexuals would keep their sexuality in their own bedrooms it would be marginally tolerable, but they do it right on the street, kissing and holding hands and displaying blatant affection for each other in parks and museums. They are doing irreparable, irreversible psychological damage to innocent gay children and impressionable gay teenagers, like myself, who ought to have the right to be protected from these odious displays of so-called affection."

Travis took a deep breath, one he deserved for showing the greatest courage of his young life. There, he was out. He watched his father's face contort, first in shock, then disdain.

"And, to think, I'm feeding and clothing you," Robert replied with a vengeful wrath that was on the verge of strangling his own son.

"Give me the word and I'll move someplace else," Travis countered.

The senator brought out an e-mailed copy of his essay from that afternoon. "Do you really expect to get into Stanford after writing this kind of smut? If I may quote you, "'He says he has never before seen such beauty as mine—eyes that drain my soul into his like water into the holy cups of Delphi. He says my V to my loins is the perfection of his temple. When he puts his mouth on my neck I can feel the wind of the south on his breath. When his mouth is on my nipples I shiver as though the wind of Mercury has slid down my spine through the crevasse that tightens with the breeze. And when he puts his mouth on my sex I feel the center of a black hole in my heart and my love for him pours out of me.'" Robert slammed the paper on his desk.

"It's how I feel. Mrs. Campbell said to be as expressive as we wanted. She's not a priest."

"Your teacher said expressive, not pornographic! How do I explain this to the board?"

"Tell them you have a gay son."

Robert stood up. "Damn you! I will not! This is nonsense that one of the Kings' boys has convinced you that you're gay. I will destroy that school if it's the last thing I do!"

Maria heard her father all the way down the hall, but she still knocked. "Dad, there's a man at the door who says he's the Stanford swimming coach. Are you expecting him?"

Robert glared at his son, then his daughter. "Tell the man we're not interested. I'm sorry for wasting his time."

Maria nodded and left.

Travis began to cry, stood up and fought through his tears to think of the meanest thing in the world to say to his father. His own goodness prevented the words from sliding out. He ran to his room and wiped his face with his T-shirt, then saw the rope ladder by the foot of his bed. Out the back window he ran around the house and barely caught this coach before he started the car.

"Sir, I'm Travis Franco. I really want to go to your college, if you want me. My dad isn't in a very good mood and won't support me on this."

Coach Benson gave a puzzled expression and one of deep concern.

"Son, I came here to talk to your parents and you about our program. Your dad is a powerful person, being a senator, and you're only seventeen. You can see my problem."

"I promise you my mother supports me completely and I'll be eighteen in a few weeks. That must mean something."

"Yes, you can sign your own scholarship, but I'm concerned about your mental state without the support of your father."

"Coach Benson, I'm going to be square with you, though I might end up losing this opportunity. I'm gay and my father has issues with this."

A smile creased the coach's face. "This I can deal with, and I certainly empathize with you, Travis. Look, son, we have several gay swimmers on our team and on our other athletic teams. This isn't a problem at Stanford. If you still are interested I'd like you to visit our campus and make a decision based on that. I'm sorry your father feels the way he does, but give him time."

"Yes, sir, I'll do that." Travis watched the coach's window rise, then realized he'd forgotten something really important. He knocked on the glass.

"Coach, there is one more thing. This might also cost me my scholarship, but I know I want to tell you. There's a boy in an academy outside of Dallas who is faster than I am. It's called Kings Academy."

"Why haven't I ever heard of this school or the swimmer?"

"I don't think they advertise too much and it's a real special school for great kids. Actually, Coach, you better be prepared for the shock of your life. If you had a dream about how you'd wish to live your boyhood, you'll about to see that dream come true. Course, I doubt if you're gay, so just pretend you're a gay guy and you're in heaven. That's the best way to describe it."

Coach Benson laughed with this teenager's conception of this academy. He was intrigued, nonetheless, by the boy's perception. "Are you sure about this, Travis? I mean, I'm not aware of any academies that are just for gay kids."

"Coach, I'm in love with one their boys. Trust me, I know gay boys when I see them and my boyfriend holds the fastest time in the

252

nation this year. Don't forget, he beat me. The third place finisher was right on my shoulder, and he was a Kings' boy."

"Travis, if you're right about this…..just don't worry about the scholarship. You're an honest and talented young man. Stanford would be proud to have you." As Coach Benson drove away he had the broadest smile on his face in anticipation.

Back in his room, via rope ladder, Travis pumped his fist into the air and eyed his erection in thoughts of Alex. "Yeah!"

Chapter Twenty

Toy and Kami rested in bed after an hour of rambunctious sex; their love and relationship after sixteen years had only gotten stronger as time went on.   To the side of this king size bed stood a hundred gallon saltwater aquarium and both men snickered with the recognition that the four clown fish had enjoyed watching the antics of their sex play.   The clowns began to pair off in a romantic feistiness of chase and rub, as if the male sexual escapade had incensed their arousal.

"Amatsu sora naru," Kami blurted out in his second language, which meant, 'That is what it means to love.'

Toy laughed, for he, also, had learned the Japanese language throughout the years.   The two men were alike and different, if not out of necessity.   They were definitely an odd sort, both aspiring teenagers when they met and had two children each by the time they were both eighteen.   Combining their talents and ingenuity they were in control of an academy that bordered on experimentation.   In control of almost two billion dollars was a major undertaking and a tremendous responsibility for two young doctors.   To keep an academy of this size and afford the daily living expenses of an abundance of teenage energy was an enormous achievement.

Designing the academy had been the easiest part; developing a structured program of sports and academics to accommodate 29 children of historical importance took on a whole other meaning.   Combining the rigors of both mental genius and physical mastery just seemed appropriate and right to the two young men who did massive research in every element of raising children from start to finish.   What they had come to realize, there is no finish.

Toy rode up on his elbow and looked face to face with Kami.   He moved his index finger down Kami's chest and around a nipple.   "We might have a curious senator on our hands."

"You're speaking of Senator Franco, I presume.   I saw his charisma at the last competition, though I think he's full of himself," Kami said with certainty.

"I've looked into his career: A stubborn, narrow-minded man, very inflexible, touchy in matters of pride, and excessively loyal to those who can pat his butt and inveterately hostile to his foes, regardless of circumstances. He's been asking around about our academy, so says my father when he called today. Dad knows too many people in politics around the state. Then I received a call from a Phil Masters, the headmaster at St. Bonaventure. He asked a few personal questions about that incident with Wendy and the boy from their school. It appears the boy's father is a lawyer and might pursue some legal action."

"Some people's kids," Kami said.

"Some kids' parents!" Toy countered. "They think their child is perfect or they see a fast buck. Lawyers! Never can figure them out. Their so-called court of law is so sacred it must be protected by a bodyguard of lies, and they're just the actors to make it happen."

"Could it be, Toy, that we've just kept our heads in the sand in hopes that nobody would notice that two thirty-somethings are bringing up almost three-dozen children? I know we've created our own utopia here, but this senator is teaching us a lesson in dealing with reality. Speaking of the senator," Kami paused, "it appears a few of our boys have fallen in love. We never considered this aspect, as least with outside kids, but the more our boys explore their environment this was bound to happen."

"I agree, Kam. They're beginning to explore their independence, which I don't blame them. One of the things we've been adamant about is presenting these boys as boys, not geeks, nerds or elite kids, born with a silver spoon in their mouths. They're born-again Christians who are home-schooled; okay, very liberal minded, hyperkinetic and exceptional, but I don't believe our boys think they're better than others their age, just more disciplined."

"There is the inevitable," Kami said and moved up to a sitting position. "I suppose we're at the stage, well, like, coaching a potential Olympic athlete or Rhodes Scholar. There comes a point where we have to exert more effort to stay one step ahead of this talent or else the child will feel stunted."

"Do you think we should confront them with their escapade of leaving the Franco house? I'm wondering what that was all about," Toy surmised.

"Would you want your parents questioning your every more when you were a kid? I think one of best things we ever did was allow our boys, when they wanted, to sleep with us, either by themselves, or with one of their brothers. They really opened up to us with their thoughts and feelings. Those were some of my finest nights, cuddling with our sons in bed and sharing the love we have for each other. I kind of miss that since they've gotten older. It seems when they turned fourteen it has given them a new independence."

Toy chuckled with the memories. "It's too surreal, Kam. Here we are, barely out of our twenties, just big kids ourselves, at times, and we're raising over thirty teens. Have we screwed these kids up, or are we the quintessential parents of the future?"

"I've wondered myself, at times, Toy. To think over the years what behavior we've instilled in our boys and your daughter. We've made them what we would have liked to have been at their age. Can you imagine either one of us being like Luke, when he slept over at thirteen and couldn't wait to show us that he'd hit puberty? Or last year when Bode and Brett slept over and they told us they didn't mind if we made love, because they were going to themselves right between us? Our kids want us to mirror and compare, as well as share their companionship and apprenticeship."

Toy smiled with these stories. "Trust me, I'd never considered sleeping with my parents and masturbating in front of them to the celebration that I'd discovered my adolescence. I'm only concerned what can go wrong when a boy's identification with his father is over eroticized and is transformed into a perversion by a trauma of some kind."

"Okay, let's examine that. Take Tad, for instance. Last year we asked him to spend the night because we knew he was slipping in class and in athletics. For some reason he was questioning himself and his identity. We had dinner together, like we do with all the boys, searching for conversation and openness. He appeared solemn at the dining room table and while we showered together. In bed we watched the Tonight Show, waiting for some reaction that he wasn't happy, and then he rolled over on top of you and rested his head on your chest."

"That wasn't surprising, but I knew we had a breakthrough," Toy remembered. "I felt his erection and I couldn't stop my own. He ejaculated within seconds, looked up with his silly grin and said, 'I love you, Dad.' I laughed and said, 'I get that impression,' and we laughed, though you weren't quite sure what was going on. When he fell asleep on my chest I rolled him over to go and get cleaned up, and he was in your arms when I came back."

"He needed one of us to hold that night. But the boy turned a total one-eighty from that day on. His attitude skyrocketed and his performances improved; all because of a love in which the person seeks to find in the other an ideal image of himself."

"Is it that simple, Kam? Whatever happened to the oedipal scenario where one must negotiate between their perception of the father as a playful and exciting figure of liberation and as a forbidding and frightening figure of discipline?"

"Because these boys have never known a mother figure, Toy. Sure they've relished the appearances of our sisters and parents, but they're gay and we're their examples of masculinity. Our sons haven't had to endure homophobia to any extent; they haven't had to play games around their feelings of gayness. Other boys discover in certain sports a masculine universe where they can enjoy the company of men and the spectacle of their bodies—as long as it is framed within competition, a struggle for dominance. The game itself becomes the phallus, something to be forever pursued and worshipped, and something that bestows manhood. These boys don't need sports for that, they have us and each other. We've never pursued sex with our sons, but they look to us to challenge our boundaries and to define their own sexuality. For you to have ridiculed Luke that night would have destroyed his ego and hope. When we shower with our boys it shows our acceptance, that we're no different or above them. Why should we be?"

Toy kissed this man of his life, a man he had watched grow from a boy not much older than their sons now. "No argument here.

Remember when we'd have to have sex in the afternoon because we were inundated with boys coming over every night? I couldn't wait until they were older and we had our evenings back. Now that they're fifteen and we don't have them relishing to sleep in our bed, I miss that. Those were the days."

"I had the same frustration, at times. Can we really see those times as something we didn't love and look forward to? Like when Marc was frantic to sleep over because he just had to ask us if Tad was right, and there wasn't really a Santa Claus. And Spirit had to role play with us because he knew he was in love with Erik, but didn't know how to approach his brother with this. And Aki wanted to improve his swim stroke, so we spent a night with him in the pool. Brett wanted us to read him anime books instead of the classics, in contrast to Tommy, who loved The Man Without a Face and The Huckleberry Pirates."

"Remember when Wendy wanted to have a pajama party? That lasted an hour and when she whipped off her pajamas because she wasn't used to them, all the boys took theirs off and pajamas were flying every which way. Then the pillow fight started and there was foam rubber all over the place."

Toy and Kami broke up with continuous laughter over their recollections. "Don't forget there are several of the boys who still enjoy their nights over. We'll have to remind them we're still available to share our bed. Tad wants to sleep over tomorrow, and Randi wants the weekend. Zach is talking about next week. Come to think of it, I think the boys still like the time away from their brothers for an evening. I'm not sure we should investigate them for any secrets floating around."

"Remember when they were younger? They loved to sleep on our chests and cuddle. I miss those times. Yes, I'd like to know what they were up to in San Antonio, but have we ever told them they all have a GPS chip in their heel? It's a tough call. I think we keep reinforcing their awareness of STDs, accountability and responsibility to each other. The China trip is coming up, so they have to stay focused and be able to separate love from the task at hand."

Toy rolled over on top of Kami. "That's why I married you, your common sense and insight; plus, you're sexy."

Nonetheless, Toy and Kami did decide to invite three of their sons and one daughter to their home high on the hillside overlooking the broad acres of the academy. This wasn't that unusual since the men often invited different boys for dinner at their own request to chitchat and for the child to learn social graces. Dinner was no less extravagant, and as soon as Alex, Dane, Lance, and Wendy arrived, they sat down at the exquisite dining room table, surrounded by glass, with the academy's night lights sparkling like tiny stars on the earth's surface. Kami had the telescope lens on the heavens that projected a constellation onto the wall beside the table.

Alex humored the moment. "So do we get to shower together and drink hot chocolate before we watch the late show?"

"I remember when you couldn't wait to do that with your fathers, young man," Toy scolded.

"Dad! I wasn't mocking that. I just wondered why you invited us four."

The laughter and relaxation didn't predispose the students that something was terribly wrong, though it was odd that just the four of them had been invited. Actually their fathers appeared in good moods. Toy finished his glass of wine and inched back in his chair enough to address everyone.

"Good to have you here tonight. I put a great deal of trust in to every child here at the academy. If these expectations are unrealistic, please speak up." Toy waited and received only blank stares. "Okay, without getting into the where-with-alls, you three boys went out on a little excursion when you visited the Franco's for the birthday party. This is not confession, but, may your fathers feel confident that you did nothing to bring embarrassment or shame to the academy? It appears you were in a hurry and went all over the neighborhood."

"How do you know all this, Dad?" Alex asked.

"Son, you know how I am with my son asking a question to a question. Get to your answer."

The four teens glanced at each other before Alex took the initiative. "No sir. There shouldn't be anything that will ever come back on the academy, Dad."

"Good! See how easy that was? I also noticed we had brother and sister in the same bed. May I assume you had company? Not so strange, but now we have children who aren't ours probably engaging in sexual activities. Are the participants being cognizant of STDs?" This was just a guess of Toy's, but he pursued it anyway.

Four nods followed the string of questions. "Do you have any questions, Kam?"

"I'm not going to give you my sermon on love. Are we all on the same page in your focus for school and sports? No dreamy eyes? Fuzzy thinking? Perpetual hard-ons?"

This received laughs, even from Wendy. Dane presented his view. "Dad, I think I'm in love with Niqui. I mean I love all my brothers but there's something special between us. I hope we can take him to China with us."

"Is Niqui gay, Dane?" Kami asked.

There was a pensive hesitation. "Probably not, but he enjoys the sex."

"Dane, boys your age have urges that would invite Genghis Khan to their bed if the guy was acceptable. I'm afraid you're setting yourself up for disappointment. Sure the boy is probably even in love with you, at least the sex part. He loves your company, your humor, your personality, but, in the end, when he finds that particular girl, his passion will go to her. Do you really want that downfall?"

Dane thought about this. "I've thought a lot about that. I think I can share love with Niqui and know that it will end. He's a cool kid and we have great chemistry. It's not all sex since we talk about politics and our future. We do things because we enjoy giving each other pleasure, and he even admitted to me he had kissed and masturbated another boy his age last year. Though he thinks he'd like to get married someday, Niqui knows he fantasizes about us two being together and masturbates to these. He even has become very much the gay activist."

"That's a mature viewpoint from both of you. There's something else to consider with this family. Senator Franco has taken a vendetta

260

against the academy, I'm afraid," Toy informed his children. "How much his family is affected by this resentment is something you might surmise if your visit to San Antonio is still on next week."

Wendy perked up. "I spoke with Maria last night and their whole family is in turmoil. It appears their father has gone over the edge. Travis came out to his father and the man completely rejected him."

"Welcome to humanity, kids," Kami said. "Feel lucky, because other boys don't have it so good."

"Dad, I have a question," Lance said and glanced at both fathers since he considered them equal to his upbringing. "Our penis sizes are much longer than any of the Franco boys. What's going on?"

"We mentioned this when you guys were twelve but nobody seemed to care what other boys looked like. Lance, as a three year old, you saw the other boys swimming in the lake and wanted a penis just like theirs. Well, you've seen Kami's enough times to know you have nothing on him. As a matter of fact, that's where you got your gene. Yes, we have eliminated health problems, depression, and other defects that people inherit at birth from their ancestors, but this was a special gift to my son. Dr. Jolson did the same for the other twenty-nine boys. You are all aware that genetic research has advanced to the point we can design the human body, internally and externally, like an architect building a house. Your appearances have not been altered, but your teeth are genetically improved so that cavities are rare. You won't have hair growth below your eyebrows and you are all aware that from a nutritional viewpoint we treat your body like a well-tuned Porsche."

"Ferrari," Lance countered.

"No, a Lamborghini," Alex replied.

"Okay, okay, okay. Your dads have Porsches, so they rule," Kami said and then had several carrots thrown at him. There was a beep from a security panel and Toy activated the voice response.

"Dr. Kerho, this is Roger at the front gate. A Coach Benson is here from Stanford. He says he doesn't have an appointment but would like to speak to either the head swimming coach or the director."

"What's it concern, Roger?" Toy asked.

"Something about your son, Alex."

261

Toy glanced at Alex. "I didn't do anything, Dad."

"You've been saying that for fifteen years, young man." Toy turned to the panel. "Send the man to my home, Roger, and call Coach Hawkins to report, as well. We have plenty of dessert."

There were introductions and pleased fathers when they heard the coach was there on a recruiting trip. The Stanford coach was amazed at the ambience of the home and the fact that the entire décor was set in the architecture of Japanese. His kimono, a variation of what everyone else was wearing, was a gift he could take home.

"I'll never tell my wife I had to take off my shoes. She'll think it's a great idea," Coach Benson said with a smile. "Is this a government secret? I can't believe the beauty of your campus and it's night time."

"We're not a public institution, Coach," Toy said. "If you'd like a tour we'd be glad to accommodate. For what purpose do we honor your presence?"

"There's a boy in San Antonio, you might know him, Travis Franco, who said there are boys up here who are faster than he is, at least one. I'm here to tell you I'm interested in giving this boy an education, if he is valid."

"They are, Coach. Meet Alex, my son. The other is Alan Marshall, also my son. Oh, here's Coach Hawkins. Coach, you know where the kimonos are kept."

The men talked for half an hour and Coach Benson knew he was sitting on a gold mine. This Stanford man listened closely and comprehended that two of these men were married to each other and taking care of 33 children.

"Can I have the boys and, of course, their fathers, come to Stanford for a visit?"

"Plan on that," Kami said. "There's only one thing you should know. The boys are only fifteen, so you're going to have to wait a couple more years."

"You're kidding?"

With a mere command on a digital keyboard the swimming race between Kings and St. Bonaventure shined on the wall. The coach watched with amazement at the race and the times. Other races were equally as stunning and most of the times by the Kings' swimmers were equal if not better than most high school swimmers.

"Can we keep this a secret?" Coach Benson asked with a grin.

"We'd appreciate that," Kami said. "Come on, we'll show you our campus."

The main tour went exceptionally well, though venues were quickly lit up and security had to hustle to prepare. He was shown advances in the sport of swimming, including the silicone suit that had his mind spinning. The future of sport was right here at his fingertips, yet he couldn't possibly remember everything he was witnessing in such a short time. Coach Benson scanned a wall display of muscle fibers, both fast and slow twitch, which he was quite familiar with. He was just about to ignore this student experiment when he saw a third type that he wasn't familiar with.

"Excuse me, but is this one of your teacher's projects?"

"Our boys do their own research with the assistance of their professors. What you're looking at is one our more creative student's work. It's a kind of muscle fiber known as type IIX, which falls between the typical fast and slow categories. The fiber is a combination of speed and endurance. Bobby found a master gene, called PGC-1beta, that when switched on can convert almost all the muscle fibers in mice to IIX. Bobby found that when the IIX mice were put on treadmills, they ran for 25 percent longer and covered 45 percent more distance than their normal littermates."

"Do you have any idea where that could lead?" Coach Benson asked in disbelief.

"I know what you're getting at, but we're directing this research away from making super athletes to more medical uses, like treating muscle-wasting diseases like muscular dystrophy."

"So you have graduate students here?" the coach presumed.

"No. Bobby was thirteen when he discovered this with Johnny, another thirteen-year old at the time."

"I have to meet these boys, if you don't mind. What if I brought my team here to swim with your boys?" the coach thought out.

"Fine with us, Coach. Bring 'em for a week and let them train with our sons. They'll never be the same," Toy said and winked at Kami.

"Where are the kids?" Coach Benson asked.

Toy led this Stanford coach to a double door that opened to a massive dojo. Inside Oshi was fighting two other boys in the sport of Kendo. Tatami mates were set along the sides of the dojo, where all the other boys sat respectfully as the match commenced. The men watched intently as Tad and Rom were giving their best to outwit this expert judoka from Japan. The kendo shiai (match) was fast and furious, and Coach Benson was amazed with the vigor and speed of the artists.

"Yeah!" Toy shouted, but the glee of surprise wasn't for his best friend but for the strike from one of the boys that ended the match.

The three kendoists bowed and took off their head guards. They all turned to the applause and eyed this surprised guest standing by the door.

"Gentlemen, I'd like you to meet, Coach Benson, from Stanford University," Toy announced. At once all the boys stood and bowed in unison. Coach bowed in return in mutual respect.

The coach followed the directors' lead and took off his shoes as they entered the confines of the dojo. Kami congratulated the boys for their victory over Oshi.

"Must suck getting old," Kami said and received his expected laugh from the boys but a frown from his dear friend.

"You've taught them too well. Gone are the days we used to take on six at a time and swat their pink asses," Oshi said with pride.

"Soon they will be taking on us two, one at a time," Kami said.

"Not you, my dear friend. You can still handle three at a time," Oshi reminded Kami.

"I'm not so sure," Kami replied.

"A man who questions himself has succumbed to the first hint of failure," Oshi said.

"Not me to succumb to my sons," Kami said and pointed out three of his finest kendo fencers from the group: Dow, Tommy, and Jacob. He removed his kimono and was immediately given a keikogi (jacket) and the hakama (skirt like trousers). Bobby hustled over and gave his father a tare or waist protector, and then the do, or chest protector.   A hachimaki was wrapped around his head to keep the sweat from dropping into his eyes.   The men, or head guard, was fitted on and firmly tired in the back.   The last items to be worn were the kote, the arm guards.   Dane brought his sensei's favorite shinai over and Kami was ready for the match.   A few calisthenics to prepare him and the three boys bowed at the same time as Kami.

"Ki o tsuke!"   Oshi yelled to the rest of the boys, which meant to sit up straight and pay attention.   No one dared to question or delay Oshi's orders.   His command of "Rei!" had the four fencers bow.

At the corner of this arena Coach Benson stood with Toy; the two men were both excited with seeing a competitive match in preparation.   Toy relayed to the coach a great deal of information on what was transpiring and especially about the love of his life.

"Kami was the Japanese All-Japan High School champion when he was fifteen.   His sensei, Mr. Kamito, was Japan's champion for six straight years.   Kami was the first non-Japanese to hold this honor."

Coach Benson nodded and was completely intrigued by all that was presented to his eyes.   He was surprised that the director had disrobed to his skin in front of the boys and was dressed by the students.

"Do all the boys participate?"   The Stanford coach asked as he eyed the many faces of attentiveness.

"All the boys have taken martial arts training since they were five.   They all hold black belts in three arts, including Kendo.   Kam told me the other day that half these boys are better fencers at this age than he was.   That's saying a great deal.   To take on three boys at once is a feat that revels the likes of Toshiro Immune."

"I have much to learn," the coach admitted and then heard the word "Hajime!"

Kami and the boys had bowed, taken the sonkyo position, risen, and assumed the on-guard maai position, and after a brief moment to permit them to compose themselves, Oshi had commanded: "Hajime!" (commence!)

The match was like watching samurai attack, with parries and fast footwork that amazed their visitor.   Toy was describing the match and the confines.

"The floor area is twenty-seven by thirty-three feet square. The boundaries are marked with that white line four inches wide. In a match there are several points that must be scored; here only one hit removes the opponent. The shinai or fencing foil is the same weight as a regular sword, less than three pounds. Kam used to take on the boys blindfolded, so great is his perception and intuition. Those days are gone, I'm afraid. The boys would be equal to many an ancient samurai, so great is their skill."

Kami's footwork and skill with the blade appeared so effortless the guest barely saw the slashing do-giri stroke that eliminated the first boy from the trio of attackers. The second boy was felled with a kesagake movement, a whip-like slash of the blade. Though the final boy hesitated with his two colleagues removed, the boy attacked furiously and was parried with ease and lightning fast defensive moves.

The coach was awed by the dexterity and speed of the fencers; the swiftness of bamboo and the lightning fast avoidance of attacks and feints. Wrists snapped these wooden creations with precision, only to have the fencer make a rapid retreat or a mere boxer's dodge to avoid a deadly slash if these weapons were truly made from steel.

Kami gripped his sword in his left hand and made a half circle to the left. His left foot pushed forward and stepped about five degrees to the left, while the right foot was stretched and planted firmly in position on the dojo floor. As the attack was made, Kam called out: "Yokomen!"

"In kendo the attacker calls his hit as he attacks," Toy relayed to their guest as contestants bowed at the conclusion.

The boys applauded this skill and they all bowed to their sensei, who showed them that his mastery was unequaled by even the number three. Kami wouldn't directly admit it, but these days weren't for long when the boys would get the best of him. He stood before his kendoists and exclaimed: "What have I taught?"

"The eyes first, the footwork next, the courage third, and the strength fourth," they all said in unison.

"Hai!" Kami proclaimed. "You think your courage and shrewdness are enough, but the lesson is always?"

"In order to learn the techniques, exercise your footwork first rather than your handwork," they again repeated this together.

"Mr. Lincoln! Name the five postures," Kami asked.

Tad stood and called out: "Chudan no kamai, jodan no kamae, gedan no kamai, hasso no kamae, and wakigamai, sensei!"

"Correct! Tea and hot bath! Get to it!" Kami shouted and returned the bow of the boys and Oshi.

"I'm impressed," Coach Benson told Toy. "This is beyond what we can teach at Stanford."

"Stanford is a great university and would serve our boys well," Toy admitted. "Perhaps you will join us for tea, sir."

In minutes the coach was sitting with all the students; the boys dressed in their yukata and partaking of this ritual with seriousness and listening to this coach praise them for their politeness and discipline.

"Perhaps I may see where you live," Coach Benson suggested.

"We'll head there next," Toy said and they were soon at the entrance to the dorm. "You'll have to strip, Coach."

"I'd ask if you were kidding but you never are." Coach Benson took off his clothes like everyone else and was soon parading through the living quarters in his birthday suit. The vastness and ambiance of the boys' dormitory had him wide-eyed and in awe. Seeing the museum first the coach caught sight of a picture of the boys in choir at a concert hall.

Kami explained this. "Every Christmas Eve, since the boys were seven, we give a concert to the homeless at the Dallas Convention Hall. It's grown to quite a spectacle that the public became more involved than the homeless. Now we reserve half the seats for the two-thousand homeless and street people, and sell the other half to support the program for the homeless. The tickets sell out in January the previous year."

"Marvelous!" Coach Benson spoke with admiration. "It's amazing how they've grown from these darling little boys to such a handsome lot. You must be proud."

"It's just important the boys remember what Christmas is all about and giving something of themselves to those less fortunate. It makes

Christmas morning that much more meaningful for the boys. I can't tell you how much it means to those homeless in Dallas and the inspiration it gives to the city," Toy remarked. "Come, I'll show you the rest of the dormitory, but I must warn you, this is the boys' domain and we don't restrict their openness with each other.

Coach Benson didn't really catch what this meant until they were entering the pool area and the coach saw two boys having sex on top of a gigantic stuffed elephant. Since this didn't appear to faze their fathers, Coach Benson didn't blink, but he sure did a double take with the length of the boy's penis entering his peer. If the elephant wasn't complaining, neither was he going to remark about the blatant display. What was even more amazing, the nubile young lady in the pool didn't even give the two boys a second glance.

Mr. Kerho called for his daughter to meet their guest and introduce her brothers as they went from area to area. For sure Coach Benson was never going to tell his wife that a beautiful fifteen-year old girl had shown him around, but, if he did, he sure wasn't going to mention that she was naked and had him talking to himself not to glance for too long at her gorgeous body. His curiosity grew from viewing penises that would make most males envious. He couldn't exactly admit to this voyeurism, so he didn't say anything.

They stopped by the music room and listened to three boys practicing for a concerto for Sunday morning's church service. They played brilliantly with two other boys singing a duet. As a licensed pilot himself, he had to climb in the F-22 and give it a try. He crashed on take-off and felt like a young boy who wouldn't surrender to defeat. Tad Lincoln climbed in the cockpit of the stealth fighter and the two males had a dogfight, which Tad won easily. A few practice swings on the virtual golf apparatus also humbled the Stanford coach.

As he was escorted to the different venues within the dormitory the coach paused to watch a couple of the boys dancing, while others hung onto each other in friendship and in laughter of their peers in a type of African dance. Only these dancers had erections in their tantalizing moves with each other.

"They have no reservations, do they?" The coach stated to Dr.

Kerho.

"To tell you the truth, Coach, we've learned as much from our sons as they have learned from us. I don't have to remind you of the anxieties and worries we all faced as adolescents when we were growing up. Be it sex or socializing with our peers, there was always some concern if we really fit in or met with others' expectations. By allowing our sons to be who they are we've excluded their anxieties and worries as by-products of excessive linear analysis. The boys have learned more about their most profound inner experiences by turning away from analytical intelligence and affirming their simple "primitive" responses that often characterize the animal world. One of our basis impulses is to sexually act out. These boys have no reservation, as you say it, to experience their arousal and love of sex with each other."

"Fascination," Coach Benson replied. "And you say this affects their athletics?"

"We train our sons in all areas of life: music, social skills, sports, art, and science. Whether they're playing a musical instrument or using their computer typing skills, we teach automatic thought; the ability to bypass the analytical thinking faculties and move directly to the motor memories. We don't stress on winning, per se, but that the best performances involves "instinct," "letting your body do the work," or "getting your mind out of the way." Why do our boys excel, you may ask? They know the secret of visual perception plus muscle reaction. When an athlete thinks consciously about physical mechanics or movements, his or her performance usually deteriorates."

"In the zone, we call it at Stanford."

"Yes. It's an overall quieting of the brain; yet at the same time, certain areas of the brain, such as the attention centers and executive-control and focus functions, become active. In addition, the boys know that distinctive biochemical changes occur, such as the increased release of nitric oxide and neurotransmitters, including endorphins and dopamine. Our boys play sports with humor and excitement."

"And how to do get them that relaxed, Doctor Kerho?"

"We show them movies of the Three Stooges and the Marx Brothers before competition." Toy laughed but he was dead serious.

A tour of the boys' living quarters had the coach in awe of this advanced electronics. He was introduced to a robot, of all things, with customizable software and one that could do a facial-recognition scan on anybody who enters the dorm and take pictures of and deliver warnings to uninvited guests.

Toy stepped into one of the boy's rooms and exhibited an LCD with touch-screen features for controlling temperature, lights, music, and ambience. The plasma display showed content that was streamed wireless from the media hub in other buildings. The boy's tablet PC with speech recognition could be used for dictation, or a portable videoconferencing with any professor in the academy.

The coach was laughing to himself and having the time of his life when they left the quarters; his nudity was only remembered when he departed this wondrous structure carved into the hillside. Coach Benson now knew exactly what Travis Franco had meant.

On the walk down the grass path, Coach Benson spotted the thirty-three lit statues surrounding the track stadium. "May I?" he asked and received an escort to this most unique exhibit. He read the poem by Stephen Spender at the entrance to this collection of artwork.

Human bodies—proud and magnificent, radiating glory,
honed and powerful, every muscle in clear
relief even when inactive. Each
figure shining as if dipped in a "dazzling whiteness."

After his own experience in the dorm the nudity was not even considered offensive or disturbing. It put his mind in a state of relaxation that he'd never quite experienced before. He turned toward Dr. Kerho and Dr. Marion.

"There is something eternal and absolute about these figures you've done for your children. I can only believe this is an ultramodern vision of the future and a sense of individual identity, so intrinsic to our own age. I can't wait for my next visit when I can photograph this for memory."

"Thank you, Coach. It's been our pleasure having you join us tonight. Next time you might consider bringing your wife. She would enjoy meeting our staff and our sisters."

"Thank you, Mr. Kerho. I'm not sure I could convince her to walk around naked with so many cute boys in the vicinity."

"Tell her all our boys are gay and she'll have the time of her life," Toy said.

"For some reason I think you're right," Coach Benson replied with a laugh. "I only dread she'll want me to get one of those surgeries to elongate my penis."

The men enjoyed the humor and promised another meeting in the near future.

When Coach Benson departed he was on cloud nine. Never in his life had he expected to partake of such an experience or come across boys with such gifts and support of loving adults. It wasn't something he could describe and make sense without expecting judgment or embarrassment. Even the boys and one girl were too well-behaved to the kids he often recruited or coached. He was wondering if these boys would even fit in to an environment of a university like Stanford. The coach wasn't sure if he wasn't leaving the future of sport in America, an absolute gold mine of talent and academic success. One of his immediate troubles, he never got the director's phone number or address. Only a gasoline attendant in Dallas had told him to follow a highway, off to a side road, then another winding two-lane road through a valley until you run into a gate.

"Keep your hands up and don't make any fast moves," the guy had said very seriously. Coach Benson had thought he was kidding.

This successful university coach had a quick inspiration to tell the press about this academy, or even the Olympic committee. Why should such an institution be kept secret? A few seconds of pondering this idea and the coach thought better of such a revelation. 'That would mean every Division One college in the nation would be privy to this pool of superstars,' he admitted out loud to himself. The coach decided then and there to put that idea in the closet.

Chapter Twenty-One

Niqui felt like a new gymnast, given the mental exercises and technical work by Dane. His confidence restored he was also challenging his hubris by smiling back at girls and being more sociable. He was even more surprised when a senior girl approached him in the hallway and asked to take the sophomore to a movie that evening. Niqui fought for words, though his initial words were, "Are you sure you have the right boy?" She nodded and the date was set.

The mere thought of going out on his first date ever didn't include telling anyone about his new step up in life. Possibly this would be his breakthrough, the spring hormones he'd heard whispers about were overflowing in every direction in his body with the anticipation of maybe a kiss. Dressed as cool as a rapper Niqui casually mentioned to his mother that he was going to the movies with a friend. He patiently waited at the curb and the girl still parked a half-block away to force the boy run to her car.

"Didn't want to make a scene," she told Niqui. He didn't get it but smiled anyway.

They were at the ticket window when his date held out her hand. "Give me fifteen dollars."

Niqui was prepared to spend his own money; it was the approach that seemed a little awkward to him. Considering this was his allowance for the month he watched the hard earned bills passed to the teller. In his mind he remembered the hours of sweat in mowing the lawn, cleaning the pool, and trimming the hedges. Another ten dollars on snacks and Niqui wasn't too sure he was ready for this dating thing.

The movie was a girl flick; she thought it was cute; Niqui was bored silly. Trouble was, the girl allowed Niqui to place his hand on top of hers, but it just laid there like a dead fish. He prayed that the girl wouldn't drive them to a restaurant or even a fast food chain. Ready to make up an excuse of a stomach ache Niqui was glad when his date pointed her car to the outskirts of town and parked on a remote dirt road that had the reputation as a making-out spot.

"Okay, handsome, let's get in the back seat," she ordered and they nearly knocked each other into the seats when they collided.

Finally this senior gave Niqui a push with her hand to the far corner of the back seat, then moved to where she was lying on her back. "Know how to kiss?"

Niqui felt comfortable with this; after all, Dane said he was a great kisser. "Yeah, I'm not too bad," Niqui said with confidence.

He brought his mouth close like Dane had taught him, tender, at first, then soft to torment and bring the tongues together. As the pressure ensued Niqui found the girl's mouth mushy and way too big. Her tongue swished back and forth and practically sucked Niqui's lip from his jaw. Niqui was convinced kissing a snake might have been more fun. Practicing on a washcloth would have been equal to this pleasure.

When this tongue lashing didn't turn him on he moved his hand to one of the voluptuous breasts squeezing into his rib cage. Not even his mother had tits this huge. With skillful ease he was under the girl's sweater and discovered that bras have metal bars—either that, or this was a cage to prevent this type of attack, Niqui wasn't sure.

Not being able to maneuver under or through this device, his date sighed with disgust and raised her sweater, then unhooked her bra with even more impatience. The mammoth tits hung like balloons below Niqui's raised chest and all he could think of was that they looked like his mother's boobs--and this after four children. It was a total turn-off and not the type of perky breasts like his sister's or Wendy's. The feel wasn't much better with this large mass of tissue in his palm and the smell reminded him of his wrestling suit after three days of use.

"Get your feel in?" She asked, like this was a major inconvenience. Niqui had to wonder what he was doing wrong.

She pushed him back as she swept off her skirt and panties, while pushing him to the side. A dark bush stood out in the night with a greenish glow from the light of the radio, barely audible music that Niqui distasted—rap. While he wondered if this is what he really wanted, his pants were being dragged downward with the dexterity of a wild animal. She all but ripped his underwear in pulling them down.

There wasn't a whole lot of excitement generated by this simple act of undressing. She began to fondle his sex roughly before jerking her hand away.

"What's wrong?" Niqui asked.

"Shit, kid, haven't you even reached puberty?"

"I shave once in awhile. Like swimming, it's a gymnastic thing, too," he said but almost laughed. She bought it.

"Don't I turn you on?" The senior pondered as if this was a given.

"Yeah, sure. Guess I've been concentrating on other things," Niqui answered but knew he'd rather be studying advanced calculus than on top of this corpse.

This girl wasn't at all what he envisioned. A vision of his sister flashed before him: slender hips, narrow waist, tits that were firm and pointy, and a crotch that had a slight trace of hair up its crack. In contrast this girl was soft like a pillow and her thighs were rounded; plus, her sex had a strange smell of a fish that had been left on a boating dock to rot.

Niqui was bothered by his clothes around his knees so he kicked off these and brought his underwear up to rest over the front seat in case someone approached.

"Nice underwear," she complimented. "At least you wear clean stuff. I've had a few boyfriends in my day who were real slobs, if they wore any at all."

Using saliva Niqui tugged on his penis with his head arched back, thinking what this would be like if Dane was below him. Close to orgasm he pointed his erection downward.

"Not that hole!" she yelled and nearly scared Niqui from the reality that he might actually have intercourse with a girl. She continued with her roughness in guiding his member into her vagina. Even then Niqui wasn't sure he had penetrated.

"Am I in?" he asked.

"Franco, if you're not, there's another guy here in the back seat with us."

For a mere second Niqui thought about the time he and Jay tried different fruits to stick their erections in. This one felt like that orange

274

which had become pure mush after he'd shoved his erection through. At the same time his neck was being abused with the suction of ten leeches. Far better than his mouth being slobbered on, Niqui tolerated his vampire bite.

This senior girl interrupted the reflection. "Well, are you just going to lay there?"

"I've…already, you know," Niqui confessed.

"Shit! Fuckin' beginner!" She pushed him backwards. "Get off of me, it's getting late."

Niqui felt no difference out than he did in. He began rounding up his clothes, but he was now agitated at spending so much money and being verbally abused by this girl who he thought liked him.

"Don't you like to kiss ears and neck and run your tongue down the person's chest and around their nipples? You know, lick all over and say tiny whispers and jokes, like sex is fun? We didn't even laugh." Niqui recanted the experiences he'd had with Jacob and Dane.

"Kid, you need a pet Chihuahua, not a girl."

"I'm only trying to tell you there's more to sex than what we just did."

"Check this out, Franco," she started while snapping her bra back on, "it's not like I have a sign around my neck that says Smorgasbord, nineteen-ninety-five. Forget it, if you find someone who does all that, I want to marry him."

"I have, but you'll never have him." Niqui hadn't meant to disclose this much but his focus was being perturbed that he'd put his underwear on backwards.

They drove back in absolute silence. Only when Niqui switched the radio station to a baseball game, did she react.

"What do you think you're doing?"

"For the money I spent tonight I could have bought a ticket to the game. I want to find out how the Texas Rangers are doing."

"You're strange, Franco."

"Strange, as in dating you?"

＊＊＊＊＊＊＊＊＊＊＊＊＊

Niqui showered and soaped his crotch three times when he got home. Something just felt dirty about the whole thing, but he'd lost his virginity. This aspect puzzled him and he'd have to ask Dane if a guy could lose his virginity twice. Back in his room he really wanted to tell Jay about the evening.

"How was the movie, bro?" Jay asked, no longer feigning sleep.

"A chick flick," Niqui answered. "I kinda went out on a date tonight. What a waste of hard earned money."

"Good for you. Spending money on a girl can get real expensive. Hey, I heard at school today that there's a group of senior girls who have bet each other on who could screw an underclassmen first. Keep your guard up."

Niqui faked a smile and knew he'd been used. It had cost him twenty-five dollars to be a sucker.

"I won't fall for that trick. Women take up too much time and they're lame, to boot."

"You got that right, Niqui. Have you heard from Dane?"

Niqui sat down by his brother. "Can I sleep with you tonight? I feel bummed out."

"Sure," Jay said and slid his bed over. "Man, I haven't seen you this wiped in a long time."

"I've never had a friend like Dane. He's like, my best friend." There was a pause, then, "Our sister has a great body, doesn't she?"

Jay giggled. "She's my twin sister, so she has to be hot."

Niqui tickled his brother and faked being the vindictive older sibling. Not only was he disgusted with himself for being manipulated but he was scared now. Tears came to his eyes as he turned away from Jay.

"You okay, bro? Was it something I said? Jay placed his hand on his brother's shoulder.

"No." Niqui began to cry and wrapped his brother's arm around his chest. Jay wasn't sure how to proceed, so they fell asleep with the closeness they now shared with one evening at the Kings Academy.

Chapter Twenty-Two

Niqui had recaptured his dignity in the days that followed his first experiment with the opposite sex. He'd managed some revenge when a couple of seniors came up to him the next day in school and blatantly asked if he'd fucked Linda Welch.

"It was more like being a spelunker," he said and walked off.

His birthday couldn't come soon enough. This time Debbie met the boys at the airport and took them on a tour of the city with the emphasis being at the Alamo. Throughout this excursion Niqui was both delighted to have Dane next to him and antsy to ask the boy a question that concerned him immensely. Wandering off inside this old mission, Niqui finally had Dane to himself.

"Dane, don't be disappointed with me but I went on a date-from-hell. There's a mark on my penis that wasn't there before."

"We'll find a pharmacy and check it out," Dane assured him.

The boys ran from the Alamo, down Paseo del Rio until they found a drug store. Problem was, Niqui knew the pharmacist as a man who lived close to their home. They found the Health Aids section and Dane did a pirouette in the aisle, with a split, then rode up with a STD detection kit in his hand. Niqui roared with laughter at his friend.

"You have to teach me that, Dane," Niqui said and felt a hand on his shoulder.

"You boys need something?" the pharmacist asked. "Oh, Niqui, I didn't recognize you at first. Odd seeing you this far downtown. Where's your mother or father?"

"Oh, hi, Mr. Thompson. My mom is over at the Alamo. We just stopped in to buy something."

"A STD kid? What have you guys been up to?"

"It's a class project. You know, health class, and all. It's okay, isn't it?"

"You go to a Catholic school, don't you? It's not like the Catholic school I used to go to, but this isn't a prescription item, so it's not like you can't purchase it."

Dane used his own money to buy the kit and carried the bag himself as to have it appear it wasn't really for Niqui. Arriving back at the Alamo they were lectured for going off on their own. If it wasn't Niqui's birthday his mother might have been twice as upset.

As a group they went to the annual Fiesta San Antonio, a ten-day festival held on Paseo del Rio center, also called the River walk. It was a gala affair and everyone had a great time. The river area was San Antonio's biggest draw these days. Stone steps lead from street level down to the footpaths, gardens, and restaurant patios that lined the banks of the San Antonio River, where tour barges floated lazily along a two-and-a-half-mile stretch of water.

Debbie wasn't sure when her husband was expected home, so they all stayed at a hotel for the night, the ladies in one room, the guys in another. Only after dinner and everyone was enjoying the relaxation by the pool did Dane and Niqui run to their room and open up the kit. They both read the instructions carefully, then Dane pricked Niqui's finger to drip a drop of blood for the needed sample.

The boys waited the few minutes for the result before Dane held up the circular disk which changed colors and denoted, within the kits' limited assessment, the likelihood of the test results.

"Tell me, Dane," Niqui pleaded with anxiety. "No problem, right?"

Dane examined the results twice, then another time. "Okay, this is what we're going to do." Dane stayed calm and made eye to eye contact with his best friend. He wanted to reduce any panic or total despair. Empathy was the foremost on Dane's mind. "This is excellent in that we caught it early. Ten days and you'll be as good as new. Think of it as a learning experience."

Niqui kept nodding, his face sinking lower by the second. "Bitch!" was his response to the senior girl who gave this to him.

"It took two to dance this number out," Dane reminded. An arm around the shoulder helped Niqui deal with the frustration. There were more tears and self-loathing, but Dane had intentions to save the night.

Morning came quickly. The best thing, it was also Easter vacation and there was a promise of no school the following week. Niqui was up before anyone and went to see his mother because he knew she'd rise early to get her cup of coffee.   Sure enough his mother and Brittany were in the downstairs restaurant enjoying the early brew.

"Mom," Niqui greeted.   "Hi, Mrs. Makawa."

"Well, hi, honey.   You are sure up early.   Want to have an early breakfast with us or wait for your siblings and friends?"

Niqui was ready to break down in tears, but having Mrs. Makawa there didn't make it easy.   The ladies picked up on this early.

"Niqui, would you feel better if I left you and your mother alone?"

"It's okay.   I mean, you have children.   I screwed up."   Tears began to flow and the women put him between the two of them with their hands on his shoulders.   "He barely blurted out after a deep breath. "I made a mistake a week ago.   I thought this girl liked me but she had a bet with her friends just to have sex with me."

"You think you got her pregnant?"   his mother asked.

"No.   She told me she was on the pill.   I have gonorrhea."   Niqui waited to see his mother's reaction.   It was somewhere between disbelief and hysteria.   "Dane said I can take oral tetracycline and get rid of it."

"Dane is just a kid himself, Niqui.   What does he know about venereal diseases?"

Mrs. Makawa spoke up.   "The boys are taught a great deal on this subject, Debbie.   Dane is right.   Niqui will recover in no time."

"See, Mom.   Our school should have a health class to teach about condoms and things."

Debbie took a deep breath and knew her son was right.   "Let's relax here."   This was more for her peace of mind than her son's.   "I'll call the doctor when we get home.   You've been safe since, right?   I mean with other children."

"Mom!   I wouldn't jeopardize anyone else.   Don't embarrass me by asking me anymore. I feel bad enough having this."

Mrs. Franco didn't want to sound like a hypocrite.   She knew now she had an older son who was gay, a middle son who was discovering sex through trial and error, and a younger son who was way over his head with a sophisticated young lady, as she saw it.   With her moments of reflection and silence she reached out and hugged her middle child, assured the boy that Dane was right about all this and this was their little secret.   Niqui agreed.

Niqui felt really relieved and kissed both his mother and Mrs. Makawa.   "You guys are the greatest," he said and darted off to wake his peers up.

Chapter Twenty-Three

The Kings Academy was presented a summons to report in Austin, the state capital, for an inquiry into the future of their license as an educational facility in the State of Texas. Dr. Kerho had no more been presented this document when Social Services showed up at the front gate. There were two representatives of this special investigative team into allegations made against the academy by unnamed individuals.

Dr. Kerho and Dr. Marion greeted a stern looking woman, Ms. Wilma McGillicutti, and her cohort, Mr. Theodore Rank. Neither one of them looked like they'd smiled in years. This pairing of state servants were shown around the facility and introduced to the teachers, coaches, and administrative staff. They also met with various students of their selection. The final meeting was in Dr. Kerho's office and Ms. McGillicutti did not mince words.

"It says here that you adopted thirty children; yet, I only saw twenty-nine boys, minus your own and Dr. Marion's. Where, may I ask, is the thirtieth boy?"

Dr. Kerho felt like he'd just been punched in the stomach. "Ma'am, my father actually took care of the paperwork, but I've been under the impression that only twenty-nine boys were on the original manifest on the plane that brought them here. I am not aware of another boy, if there's one at all."

"Dr. Kerho, our department does not make mistakes. I have the application right here in my hand that Dr. Jolson requested immigration papers for thirty boys. Certainly he wouldn't make a mistake like that."

"This is going to take time to sort out, ma'am. I'm assuming everything else has met your approval," Toy said.

"My full report will be available in a few days. Whatever discrepancies I've noticed will be forwarded for your review in Austin. I expect an explanation on where this other boy is by that time. Let me remind you, Doctor, this is a serious matter. I don't want to have to bring in the Attorney General to investigate a possible missing boy.

This is exactly the type of publicity that could close the gates of this academy."

"Jumping to conclusions appears to be one of your foibles, ma'am. Before you accuse this academy or myself of any criminal nature I'd expect time for us to find the mistake that apparently has been lost in the paperwork. I highly doubt that we've misplaced a child along the way."

"This is no humorous matter, Dr. Kerho."

"I was hardly being funny, given the present company. Security will see you out."

As soon as Toy had gathered his wits he called his father and explained what had happened. "How could we have missed this, Dad? I mean, Kam and I have gone all through those boxes and have never seen evidence of any thirtieth child."

"There is one more box down in the basement, Toy. You've left it alone because of the radiation sign on the side, but I had it checked out at the airport for any radioactive ingredients and they assured me it had no danger. I've never had a reason to open it up."

"Kam and I will be there in about an hour. Maybe Dr. Jolson was playing a trick on someone."

"Dr. Jolson has long played his trick on the world," Dr. Kerho told his son. "He knew he had the last laugh before he took his own life."

"I didn't know that, Dad. How'd you find out?"

"One of his nurses told me. She said the doctor knew that his cancer was a painful end and he only had a month to live. Dr. Jolson's last words were to this nurse. He knew his genius would survive in you two boys. You know what? I think he was right."

"Thanks, Dad. Genius is pushing it; maybe having great parents and a Japanese mentor made Kam and I what we are."

The trip to Dallas took just under an hour in a speedy Porsche. Kami and Toy brushed a few cobwebs away as they marched down the stairs for the umpteenth time in their lives. As teenagers they all but lived down there in their research and reading of Dr. Jolson's memoirs and research. It took a crowbar to pry open the wooden crate. When the lid creaked back and was lifted up there were multiple canisters with labels, volumes of paperwork, and instruments that were all but archaic to the scientific world sixteen years later.

Each canister was taken out with care, the contents were follicles of hair, bone, teeth, fingernails, and even particles of clothing with hair samples on them. Names were attached to each jar with the location from which it was obtained.

"Well," Toy sat back and took a breath, "let's line these suckers up and see if there are twenty-nine or thirty. Across the cold cellar floor the see-through jars were lined up alphabetically. Names appeared that were now common knowledge with faces of boys, close to resembling their forefathers.

Stacked in rows of ten, Toy counted the first group out, then went to the second row for ten more and expected nine in the third. He counted ten. "Kami, you count them."

"I did. There are thirty. You call off the names until there's one we don't recognize."

The fourteenth name had two faces staring at each other—Charles the Twelfth.

"Where is Charles?" Toy asked with a chuckle.

Kami shrugged his shoulders and shook his head while smiling at the same time. This was funny but for Social Services thinking that this boy was supposedly living at the academy.

"Possibly your parents wanted to raise one more child," Kami humored.

"Dad! What did you do with Charles?!" Toy yelled up the bottom of the cellar stairs.

Dr. Kerho wobbled down the rickety steps-- age had slowed up his walk and made climbing or descending not his favorite method of getting anywhere. "I signed for twenty-nine."

Through the paperwork the three men began to dig. They knew the remains of Charles the Twelfth were found in Riddarholm Church, Stockholm, Sweden. As each paper was examined new information became part of a puzzle that had long been forgotten because of so many missing pieces.

"This is it!" Toy explained.

"You found out where the kid is?"

"No.   This explains how the doctor did it.   He used nuclear transfer with a super computer.   More precisely, molecular simulation with atomic particles and genetic sequences.   I believe the process is more for a quantum computer, similar to magnetic resonance imaging scanner," Toy added.   "In reading this paper it appears he used a Beowulf cluster, a popular way of compounding super computers in his day.   He could bring forth a DNA sample with a human hair of only one hundred microns. Amazing!"

"Explain.   I'm only your average family physician," Kami said.

"The problem with cloning dead people is that it is highly unlikely that the complete sequences of DNA will have survived intact.   There may be thousands, even millions of pieces missing, so it may well be impossible to put it back together.   Dr. Jolson found a way by using nuclear energy to figure out how to put these pieces together in the right order.   All a Beowulf cluster is, is a combination of super computers working in harmony.   Even though bits of DNA can quite plausibly survive, you are going to need millions of such pieces to make a whole genome.   Like solving a huge puzzle without having all the pieces. The computer helps to solve this by exploring all the possibilities and variations in different pieces of body parts.   I dare say Dr. Jolson was fifty years ahead of his time."

"I always assumed that was the case," Toy's father explained. "Here, boys, this answers our problem."   Dr. Kerho handed a single sheet of paper to Kami.   He read it, then handed it over to Toy.

"I can't believe it," Toy mumbled.   "One of the surrogates refused to turn over the baby after it was born.   My God, we have Charles the Twelfth out there in the world and he has no idea who he represents. This is too mind-boggling!   I still want to take all this material and place in our computer to analyze the boys' current DNA to these samples.   We can verify and make positive identification which, I'm sure, Dr. Jolson has already done.   Install Charles' DNA in our files with the rest of the boys, as well.   God only knows if we'll find him someday."

"How do we explain this to Social Services?"   Kami asked.

"Dad, you want to handle this one?"   Toy asked with tongue in

284

cheek.

"Look, boys, I have the original transfer of twenty-nine babies. Let Social Services find out what happened at the point of departure," the elder statesman said with determination and a don't-blame-me posture. He passed a few more documents to his boys.

"Figure you'd like to see these. Abraham Lincoln's letters and his sexuality revealed in his own handwriting."

Chapter Twenty-Four

Leif and Sveinn arrived in Springfield on a rainy spring day, but this didn't hamper their motivation to finish this assignment and head back to Florida to renegotiate a new contract for all they've found so far. This was big, bigger than any story in the past ten years and this tabloid was making a mint in the last few weeks with what Leif had given them.

They went right to Oak Ridge Cemetery outside the city. They'd long learned that viewing the actual grave site accomplished very little. It was the caretaker or funeral director at times that gave them the information they needed.

At the cemetery the caretakers, grave diggers, and chapel director weren't old enough to be suspicious characters in this investigation. Once again the ploy about Leif being a local high school student in need of information from sixteen years ago worked like a charm. They were given the name of a man who had worked at the cemetery for forty years but had retired ten years ago. The man had apparently inherited a great deal of money and retired to travel around the world. The worst thing, the man had moved to a warmer climate, somewhere in Florida.

"Where?" Leif pleaded.

"Don't rightly know," the oldest caretaker there replied and turned to one of his cohorts. "Wait! Didn't he visit the grave site of his old friend, Wilbur, when he died a few years ago?"

"I do remember that," a young man who watered and mowed the lawn agreed. "He signed the registry book in the chapel, I believe. Check there."

Leif and Sveinn hustled to the chapel and asked for the last few years of logs. They were handed two boxes. It took four hours of tedious work to find the man's name. The best thing, the guy had put under **Address**: Cocoa Beach, Florida.

It took three transfers to reach Orlando, Florida, then a rental car to travel the sixty miles to the east coast of Florida. A check in the phone book was all that was needed and they were on the man's doorstep within the hour.

The elderly man, almost ninety years old, hadn't had a visitor in eight years. Surprisingly he greeted them with a most shocking introduction. "I've been waiting on someone to show up. I sure didn't expect two young whippers like yourselves."

Leif and Sveinn sat down in a living room smelling of cigars and old age. They had come to accept odors that seemed to go along with old age. Mr. Stevenson introduced himself as an admirer of blond-haired people, women or men, he didn't care. When he sat in a recliner that had seen its day he pulled out three issues of a tabloid the Icelandics knew only so well.

"You've done your research well, gentlemen. Ralph called me from Springfield, said two young men were snooping around looking for me. Guess I could've run but what's the use? Neither one of you look too dangerous."

Leif took the lead. "Sir, we're not here to wreak havoc on your life and not a word of what is said here will go anywhere further. We came to ask if you gave any artifacts to a doctor in exchange for a large sum of money, but that won't be necessary now."

The old man gave a hearty belly laugh. Leif reminded him of a grandson and asked the boy where he was from, how old, and what interest a handsome boy would want with such information. Not really satisfied with the answers, the elderly man only grinned.

"I'm an old man, boys," he started and lit a pipe that made the room smell like wood burning. "What are they going to do to an old man? Arrest me? Take me to a senior village? They're welcome to do that. I'm tired of taking care of myself. Lousy government won't even give us proper care or cheap prices for my medicine. It's not right I tell you. It's true what you say, son. I spent several nights figuring a way into that coffin--had to fake an excuse to close down the tourists for a few days until I could repair all the damage. It rained terribly that night and almost killed me. Being in my seventies at the time, why, that's not the type of work an old man should be doing. Not sure I have the address anymore but this doctor lived in Israel; at least that was the address where my extra money came from. I told him my troubles were worth more than the fifty grand he offered.   Damn if he didn't pay me the extra fifty.   The codger was as old as myself and a few weeks later the doctor showed up at my door.   We hit if off right away because we spoke the same language--senior citizen complaints.   Reckon the doctor liked me 'cause he told me what he was up to: making clones of famous people.   Now doesn't that get it all?   Modern science thinking they can do God's work.   Why, I was fit to be tied until he explained how this all could serve mankind with intelligence and strong leadership to get these wannabes who do nothing but raise our taxes and then complain that our benefits are too high.   Just ain't right, I tell you."

Leif smiled and liked the old guy.   "Sir, did this doctor say anything about the man's sexual orientation?"

Mr. Stevenson jovially chuckled with the best roar a ninety-year old could render.   "Funny you should ask that.   The dear doctor said the only reason he was paying me more was because these cells…..whatcha call them?"

"DNA, sir," Leif volunteered.

"Yup!   That's it!   Well my samples had them, so he says, and I had proof of the deceased habits in letters.   Nothing against them kind; actually I thought this doctor was one strange bird.   Anyways, I gave my new friend a few papers that my father had given me.   See, my paps had the job at the cemetery long before I ever took over.   He knew Robert Lincoln when the man worked up yonder in Chicago for Pullman Railroad.   Dare say Mr. Lincoln had all sorts of his father's paperwork but didn't dare give all of them to the government 'cause it would've made his pa look bad.   So, Robert gifted my paps with these documents and I'd been stuck with 'em after my parents died.   Reckon this doctor would find some use for them to explain all this."

"What were in the papers, sir?"   Leif asked excitedly.

"You're a tad too young for me to explain all this," Mr. Stevenson said.

"Leif is gay, sir, if that relieves your conscience," Sveinn said.

"Well blessed be!   Now's I sees why you're his companion."   The old man patted his knee with the most fun he'd had in long time.   "Reckon you got yourself a might pretty boy here.   Just take care of 'im and watch out for t' friggin' law. Reckon they just don't figure proper how a boy can satisfy a man's loins."

Sveinn felt guilty for being honestly portrayed. He smiled weakly and decided not to try to defend himself. Hell, the old guy was right!

This senior citizen rubbed his jaw. "Now let me see here, there were a few poems and letters to what I might think are men you just don't write these things to unless you're ready to come to bed. There was one or two soldiers, a David V. Derickson, if I remember right, 'cause I'm a Civil War buff and this soldier slept more with Abe than Mary did, I reckon. What the two of them wrote each other could firm a lad up real quick just reading them. Now a-days, I hears they got pills to straighten things up that is too tired to make the effort any other way. I tried 'em a few years ago and surprised myself with something I haven't seen since my mid-seventies. Damn if I couldn't find a woman or a newspaper boy around to make it worthwhile." His laugh was contagious and had Leif and Sveinn chuckling.

"I would've let you love me," Leif said with compassion.

"Laddy, you would've killed me with your beauty. I've forgotten what pleasure is, to tell you the truth. I forgot to tell you about another lad that took a-likin' to old Abe, though I bet it was young Abe at the time. Colonel Ellsworth was quite the looker, a short man, but had a hearty pistol, according to the letters they exchanged. Joshua Steed was Abe's first love—would have been real jealous to know how his friend spent many a night in Washington in the arms of other men."

"Didn't his wife know?" Leif asked.

"Mary spent quite a few nights visiting her folks. A whole lot of northerners say she was spying for the South because so many of her kin were from there. Abe couldn't stand the woman's mouth anyway and had a bedroom to hisself as much as he could."

Sveinn wanted to redirect the conversation. "Sir, did you ever hear back from this doctor after his visit?"

"Poor soul. We wrote each other quite often. Reckon you boys will find out sooner or later but he once admitted having thirty boys in the hopper. He paid a tidy sum to find the women to bear these for him. Even admitted that one woman had given him a tough time after the birth. I can understand why a woman might do that, going through nine months of labor, then giving the kid away. Damn, I couldn't go through the pain and just give my blood up; course, I'm not sure it was her blood to give up. The good doctor said he'd discovered he had cancer a few months before we'd last spoke; all too late because he'd devoted his life to doing this research and neglecting his own health. I'm pretty sure he died. Sure would like to know what happened to all those children."

"If we find out, sir, we'll make it a special trip to let you know," Sveinn said.

Mr. Stevenson grabbed his cane and waddled his way to a closet. He rummaged for a few minutes, then moseyed back with a stack of brownish papers sealed in a clear, zip lock container.

"I must apologize for my lack of honesty with the doctor. It appears I have even more paperwork from the collection. The last, I promise."

Leif held them like they were a treasure. In their own right the papers were invaluable, all but verifying that Honest Abe was honestly gay.

Chapter Twenty-Five

The Texas Board of Education and the Department of Licensing called it a review; Dr. Kerho saw it more as an inquisition in front of a state appointed judge. Dan Howard was representing the Kings Academy, a lawyer for over thirty-five years—fourteen of these with the academy. Dan had been a longtime friend of Toy's father, and though not the most brilliant of lawyers, he was devoted to the Kings' philosophy and goals. The man did know how to run a non-profit organization, but inside a courtroom was an entirely different matter.

Dr. Kerho was representing the academy as its primary director. He sat down and scanned the courtroom, thinking that it had been only three weeks earlier when he received the papers questioning his academy's qualifications to be licensed as a private school in the State of Texas.

There were several complaints being filed, all by associations with St. Bonaventure, though they had the earmark of one Senator Franco written all over the accusations. Across the aisle a man with tanned skin and a marine crew cut sat at another table. Toy figured this must be the boy's father who had fallen under his daughter's wrath. Apparently the man had quite a reputation as a trial attorney for lawsuits and torts.

The courtroom was small, no more than fifteen by twenty feet, and there was no jury box or large spectator gallery, only an elevated bench for the judge and a couple of tables facing it for the parties and their lawyers. There was no court reporter, either, but only a tape deck set up on the bailiff's desk.

The witness stand was toward the rear between the two counsel tables, facing forward, while the lectern for examining counsel was up front, where the witness stand would have been in any other courtroom. There was a reason for all this: the judge was the only fact-finder in these proceedings and his viewpoint was the only one that mattered.

Judge Trayson came through the door from the private corridor, dressed in black robes and with a no-nonsense frown.   Dan Howard remained on his feet after the judge took the bench.

"Good morning, Your Honor.   We're here today to answer the complaints of said petition…"

"I know why we're here," Judge Trayson barked.

"If I may make a brief opening statement, possibly we can avoid…"

"You may sit down, Mr. Howard, and let me run my courtroom my own way."

Arch Johnson exchanged a glance with his opposing counsel, both of them thinking the same thing: this old-timer who had been so mild-mannered and jocular in his judge's chambers, turned into an asshole in his black robes on the bench.

"Of course," Dan replied and slowly lowered himself.

Everyone waited in a deadened silence while the judge fingered through the file.   Toy just sat motionless while taking all this formality with disgust.

"All right, let's get started.   The court documents before me insinuates that there is a missing boy.   How does an academy lose a boy trusted to their care?"

Dan wasn't sure if the question was rhetorical, but he stood and waited a fraction of a second before he addressed the bench.   "Your Honor, I have a manifest here in my hand that denotes twenty-nine boys were delivered per the request of a Dr. Jacob Jolson.   If there was a thirtieth boy, the people on St. Barts never put him on the plane."   Dan gave the proof to the bailiff, who handed it to the judge.

Judge Trayson examined the document closely.   "This appears to be official, so I'll take this under advisement to dismiss this accusation.   There's no use involving the State of Texas in a mix up outside of our country.   Let's proceed.   Mr. Johnson, I presume you have a witness."

"Yes, Your Honor.   I'd like to call Dr. Joyce Benjamin to the stand."

Dr. Benjamin was ushered in.   She was a short, heavyset woman with a briefcase that looked like a purse compared to her size.   She professed to have a doctorate in psychology and fifteen years of experience with child psychology and sexual abuse victims. Mr. Johnson walked the witness through her credentials and qualifications, then her review of various file materials. It was twenty minutes later before he asked her to describe her interviews with St. Bonaventure students and what they had witnessed while visiting the Kings Academy.

The doctor gave examples of what several students had confirmed: a masturbatory project of graphic pictures depicting an erect penis and testicles, including an explanation of this act; various nude statues and paintings; a written short story of erotic, homoerotic! mind you, content; athletic clothing that cannot be described as cloth, but a second skin that was quite revealing; kissing and hugging by Kings Academy athletes; and other sexual innuendo.

"Would you describe much of this material as pornographic, Dr. Benjamin?" Johnson asked.

"Objection! Calls for speculation," Dan protested.

"Overruled. Answer the question, Doctor," Judge Trayson ordered.

"Absolutely! I mean to the pornography. Rather blatant pictures of male and female genitalia. The impact of pervasive pornography is transforming sexuality and relationships for the worse, especially among our youth. Children who frequently view porn may develop unrealistic expectations of women's appearance and behavior, have difficulty forming and sustaining relationships and feeling sexually frustrated. The image of a lonely teenager masturbating to porn is a sad metaphor of our decade, I'm afraid. Such displays only trigger these fantasies and unhealthy behavior. I also believe the erection in question was possibly a dildo, but, nevertheless, it gives the wrong impression that all males have ten-inch penises, which could cause an inferiority complex in other males."

"And about these displays of near nudity, Doctor," Johnson directed for the lady to continue.

"Objection, Your Honor. The counsel is leading the witness. No definition of clothing has promoted this as near-nudity and it calls for speculation," Dan protested again.

"Overruled. You may continue, counsel."

The doctor didn't wait for the question to be repeated. "Seductive nudity can affect young people at an increasingly young age. It offers a persuasive interest in voyeurism."

"Objection! Your Honor. Argumentative and nudity, in and of itself, is not a crime in the State of Texas." Dan stood with a solid objection that had little persuasion in this court.

Judge Trayson lowered his glasses and peered over them with a look of aggravation. "Mr. Howard, I don't plan on having this review last for weeks because you object to all the questioning! Overruled!" He glanced back at the witness. "Ma'am, be careful of your wording, such as seductive."

"Yes, Your Honor," Dr. Benjamin replied so sweetly and with a weak smile that truly defined a fat lady hitting on someone she had no chance of impressing. "Nudity is a way of communication. For teenagers, it's a sexual invitation. Now I realize these were sporting events, per se, but this just creates a larger audience and promotes exhibitionism, if not an invitation to something more seductive....oops! I meant a sexual liaison. It's surely giving young people the wrong message."

Arch Johnson turned his back on the witness in a form of a dance step, then remembered there wasn't a jury to impress or act for. "If you knew the Kings Academy practiced nudity, even coed nudity, would this increase your denunciation of this school to be licensed by the State of Texas?"

"Of course. Such an environment only enhances sexual immorality, encourages unsafe sexual practices and creates confusion to young people who are, more than often, ashamed of their growing bodies and sexual functions. I can bear witness that such a policy confuses youth on how to treat others, how to cope with relationships and sexual frustration, and what "normal" or "healthy" relationships are like. Deviants like specific environments they can control."

"Objection! Lack of foundation and personal knowledge."

"Sustained. Dr. Benjamin, it is not necessary to label the defendant," the judge corrected with a mildness that made Dan sink in his chair with grief.

Dr. Benjamin elaborated on her interviews with various students at St. Bonaventure, especially one of the girls who was currently seeing a psychologist. The child had been traumatized by a boy at Kings, a boy who was on the gymnastics' team, she remembered, had persuaded the girl to tour the building with him, then took advantage of the girl. This young lady did admit the sexual rendezvous had been consensual. It appeared, in all cases, the students at Kings Academy were very aggressive and out-of-control.

It was two hours before the judge allowed the witness to state the conclusion set out on the last page of her report; it was in the state's best interest to rescind the license of Kings Academy as a private institution of education.

Another hour and a half for the lunch recess, and then Dan Howard was up behind the podium to ask "just a few" cross examination questions. Toy wasn't as confident of the senior citizen as his father had often given great praise to. Toy didn't remember his father ever having to go to court and be threatened with having his very livelihood and credibility, if not character threatened-- let alone, losing his children. This was a different ballpark than filling out corporate papers and non-profit tax reports. Toy considered Mr. Howard was way out of his league with this one. Dan's demeanor was blatantly too nice, sort of the down-home, Texas tradition of good ol' boy.

"Dr. Benjamin, you've been conducting therapy evaluations and practicing for quite a few years, right?"

"That's correct, sir."

"Can men be objectified?" Dan casually asked, which confused more than the witness.

"Well, yes, of course," the doctor answered.

"Hmmm, just wondered. You only mentioned women when you were discussing pornography. Are you homophobic, Doctor?"

"Of course not," she replied with a huff.

"Ever worked with a homosexual? Been a therapist to one? Had one as a friend?"

"Objection!" Arch protested. "Irrelevant and compound questions."

295

"Where exactly are you going with this, Mr. Howard?" the judge requested.

"Your Honor, as long as we're talking about sexuality, it's best to know this psychologist's personal viewpoints to get a perspective on bias."

"Okay, I'll allow this question only once."

Dr. Benjamin didn't hesitate. "I am sure I have, I mean worked with one. I don't remember…wait, yes, I have had a few in group therapy before."

"And how did you categorize them? Mentally ill? Sick? Confused?"

"Objection. This is not about the mental state of children."

"Your Honor," Dan butted in, "the doctor referred to my client as deviant, the Kings' students as potentially ashamed. The defendant in this case just happens to be a gay man. I'm just trying to assess her opinion."

"Overruled! Get with it, counselor, but don't dwell on it."

"Do you want me to repeat the question, Doctor?"

"No, that's okay. I'd say they were mostly ego dystonic homosexuals, which, yes, I'd say is a mental disorder."

"Ego dystonic homosexual." Dan let the words hang for a few seconds. "If I'm not mistaken that is psychological jargon for, it doesn't feel comfortable. Am I correct in this, doctor?"

"Yes, but it means more. It means the person has trouble feeling compatible with his orientation because it interferes with his morals or upbringing."

"Fascinating," Dan said with head lowered. "I find it interesting that homosexuality, per se, no longer qualifies as a mental disorder, yet you still classify one as a "mental disorder" classification. Why is it that there is no such category as "Ego dystonic Heterosexuality" in the manual? Actually, doctor, there isn't such a category for homosexuals anymore. Can you explain why you're years behind?"

"Huh, I'm not sure. I mean, about why there isn't a heterosexual explanation for one. I certainly believe that people have a right to choose their sexual preferences. I would never claim that homosexuality was wrong or immoral. I just think they shouldn't be in charge of impressionable youth."

"What other judgments do you believe, Doctor? I mean about homosexuals."

"I'm not judging, sir, I just don't think they have a great chance of being happy as much as heterosexuals."

"And you rate happiness based on one's sexuality? Interesting." Dan let it hang as he acted dumbfounded to the amusement of Dr. Kerho. "Who asked you to do these evaluations, Doctor?"

"I believe the school's headmaster, Phil Masters."

"You're not sure?"

"Yes, I'm positive it was him."

"Who do you think referred your name to him?"

"I have no idea, sir."

"Do you know Senator Franco, for instance?"

"Yes, I've heard of him. I believe his children go to St. Bonaventure."

"Do you have a favorable impression of Senator Franco?"

"I suppose I do."

"Favorable enough to vote for him in the last election?"

Johnson rose from his chair before the question could be answered. "Objection! We have the secret ballot system in this country."

"We have secrecy in a lot of things," Dan noted. "I believe such secrecy is forfeited when a witness offers an expert opinion and there's an appearance of bias."

Johnson was livid. "Your Honor, unless you transfer venue to another state or country, there's little way of avoiding this implied bias," he asserted. "Everybody in the State of Texas either voted for Senator Franco or didn't."

Judge Trayson appeared perturbed by this distraction. He glanced at Doctor Benjamin. "Did you, ma'am, allow your political leanings to influence you in any way in our opinions?"

"Absolutely not, sir. I was professional in my evaluations as I am in any other assignment."

"Objection sustained. Move on, Mr. Howard."

"Then you can assume it was Senator Franco who volunteered your services?"

"Objection. Asked and answered."

"Objection sustained. Maybe you did not hear me, Mr. Howard. Please move on in a different direction."

"Yes, of course, Your Honor. Doctor, could you tell me the difference between erotica and pornography?"

"I believe I can. Most professionals describe the difference this way: porn is objectifying and derogatory while erotica depicts mutually satisfying sex between equal partners."

"A matter of taste, then?"

"Ah, for lack of better words, yes."

"Are you then an expert on freedom of speech? Pornography? Erotica? Would the community want you deciding what books or art work their child may view?"

"I'd probably be an excellent choice for that," Dr. Benjamin admitted.

"To be absolutely certain about something, one must know everything or nothing about it. Which are you, Doctor?"

"Objection! Your Honor, the counsel is badgering the witness."
"I withdraw the question, Your Honor. Doctor Benjamin, would it be fair to say that pornography is one of the few ways to assess and affirm a person's sexual feelings and desires? Does pornography, by presenting the male body or male sexuality in a glorified form, go beyond depicting and defining desire?"

"I...I...suppose it can in adults, but I, we are here..."

"And, anyway, Doctor, despite the noises about "social standards," the real battle here is about freedom of speech, access to information and, most importantly, freedom of the imagination, isn't it?"

"Yes, of course, but we're talking about children, not adults, sir."

"Children by whose definition, Doctor? Your subjective viewpoint, or a clinical definition? Are we dealing with children or adolescents here?"

"Children are usually prepubescent, so, yes, I see what you mean."

"Do you, Doctor? Have you not tried to create hysterics by

denoting these teenagers over and over again as children?"

"Objection! Counsel is badgering the witness. Who cares what they're called," Arch said.

"The distinction has been made, I believe. Move on, counsel," the judge ordered.

"Doctor, you mentioned an interview with a teenage girl at St. Bonaventure, one who's been seeing a psychologist. Is her visit to Kings Academy part of this trauma, or is this an ongoing problem?"

"Doctor-patient privilege, sir."

"Which the patient is free to waive. Or, in this case, her parents are free to waive for her."

"Yes, well, they chose not to."

"And that doesn't concern you? They're withholding information from you, from us! And that doesn't mean anything to you?!"

Judge Trayson's chair squeaked forward before the man took off his glasses gain. "Mr. Howard, what do you expect to gain by badgering the witness? You've tried my patience over and over."

"My apologies to the court and to Dr. Benjamin. If I may continue, Your Honor, just a few more questions and I'll be done."

"Not soon enough for me," Judge Trayson said, but gave a nod.

"Dr. Benjamin, have you interviewed any of the Kings' students? Please speak up, Doctor."

"No, I haven't."

"Any staff? The directors? Have you reviewed the school's philosophy, curriculum, or even been to the school? How about asking for a review of all video cameras to see if any harassment really went on?"

"I haven't had the chance, sir."

"Would it be fair to say that Mr. Theodore Rank with the state's social services has had a chance to review the academy's philosophy and students?"

"I've read his report, if that's what you mean."

"That wasn't my question, Dr. Benjamin. Do I have to repeat the question?"

A perturbed and demeaning stare was like a laser beam at this

Kings Academy attorney. "No sir, you don't! The man gave a thorough accounting."

"You sound like his report was in contrast to your own, Doctor."

Judge Trayson leaned over from his bench. "Is that a question, Counselor?"

"Not exactly, Your Honor. If I may continue." Dan held up his hand, as if to tell the judge to shut up. "Dr. Benjamin, are you familiar with a University of Virginia study of recent years about adolescents and sexuality?"

"I've read the report. I believe there are a few loop holes, but it has merit."

As if Mr. Howard was lecturing a class he turned and addressed the court room. "The study I'm referring to has to do with adolescents and early sexual relationships. Such early experimentation has proven to be very beneficial, promoting more maturity and less juvenile delinquency or criminal behavior. It's quite a contrast to early expectations and forecasts of out-of-control teenagers over pre-marital sex."

"May we get to the point, Mr. Howard," Judge Trayson admonished.

"Yes, Your Honor. Mr. Rank has written in his report….if I may direct the court to page thirty-six of exhibit two. He writes that rarely has he witnessed a group of students so talented, mature, and confident in their sexuality as those at Kings Academy. The boys' acceptance of their environment and structure is evident and paramount to this degree of compliance and self-control. I was just wondering if Dr. Benjamin wished to elaborate."

A mild stutter and a clearing of the doctor's throat preceded her deep breath. "As I said, there are loop holes to that study. I'm sure there's a segment of adolescents who become quite rebellious and out-of-control with unsupervised sex before their ready. And they weren't talking about child-adult relationships."

Mr. Howard had to chuckle. "Neither was I, Doctor. That assumption is purely your own. So what I'm hearing is that when the teens are ready for sex, they need to be supervised. Interesting. It sounds to me like you're promoting adult interaction."

Even the judge had trouble not finding humor with this faux pas. Of course the only one who didn't find it funny was Dr. Benjamin.

"What I meant was, teenagers need education and to take responsibility for their actions. It's why we have so many young women pregnant and children without fathers," she reinforced and added, "I don't mean to imply that the boys at the academy were having sex with adults. They all appear to be very sophisticated in their sexual knowledge, is all I meant."

"It's so easy to underestimate the imagination of youth, isn't it, Doctor? The boys and young lady do receive a substantial education and are quite knowledgeable about their sexuality because they are allowed to share their desires with their peers. Actually, Doctor, and correct me if I'm wrong, there are some 200 studies that find teaching children to manage their emotions and cultivate social skills boosts academic performance. Any report of Kings' students being ADHD?"

"None that I know of," the doctor answered with reservation.

"Exactly, so being able to pay attention, and knowing such rudimentary math as what numbers mean and how one's body lives and how to court, matter much more than when someone loses their virginity, don't they, Doctor?"

"Ms. Wilma McGillicutti thinks the boys are over-sexed," the doctor stated.

"Ah, yes, over-sexed. An interesting concept for a woman to use about the male population."

"Objection, Your Honor! Counsel is giving personal opinion."

"Sustained. Counsel, you are beating this one to death."

Dan would have loved to run further with this but decided to change directions with an insulting finish.

"You do fund-raisers for the mental health department, Dr. Benjamin. That's not a question, by the way. Is it not true that Senator Franco was a recent speaker and he received campaign contributions for his participation from which you were in charge of?"

"Objection!"

"I'm finished with this witness, Your Honor."

The courtroom was a bit tense for a few minutes, then a high school student from St. Bonaventure entered, took the oath and began to tell of her visit to Kings Academy. She thought the science projects were blatantly sexual and disgusting. This wasn't something her own school would condone. The boys' outfits were almost obscene and one could see their penises outlined through the cloth. Dan Howard didn't dare object and allowed the girl to ramble on about how all this traumatized her.

On cross examination Dan was most polite to this teenager, dressed up in a very nice dress that she had bought for the prom, though a trifle too much for a court of law.

"Ms. Braxton." Howard started and reviewed his notebook. "May I call you Susie? Thank you. Do you have a sex education program at St. Bonaventure?"

"No, not exactly. I mean, you know, our gym teacher, she's a female, she might say something if someone asked her."

"Why didn't you just walk away from the different exhibits if they bothered you so much?"

"Ah...I don't know. I mean, you know, whatever. Like, some of them were really interesting, I guess."

"Did you learn anything?"

"Well, I pretty much know everything, but, yeah, there was some good stuff," the girl said honestly.

"The boys' outfits, they didn't reveal any nudity, did they?"

"I don't think so. I could see their butts, but I guess everyone has one of those. Their, you know, penises, were there, like, so there."

"Like a Speedo? You're familiar with them. Several St. Bonaventure boys wore them."

"Now that I think about it there wasn't any, like, real difference."

"Yet, you continued to stare."

"I've never seen that long of a penis and everyone had one."

When Judge Trayson finally broke a smile the bailiff added his chuckle and even Johnson was amused.

"Susie, would you describe your school as aggressive or disruptive?"

"Well, you know, we have cool cliques, like really cool. I'm a

302

cheerleader, so that's really in," Suzy said.

"I'm impressed," Mr. Howard tried to sound authentic.

"But there are, like, (with a twisted expression) really weird people. I stay away from these."

"Did you meet any boys or girls from Kings Academy who don't fit your standards of excellence, Suzy?"

"Ah, not really. They were, like, so awesome and nice."

"Susie, just a few more questions. Obviously a bright girl like you has the education to know what a male looks like. I'm surprised you were embarrassed. Are you Catholic?"

"Of course, but I really wasn't embarrassed, like, you know. I certainly know what a male looks like." (Suzy tilted her head, as if that was the stupidest question imaginable)

"Then masturbation isn't a taboo subject for you?"

"Objection! I think we're getting too personal here." Arch sat down and sighed.

"Your Honor, the young lady says she knows a great deal about human sexuality. I'm questioning this belief," Dan stated.

"Overruled. Continue, Mr. Howard."

"I've had boyfriends, you know, so it's something I've seen. Gee, get real here. My gym teacher thinks it's a better substitute than sexual intercourse." Susie Braxton knew she was impressing the court with her knowledge of safe sex.

"I'm impressed with your responsible actions, Susie. Let's say a male masturbates, when or how long would be his refractory until he can ejaculate and potentially get a girl pregnant? What I mean by refractory is the time until he's capable of being aroused and capable of ejaculating again."

"Oh. Well, I'd say a day, or so," she replied with certainty.

"One more question, Your Honor. Susie, if a girl is on the pill, can she get a venereal disease?"

"I don't think so," the girl answered, now a little unsure of herself.

"No further questions, Your Honor." Dan dismissed himself and Johnson saw no reason to continue with this student.

"I believe we have time for one more," Judge Trayson spoke.

303

In through the double doors came a polished, uniformed cadet. The boy's hair was swept back with some form of gel and the lad appeared confident, if not brash.

"State your name," the bailiff ordered.

Ralph pointed at his dad. "That's my father there."

"Son, I asked you to state your name," the bailiff repeated.

A tint of temper gravitated from the boy's face but he finally stated his name, took the oath and sat down. Arch Johnson led his son through his competition at Kings and had his son admit that the exhibits were cool, especially the one on masturbation, though the dick wasn't real, Ralph stated with certainty.

Yes, he did tease the younger Franco about losing to a girl, but then the "young lady", as he called her, got all pissed and said she could beat his butt, as well. Ralph said he was just funnin', but she took it all personal and proceeded to slap and kick and scratch him, until his friends were able to pull the girl off, only because he wouldn't dare strike a girl.

Several questions followed as this cross examination centered around his school behavior. He hadn't even seen the Franco boy since they had returned from Kings, but he was sure that the three boys who were waiting for him and his friends while they played at the park—soccer, he was pretty sure of the sport—were Kings' students seeking more revenge for getting one of their students in trouble.

"Objection! Calls for speculation," Dan spoke up. "One second the boy says they were wearing masks, ninja outfits, I think he said, then he thinks they're from Kings Academy."

"Objection sustained. Move forward, Mr. Johnson. No, not you, young man, your father."

The teenager admitted to having some academic difficulties and past scuffles, so the headmaster and his father suggested a more structured academy, which was why he was wearing the uniform of a cadet. There were no further questions from Johnson to his son.

The judge glanced at Mr. Howard. "No questions, Your Honor. Excuse me, maybe I do have one." Dan stood up and approached the teen with an admirable smile.

"Ralph, you appear to be a strong, husky teenager, and, if I don't say so myself, a proud military cadet." Dan watched the boy sit up a little straighter, a hint of over-confidence came through. "Was it really necessary to have so many of your peers hold the Franco boy down while you whaled on him and tried to sodomize the boy?"

"Shit! I didn't need them. I can take care of the punk with one hand tied behind my back!" The teenage delinquent brightened with his delivery and reputation.

"No further questions, Your Honor," Dan said and noticed the disappointment from the witness that the compliments had ceased. Arch Johnson had his head down and wished he wasn't in this courtroom. "Your Honor, we would like to put into evidence the following security tape which will show the witness had perjured himself to the actual events at Kings Academy."

"I will view the tape in my office," Judge Trayson said and accepted the video. He dismissed everyone for the day.

Chapter Twenty-Six

Debbie Franco's world had been turned upside down in the last few months. From an existence of playing a senator's wife, smiling when she didn't necessarily care to, posing for those she didn't respect, and going to functions where women tried to out dress the other, she felt both an accepted mother and a frustrated one to the point she no longer wished to have her husband around her or their children. Debbie knew the boys, especially, were scared of their father. Travis was non-existent when his father was around, and Jay and Niqui had few words, unless their father spoke first. Maria had sided with her brothers and pretended that her father was this strange man living in their house.

There were other things on Debbie's mind, things she'd put in her memory to ponder and hope they weren't true. The day she had taken Niqui to the family doctor was not only embarrassing for Niqui, but it wasn't the greatest moment for a senator's wife to bring her middle son in for a venereal disease.

Dr. Foster had stayed upbeat and actually complimented Niqui for his fortitude in getting the test kit and making the best decision for his health by talking to his mother. He'd known other kids who had just hoped the problem would go away, only to begin years of complications and more serious damage that could eventually kill a person. The other hard thing for Niqui was giving up the name of the girl so her parents could be contacted.

"He took it like a sailor," Dr. Foster told Debbie when her son came out of the exam room, holding his butt. She was glad no other patients were waiting.

What distracted her more at the time was not her reason for being there, but, as Niqui was in seeing the doctor she had scanned the various magazines available for the patients in the waiting room. One of the magazines available was a tabloid. Debbie wouldn't normally find such rags to her liking; it was the picture on the front that caught her attention.

The young boy standing beside two adults was only twelve, according to the caption, but the youngster looked a whole lot like a boy she'd met during the banquet that night—Trevor Alekseyevich. The men in the picture resembled Dr. Kerho and Dr. Marion. She was so enthralled she continued to read the article, slowly telling herself that this must be the boy who was not only an exceptional swimmer but one of their star volleyball players.

This boy couldn't possibly be the once czar of Russia, she kept telling herself. Her fingers shook and dropped the magazine on the table when her son and the doctor came back. She was so much in shock that the doctor became quite concerned for her own health.

The mother and son sat in the front seat on their way home and Niqui was the first to talk from relief of his problem. "Dr. Foster asked me if I missed having pubic hair. I told him it turned on my boyfriend. He laughed. I asked what it takes to make my penis longer, and he said there's a surgery that could give me another inch or two but it wouldn't look me straight in the eye when it got hard anymore. I like that guy."

"I suppose we can ask Dane how he got his," Mrs. Franco teased her son. "By the way, has Dane ever said anything about his parents?"

"He's adopted, Mom. All the boys are adopted."

"Really. Did you meet Trevor Aleksyevich when you spent the night?"

"Sure. He's really cool and really tall. Plays tennis, too, but Jacob smashes him. Cool! I just said a pun."

"What's Dane's last name? I forgot."

"Gee, Mom, you could at least smile. Nureyev, Mom. Why the questions?"

"Sorry, I'm just a little tired. Call if a mother's curiosity. It's good to know who is making love to their child. Does he dance?"

"Does he dance?! Dane is like the best dancer I've ever seen. He does all sorts of cool stuff all the time, even in the store we went to. Dane says he's going to teach me some of his moves. I can barely wait. They had other boys there who dance really well, too. I think their names were Trent and Tomas."

"You've probably never heard of Rudolph Nureyev."

"Nope.   Sounds like a reindeer I've met at Christmas."   Niqui laughed.

"He did ballet.   One of the best ever," Debbie explained.

"Dane does that.   Boy is he flexible.   Awesome in bed."

Debbie finally broke a smile, but her mind was running wild.

After they'd gotten home Debbie made it a point to talk to her eldest.   She asked me why Alex's last name was the same as the doctor's.

"Got me, Mom.   I'm still trying to figure out a riddle he gave me." Travis repeated the riddle: "It's ten miles long and can be rolled up inside a fist.   The truth is, what I was, I am.   What I am, I never was.   I define a moral Rubicon."

Mrs. Franco considered the riddle for a few seconds and then raised her finger as if a brilliant thought had come to her mind.   "Your mother isn't as dumb as you think, young man.   If you took one of your own chromosomes and magnified it until it was the width of a very skinny rubber band it would be about ten miles long.   I'd tell you the rest, but I gave you a good clue to help solve it."

Travis kissed his mother.   "You've learned a lot from us kids, Mom.   Keep up the good work."

As soon as Travis had run to his room to look up something on his computer, Mrs. Franco stood staring at the aquarium in the family room. She'd realized that Alex had given her son a riddle that verified this most amazing academy.   By God, what had those doctors done?!   The complexity of this even had her mind spinning, but she sure wasn't going to run to the newspapers or embarrass an academy that her own children had quickly fallen in love with.

She had to lay down with the realization that she had seen the greatest ballet dancer ever, naked, then fantasized about having sex with Nureyev; at least an exact duplicate.

Chapter Twenty-Seven

By the next morning the courtroom proceedings in Austin had changed its direction. There were no more accusations of violence by any of the Kings' students. Arch Johnson had his secretary put his son on a plane with a one way ticket to South Carolina. He was already checking on the continuation of his son staying for summer school at this military academy, as well. There were summer sailing excursions for troubled teens, which Arch had his secretary explore on which one went the farthest away. Either way, he didn't want to see his son for several more years.

It had certainly come as a surprise to Dr. Marion when he was subpoenaed by the Department of Education. Almost immediately after taking his oath Kami was barraged by an intense interrogation of his past. Senator Franco had arrived and taken a chair at the table next to Arch Johnson.

There were the preliminaries of the doctor's age, current residence, biological father, etc. "You attended high school in Charleston, South Carolina. Is that correct, Dr. Marion?"

"Yes"

"Is it true you left school by tenth grade and never graduated from high school?"

"Yes," Kami agreed to a point, but he wasn't going to elaborate or explain.

"Yet you profess to be a doctor. I've checked for your name in every college in the United States over the past dozen years and all you have is an undergraduate degree from the University of Texas. Interesting."

The two men just stared at each other, except Kami remained perfectly confident and calm. Johnson lost his chain of thought for a few seconds and forgot where he left off. "We'll get back to that. Do you promote sexual promiscuity at your academy, Doctor?"

"If you mean, do we teach against sexual embarrassment,

repression, and shame—which leads to sexual inhibition, dysfunction, and violence, I would admit this to be fact. One does not engage in sex because one realizes one's responsibility to the human race to procreate. One engages in sex because it is the natural thing to do. It is built into our genes. You obey a biological imperative. We don't lie to a fifteen-year old and deny their sexuality. We give it responsibility and accountable behavior with education and appropriate direction."

Arch moved closer to the doctor's face. "And you do this with allowing children…adolescents, excuse me, to exhibit their sexuality, if not expose it?" Johnson wasn't even sure he wanted to address an issue as this.

"Our students are very comfortable with their sexuality. Not one of them is shameful about what God has given them. Where there is any attempt to separate the sexual experience from the total person, that first act of objectification is perversion. And because the sexual imagination can and does respond to all sorts of stimuli, it is not possible to protect and shelter it. Sexual images, being both blatant and suggestive, play a large role in sparking the imagination. The nude art and sculpture at the academy, the research and analytical investigation of their own sexuality may present them with images, ideas, or feelings they have never experienced before, or it may ignite dormant feelings and desires which need affirmation. Feeling sexuality and eroticism is an impulse everyone feels on some level, no matter how much they consciously support the restraints parents and society attempt to impose on the young."

Arch had no idea he was facing someone this prepared. He tried pinning the man as a violent youth who attacked his father. All he wanted to do was expose this man as a fraud, maybe a guy who got his diploma through a paper mill. He decided not to pursue this angle at the last minute.

"Dr. Marion, I've been told by several adults who attended this competition between the two schools that your athletes were rather well endowed. Is this an implant they put in their shorts?" Johnson laughed with his question, but the Kings Academy had to be putting on a bad prank for the audience.

Kami did not return the humor.   "Possibly, sir, it verifies the study by Anthony Bogaert of Brock University in Ontario and his colleagues. The results, published in 1999, showed that gay males have longer, thicker penises than did straight men."

Johnson was so amazed he wrote this study down on paper.   "So you're saying this is a school for gay boys?"

"That's not what I said.   Your question was about penises," Kami replied.

"Of course, but is your school for gay students?"

"If we ever advertise for students, sexual orientation would not be a requirement.   Everyone has one."

The bailiff laughed and received support when the judge half-heartedly gruffled, as well.

Arch was thrown off-balance by the frivolity.   He lost his focus and allowed the defense to proceed.

Dan Howard approached a man he'd known for almost fifteen years.   This was a slam dunk for him.   "Kami, how to you pronounce your first name?"

"I prefer, Ka-mi, two distinct syllables, or Kam, which is what Toy calls me."

"Okay, Kami, where did you finish your education?"

"At Meiji University in Tokyo, Japan.   I passed their exams through on-campus courses and video conferencing under the direction of Dr. Parkinson, the previous pediatrician at Kings.   I took my boards in the State of Texas and my internship at Dallas Metropolitan Hospital."

"Mr. Johnson made mention that you had sliced off your father's hand as a boy and that you teach martial arts at the academy.   Are you this violent instructor who can't control himself?"

Kami confessed that, when he was fifteen, he had returned to Japan and discovered his father was a criminal and involved with a drug organization.   He had been put into a position of defending himself when he thought that his sensei, Mr. Kamito, was in danger of being shot by his father.   Martial arts had been taught to him to give self-confidence and to build his self-esteem.   There had never been an incident of excessive force until the St. Bonaventure Academy visited the Kings and Wendy had had to defend herself. Kami answered all the disputes with integrity and had vindicated his reputation.

Dr. Marion stepped down from the stand and heard Arch Johnson whisper to the senator.

"With the knowledge that the academy does have many gay youth, why all the art, books, statues, and sexual innuendo?"

"Certain people tend to reduce sexual media to a clever game which even society cannot keep from playing. One's essay on homosexuality in theater is not his misplaced, condescending liberalism, but rather his insight. So it is with our students. A passing reference, an entire book, or even a gay character in a novel can make all the difference to a young gay person because it gives concrete form to unspoken, unacknowledged, and unexpressed feelings.

"See, in a sexually repressive school, such as St. Bonaventure's, with no disrespect, any depiction of sexuality is unusual. In a society which as a distinct heterosexual bias, any depiction of gay male sexuality is, for gay youth, a breath of fresh air."

"No further questions, Your Honor." Dan pointed at Johnson, but he thought better to cross-examine.

Arch turned in his chair to speak to the senator who had just recently arrive. "What the hell did you get me into, Robert? Your kids better come through."

Dan Howard requested a dismissal from this review without further ado. The charges were groundless and superficial, at best. Dan verified that he had handed over a DNA sample of one Tad Barrie, to verify he was not the father of a child so claimed by a girl from St. Bonaventure. Arch stood up in his humble demeanor.

"Your Honor, I talked with the girl's father last night. There appears to be an error in the girl's memory. She has admitted that the father is a gymnast on the St. Bonaventure team and not any boy from the Kings Academy. We apologize for the inconvenience."

Judge Trayson sighed with more disappointment. He looked first at the senator, wondering what was going on here, then back to Johnson. "I am reluctant to proceed here, gentlemen. I trust your next witnesses are more truthful, counselor." Both Johnson and Senator Franco nodded.

"I'd like to call Travis Franco to the stand, Your Honor."

Travis came strolling in and even nodded at Dr. Marion. He spoke his name without glancing at his father, then took the oath. A wave to his mother in the back row gave him the courage to do his best. It was true he spoke quite often to boys at Kings Academy; it was the next series of questions that had Travis suspect his father had listened in on his phone conversations.

"On the night of March 25th, a Friday night, did you talk about orgies, girl or girls naked in the shower, and playing video games with one Alex Kerho?"

"I did, sir," Travis admitted and saw his father beam with a smile. "What I was referring to was a movie I'd seen—The Dreamers, I believe, that I'd seen recently. It had nothing to do with Kings' boys being in the movie." Travis smiled with the truth that he had seen parts of that movie that night.

Arch Johnson's face went white. "Does your father know you look at adult films?"

"It was rated NC-17, sir. I'm seventeen and have to register for your draft in a few months. Sex and nudity don't bother me. I would appreciate if my father wouldn't listen in on my phone calls and, if he has bugged my room, I'll find it; he doesn't have to remove it by entering my privacy."

Arch had no retort for that one. "Did you write a pornographic essay in school, Travis, an essay that might have concerned a boy at Kings?"

"It was erotic, not pornographic. Writing makes me feel in control and have power; a viable teenager who is otherwise persecuted for being gay. I find I can express my feelings and thoughts in writing; things I don't even tell Alex when I'm with him. My teacher said to write from the heart and be romantic if we wanted to. I did."

"So, young man, did you know you were gay or straight before spending the night at this Kings Academy?"

"I've always sensed my attraction to males. Swimming and our locker room have been difficult for me to refrain from focusing on my peers and the nudity around me. Alex made me realize who I was and made me glad to be a gay teenager."

"Glad? By having sex with you? Kissing and hugging on the pool deck when you least expected it?"

"Sure, it was a surprise, but Alex made me feel alive. I've known since I was eleven that the greatest obstacle that gay boys have is their fear of rejection on account of their attraction. My father rejected me, but Alex didn't. I needed feedback from reality in terms of my real feelings and my real self, and I wasn't getting that during the first 17 years of my life."

"No more questions," Arch said with a demeaning voice.

Dan Howard stepped forward, his shoulders back and withheld his smile. "Morning, Travis. Were you able to view any of the exhibits on the Kings' campus while you were there in competition?"

"Yes, sir. I wish our school would allow this type of critical thinking and research to explore new ideas and learn the truth about our sexuality and other areas of science. I love science and now I can see why the students at Kings have the highest scholastic scores in the state."

"Were you embarrassed with the outfits the athletes wore at Kings?"

"No, sir! They were very fashionable and discreet. Our sport teams have discussed getting away from the more traditional bagginess and presenting a look that draws attention and, yeah, is kinda sexy."

"Have you learned anything from a Kings' student or in viewing one of these exhibits?" Dan was having fun with this boy, a dream witness.

"Alex has been a godsend. My burden of being gay has been lifted. It's like being given permission to like and to love myself. Being gay is in itself a forbidden fantasy. Kings wants their students to celebrate sexual experimentation and endorse sexual freedom with common sense and respect for others. The questions my school can't teach and my father ignores I learned from simple projects that the Kings' boys do every week."

Dan sensed he could torture this judge and so-called prosecution with this kid for hours. In respect for the court, Dan smiled, then said he had no further questions. Everyone in the courtroom was relieved. The gray-haired gentleman, who could have passed for Andy Griffin, winked at Toy and Kami who were trying to hold their grins.

Arch Johnson was more than chagrin. His ploys were running out, but he called Niqui Franco in hopes of discovering one straw to hang his hat on. Niqui was so well-dressed the boy looked sharper than his father. He scanned the courtroom and didn't appear as nervous as his older brother.

"So, Niqui, I presume that's pronounced like Niki?"

"Yes, sir."

"You are the gymnast of the family," Arch began with a grin. Niqui wasn't sure whether this was a question or a fact so he just kept quiet. "You have quite a physique under that jacket, young man: strong, confident, flexible, probably aggressive. Are you capable of taking on a group of your peers?"

"I'm not a fighter, sir," Niqui said. "I wasn't there."

Mr. Howard rested his elbows on the table. Johnson tried to copy Howard's method of interrogation on Niqui, but it was a poor imitation; plus, it didn't work.

"Do you know who was there?" Arch asked.

"Objection. Calls for speculation," Dan challenged.

"Not if the boy knows the truth!" Arch shot right back.

"The boy said he wasn't there, so where's the truth?" Dan debated.

"Objection overruled," Judge Trayson decided.

"Fiddlesticks," Dan replied, barely audible.

"What did you say, counselor?" The judge almost stood up.

"Sorry, Your Honor, I was speaking with my client."

"If I hear another comment from you or your client I'll have you both held in contempt."

"You may answer the question, Niqui," Johnson diffused the bantering.

"I've heard rumors," Niqui admitted.

315

"Objection, hearsay," Dan said.

"Objection sustained," Judge Trayson sided only because he had to.

"What do you know?" Johnson tried another approach and was surprised when Howard let it pass.

"I heard there were three guys in ninja outfits. Probably fathers of boys your son has bullied before."

Arch could have declared the boy a hostile witness, but he held his tongue and moved to another subject. "You spent a night at the Kings Academy. Did anything happen that made you feel uncomfortable?"

"Yes, sir," Niqui answered in a flash.

Arch's demeanor perked up. This was the break he'd been waiting for. "Could you describe what this might have been?"

"Sure. I had to run and swim over three miles in the morning. Those guys are in great shape."

Johnson sighed. "Did you notice any excessive nudity or promiscuous sexual activity?"

Niqui pondered the question. "Like in the showers, there was a lot of nudity, but nobody tried to rape me like your son tried with my young brother."

"Your Honor, I move to strike the last comment," Johnson stated in frustration.

Judge Trayson nodded and agreed. "Young Franco, please just answer the question."

"Yes, sir," Niqui replied.

Arch took off the kid gloves. "I'm told by your father that you caught a venereal disease. Who gave you this, a boy from the Kings Academy?"

"No, a girl from our school."

"Didn't a boy named, Dane Nureyev, help you cover this up?"

"Why would he do that? He helped me. I screwed up and knew I had to tell my mother."

"Niqui, a clerk at this pharmacy said you and this other boy bought a kit to detect STDs. Was he lying?"

"I don't know why he would tell you that. It's not like we needed our parent's permission, and what business is it of his? Why don't you ask me about how your son almost raped my brother?!"

Arch's patience had run out. He'd never been so embarrassed in his life with what the senator had said was a sure fire case to destroy the Kings Academy. He took a deep breath and told Judge Trayson he had no further questions.

Judge Trayson nodded and Johnson sat down after giving Senator Franco a stare that could have meant a fight in the parking lot. Dan Howard began to stand up and approach the witness stand when Judge Trayson spoke directly to Niqui.

"Son, I have an idea the questions Mr. Howard will be asking you. I'd like to save the court's time here. Is there…let me ask this another way. What are you going to say about Kings Academy?"

"I'd really like to go there. You know, like, be a student," Niqui answered.

"For some reason that's what I thought you'd say," the judge calmly replied. "Mr. Howard, can we move on?"

"Yes, Your Honor. I have no questions for the witness."

"How many more?" Judge Trayson asked with frustration.

"There's one more brother, Your Honor," Arch verified and was assured by the senator that this boy was the more susceptible of all his sons. Jay walked in very tentative, uneasy, and wishing he was anywhere but in this courtroom. He was told to speak up twice just to get his name.

"Good afternoon, Jay," Arch began with a weak smile and a best-buddy approach. "Son, on the night of…"

"I wasn't there," Jay interrupted.

"So you know nothing about five boys from your high school getting beaten up by three boys dressed in ninja uniforms?"

"Objection. Counsel is leading the witness," Howard criticized.

"Mr. Howard, I'm at my wits end here. I'd really appreciate if we could get through this and discover the truth. Objection overruled."

"The boy might have lied, sir. It sounds pretty far-fetched to think there are people walking around in black pajamas," Jay stated with a lot less frivolity than when he and his two brothers were joking on what to say.

"Did you see anything at the Kings Academy that would make you feel uncomfortable?"

"There was this three mile run and swim," Jay started.

"I withdraw the question. Did you have sex during the night you stayed at the academy?"

"Yes. I'm quite able to have sex at my age."

"Did one of the boys insist on this or was it your own decision?"

"It wasn't with a boy, at least not that time."

Senator Franco moved forward in his seat with disbelief, his mind not able to conceive his youngest child in bed with anyone.

Arch appeared confused. "I wasn't aware of the academy having girls. This wasn't with a staff member, was it?"

"I had sex with Wendy Kerho. She is the most beautiful girl in the world, though my sister is pretty, too."

Robert sat back with a stunned look. He would have given his right ball to have sex with that girl and here his youngest son got lucky. Maybe his boys didn't lose that day. He almost smiled with his thought.

"I haven't got to the part where your son demeaned me for losing to Wendy," Jay blurted out. "And the part where your son's friends held me down so he could rape me."

"Your Honor, I didn't even ask a question," Arch said and went back to his table.

"I have a few questions," Judge Trayson said and leaned toward the boy. "Young Franco, do you know how many chil....teenagers knew about this Johnson boy beating you up?"

"Everyone, sir. The two girls who were there had watched the boys strip me and hold my legs open so Ralph could stick his you-know-what in me. Well, they signed a paper stating what they saw. They gave it to the headmaster."

"Objection, Your Honor. Hearsay," Johnson protested.

"Mr. Johnson, I don't want to hear another word from you. So, if I'm hearing you right, young man, your headmaster knew about this from the start."

"Yes sir! And he told my father and he discussed it with me."

Judge Trayson looked at the senator with questioning eyes. The senator rose. "Judge…er, Your Honor, my boy had told his mother that it was a wrestling incident."

"That's not true!" Jay blurted out. "Maybe one of my brothers said that to protect me, but I didn't say it."

"So when did you find out, Mr. Johnson, about your son's behavior?"

"I believe the headmaster called me that evening, Your Honor. I wasn't fully aware of the circumstances, only that I met my son at the hospital. He had two broken wrists and was pretty badly banged up."

"St. Bonaventure brings accusations about an outstanding academy to this court, knowing full well that these weren't true, even putting your own son on the stand with the full knowledge that he was lying. Where's the damn confessions of two girls, Counselor?! Do you think, Mr. Johnson, that I have nothing better to do with my time? Does the State of Texas have a slush fund that I'm not aware of to make a mockery of justice? You bring to me adolescent nuances and act like they're destroying society's morals and accuse an educated man of improprieties because you didn't simply ask where the man received his degree! If I didn't know the senator so well…strike that statement. I am dismissing this review and I expect St. Bonaventure to fully reimburse Kings Academy for their legal expense, for the director's time and effort, and write an apology to this academy for the inconvenience.

"Furthermore, I will be writing a letter of censor to the bar on behalf of Mr. Archie Johnson for his behavior in this court of law. You'll be lucky to have a license, Mr. Johnson, by the time I'm through with you. I'm also remanding your son back to Texas juvenile detention, where he will face charges of attempted rape and sexual battery on this boy."

Johnson looked very pained. "But, Your Honor, I just…got rid of the boy."

"Maybe that's the problem here, Counselor, you have never paid attention to him to begin with. This case is dismissed."

Toy and Kami stood with their lawyer and shook hands. As Jay walked by with his mother beside him, Toy reached out and brushed the

boy's hair with a silent nod.   The stare from the senator had knives
attached.

Chapter Twenty-Eight

From Cocoa Beach Sveinn drove the two of them across the state to Tampa. They stayed the night perusing their new find and agreeing on what price they should ask for their biggest discovery.

The tabloid's chief editor met with his assistant the following morning, knowing full well what these two adventurous souls had done for their circulation and notoriety. They had a bit of good news for the guys themselves.

"We received a phone call less than an hour ago from a woman in Washington D.C. She said she had information for us on dead kings and masters being cloned. As hard as we tried to get more information from her, she said she'd only talk to this boy we've mentioned in our articles who is out there digging up the facts. I guess that's you, Leif."

Leif had a pleasing smile with this turn of events. "Before I do that, Sveinn and I have come across a biggie that will shock the nation. It's worth more than $2500. As a matter of fact I think you're making a mint off our stories and taking us for suckers."

"Hey, wait a minute! You never would have found these without our support."

"I'm sure the National Enquirer would have been interested in my fortitude. Maybe I should try there," Leif said with a new demeanor.

"We had an agreement!" the editor protested.

"May I see a copy of the contract?" Sveinn asked.

"Okay, what do you want?"

"A hundred thousand for our information."

The two editors laughed. "This has gone to your head, young man. We're not a bank. Tell you what, let us ponder your offer and you go to Washington to see what this lady has on her mind. If it's that big we can broker a deal."

Sveinn moved forward in his chair. "It's interesting that there are no contracts here. Wow, what a deal; we go to Washington and you get to see the facts before we negotiate. Do we really look that lame? We can easily sell our information to the highest bidder."

"Come on, guys. We can't just write checks without permission from the higher-ups. Call us when you get there, okay?"

"Have your checkbook ready." Sveinn finally nodded, but he was also wary of this approach. They were given an address: Lauri Fisher, 1301 K Street Towers, Washington D.C. Leif jotted down the phone number, as well. Departing the office building they both saw dollar signs and a prolonged stay in America, maybe for the rest of their lives.

"What do you say we settle down and buy our own home, Sveinn? I'll go to school and you can be my teacher. No one will have to know I live with you."

Sveinn knew this only meant trouble. "I'll think about it."

"Oh, great, you sound just like those people we just got done talking with. I'll think about it. I'll think about it. I'll think……"

Leif was lifted up into Sveinn's arms and whipped around in circles until he was totally dizzy. The two males collapsed on the grass, dizzy and giggling. They were definitely a pair again.

Leif sat in the window seat—his favorite—on the 727 to Washington. Flying had fascinated him a few weeks earlier to where every trip had his body pumped with adrenaline. Now it was becoming old hat. His mind was often on Fire Island, a place he felt defined his living as a gay person. He loved the way men had stared at him and paid attention to his every move. His ploy had worked by holding out for a big offer, and Leif knew he was lucky to have met a man like Michael Rush.

There was also a truce and an appreciation for Sveinn for being his protectorate during all this time, though the magazine offers at that hotel in Ft. Lauderdale would have been interesting to try. Sveinn had refused them all, saying Leif wasn't legal, too cute to be considered legal, and his skin was too tight and soft, a smile too quick and innocent to be a boy that would set American hearts on fire. The porn industry would eat him up. The bottom line: Sveinn was afraid of losing him if he allowed the boy to be a model.

One thing Leif knew he wanted to get on with being a kid. School wasn't so bad and he loved sports, especially surfing, cycling and swimming. He also had learned that he had a soft spot for older people, a desire to help them feel good about themselves because he had struggled feeling good with himself. His ribs got a nudging from Sveinn.

"I keep noticing this guy looking at us. I also saw him in the terminal. I'll bet you ten bucks that tabloid is having us followed," Sveinn whispered.

"Of course we're being followed. These people on this airplane are going to the same place we are," Leif said.

"Very funny. No, I mean it. We'll have to be careful where we lay our baggage."

Leif thought about it, just not hard enough.

Chapter Twenty-Nine

Senator Franco had left the courtroom that day bent on retaliation with a vengeance. He'd never been so humiliated in all of his life, and no two-bit institution was going to bring him to defeat. Worse yet, he was to meet Joe Trayson for a drink at a local pub, and Robert was ready for an ass-chewing.

The two men had taken a table to the rear of the establishment; the pub's owner had always appreciated the judge's consistency as one of his favorite customers. There was no doubt that Robert was humbled and wished to apologize within the first few seconds, but Joe Trayson cut him short with an upright hand.

"No excuses, Senator. You know, Robert, if we weren't such good friends I'd have thrown the book at you boys, playing the courtroom with the scheme you and Arch Johnson cooked up. Whatever you have against the Kings Academy I recommend to you that you drop this nonsense. It's going to do nothing but bite you in the ass. I don't care if they prance around with hard-ons, or fuck each other in the ass, if you don't have better evidence and explicit eight-by-tens of their faces and bodies in action, then stay away from my courtroom. You're damn lucky I didn't get you both for contempt and tampering with witnesses. One would think you would have at least had control over your own sons!"

"You're right, Joe. I'm having family problems right now, no thanks to that damn academy. Even my wife is being brainwashed by those people, and I sure can't put my finger on it. Okay, I screwed up, but I want you to know that I'm pushing real hard for that federal judgeship for you. Politics, you know the game."

Joe Trayson chuckled, which helped Robert feel more at ease. "You're an up-and-comer, Robert. Rumors have it that the Republican Party is keeping a close eye on you with a lame-duck president on his second term. Don't worry about this latest setback. I'll keep it quiet and it doesn't appear that this Kings Academy is going to seek retribution beyond what I gave them."

"Thank you, Joe. Unfortunately I haven't heard the same rumors about my political career. I've done a few favors for the president, but he hasn't exactly praised me within the party. Not a good time anyway, here I am preaching family values and anti-gay bills, then my own son comes out to me. It's crazy when a person tries to do right and uphold one's church, only to do battle with his own family. My boys are having more sex than I am. When I was fourteen I was stroking my cock three times a day, and here my youngest, who I was sure was just discovering a hard-on in his hand, is humping this gorgeous chick that would make your dick drip."

Joe took the edge off the moment by telling a lawyer joke. Their laughter preceded six beers each and they both left feeling rather light on their feet and a whole lot happier than when they entered.

The senator flew directly to Washington from Austin. To confront his family would only put him back in a foul mood. His loins were pulsating from lack of sex and Lauri was just the youngster to make him feel better. Robert's mind went back to college years when he met his future wife. The girl was running against him for senior class president and there was no way he wanted to lose to a female. Debbie Rasio had a great body to go with her Spanish eyes and texture. Robert had suggested to her that she would be better off running for Miss Texas than wasting her time worrying about the politics of a university. This appeared to inflame the woman that much more, so Robert tried another approach—taking her on a date.

Inheriting a substantial fortune from his own parents, Robert attempted to impress the young lady with lavish gifts and the best restaurants. When these inducements failed to persuade her to drop from running, Robert managed to spice up the girl's drink with Ecstasy. Debbie was soon giggling and dizzy when Robert took her to his apartment off campus. She was only slightly conscious as Robert removed her clothes and performed intercourse while a video camera recorded the occasion.

By the next day Debbie wasn't sure all of what happened. She was slightly attracted to this ambitious Spaniard, born in the United States with the distinguished markings of a proud Spanish aristocrat. The sex wasn't what she would have normally done so early in dating, but she was on the pill and her only concerns were with STDs. Robert assured her he was perfectly clean.

The catastrophe came when Robert admitted having a roommate who loved to play practical jokes; this latest one being the filming of their sexual encounter and threatening to show it on the Internet. Debbie was more than alarmed, scared of the notoriety in fear if her parents found out and her future in being a teacher would be greatly harmed. Robert assured her that he'd straighten things out but it would take time—a few days, at best. He did perform the ultimate miracle, so he said, and Debbie had been so grateful to keep her virtue she offered to drop out of the contest for class presidency. A year later they were married.

His children were outgrowing their father's influence. At least this is how Robert justified the recent animosity. They would soon learn who rules the family and makes the gold. He just needed to be patient and commanding. For the time being he was determined to explain to his eldest son that a gay lifestyle had few positives and even marrying a woman was an image to success and far better than what being gay could offer. The senator was already pushing a bill through Congress to eliminate gays adopting children. He might even get Jay to loan him his girlfriend.

He was surprised on his return to his office to see a message that the president wished to see him at his convenience. Convenience meant when the president had time. He called the White House and informed the president's secretary that he was in town, then called Lauri with words of love and romance.

"Every missed moment with you has caused my heart grief," he had said. What he meant was, his loins were about to explode from lack of sex. Of course he impressed her with the idea that the president wanted to speak with him when Robert had the time. Robert said she came first.

The senator spent a few hours clearing his desk of overdue business. He hastened his exit and stopped to buy a bottle of red wine. Even though he was giving Lauri an extra five-hundred a month he still knew how to please a woman.

When the senator arrived at Lauri's apartment he loosened his tie, unzipped his trousers and let his dick flip out, more than ready. Lauri loved to kiss and build up with foreplay but this wasn't on Robert's mind. He forced her head between his legs and proceeded to ram down her throat until his orgasm built with his moans and her gasps preceded severe choking. While she caught her breath Robert zipped up.

"The steak smells terrific," the senator commented and patted the girl's bottom as she darted for the kitchen. Relieved and feeling like a new man Robert sat down on the sofa which he found to his liking. Within reach were a variety of magazines and books. The headlines from one of these caught his eye:

EUROPE'S TREASURES ARE BEING CLONED

He was ready to scan through the rest of the periodicals when the picture beside the headlines had the senator do a double take. A second look, then one closer to his eyes, he placed the magazine on the glass coffee table and examined even closer the boy's photo with two men.

"Here is a glass of wine for you, Robert," Lauri said pleasingly and more daring than ever, now that she was calling him by his first name instead of sir or senator.

Without looking up or even at the wine glass Robert gulped the liquid in one lift of the goblet, stood up and grabbed his jacket. "I have them bastards now! You're finished, Kerho!"

Lauri wasn't sure who the man was talking to or who he was mad with. Before she could ask, he was out the door.

Robert was above the speed limit when he drove back to his office several blocks away. 'Fuck with me! I'll show them', he kept telling himself. His appetite for a good meal had been forgotten for the moment. With his rocks relieved he felt refreshed and more focused when sex wasn't a big priority.

The trouble with being in his office was the constant interruptions and phone calls. For some reason Robert figured he might have been better staying at his whore's apartment. Put a fifty in her hand and send her shopping was what he should have done. He tried to formulate a plan, a sure fire, foolproof way to investigate this academy and assure himself of victory this time. The judge had been right about some things, maximize the evidence and know everything before any further attempts to destroy what Robert considered his nemesis.

An important email was brought in by his secretary. The president wanted him at a State dinner that evening. Robert was both anxious and delighted that he was thought of by the Head of State.

His patience grew thin when his private cell phone buzzed. He allowed himself to inhale a deep breath, aware of who was calling, because he'd only given this number out to his new sexual outlet. When he snapped the Digiscreen open a set of voluptuous boobs eclipsed the entire three-inch screen.

"Hi, Honey Buns," Robert greeted with regret of not staying a while longer and enjoying these fruits he found so delightful and fresh, compared to his wife's who had allowed four children to suckle and drag them down.

"I was worried about you, Robert," Lauri responded. "God! You just ran off without even a goodbye."

"I'm really sorry 'bout that, Honey. I've had so much on my mind. I promise I'll be there first thing in the morning. I'm meeting with important people tonight. Could you lower the lens a fraction? Yeah, there. That's so cute how you shaved your twat to the shape of a heart." The senator sensed the growing presence in his crotch but it was too close to his appointment with the president to hurry back for a quick fuck. He still squeezed his balls and allowed himself to imagine.

"All you want me for is my body," Lauri said and giggled, then ran a finger through her slit to tease the man who promised her a successful career in the political arena of Washington D.C. "See what you missed out on?"

"Baby, you have a lot more going for you than just your body," Robert lied.

"You ruined the fine meal I had ready for you. And what was that all about, you picked up one of my magazines and kept repeating, 'Kings Academy, those bastards!' I thought your face was going to explode."

Robert tried to remember what he was thinking or said, but he'd been so focused and intense. "A lot on my mind, Sweetie. Point the lens between your legs. Yeah, excellent. You know I've been fighting this academy in Texas. Bunch of queers who need corralled, Texas style." Robert was rubbing his cock through his pants, before he eased off at the last second so he wouldn't come in his underwear and end up smelling like a used rubber while dining with the president. "Keep it hot and ready, Baby. I'll be there first chance I get."

The senator shoved his phone in his pocket, pushed his erection sideways, and realized his sons must have gotten this addiction to sex from their father.

Lauri had already found the phone number to this tabloid and read how they rewarded tips with cash. She called them instantly.

Senator Franco had visited the White House before, usually with associates or another senator. There were a few discreet security measures and, though the White House officers were formal and professional, they were pleasant. As soon as they had identified the senator he was hurriedly escorted in to the Red Room which was furnished in the Empire style of 1810-1830 genre, with a sofa that once belonged to Dolly Madison. It was lit by a gilt wooden chandelier with the walls covered with a specially woven silk in sumptuous cerise. Robert had hoped for the Oval Office, but this room was very Spanish in décor and fitted his taste of décor.

The president strolled in with his closest associate the Attorney General. Robert stood up and only sat after the president sat in an authentic piece chair of colonial vintage. The usual pseudo concerns of family and health, then the president eyed his watch and went right to his agenda.

"Robert, we're having a formal dinner with the President of Peru tonight. I want you there by my side. We've been keeping an eye on you, Senator Franco. I like your voting record and your support for my family values' program. Your father was quite the up-and-coming "progresista" Catalan, until his unfortunate plane crash. I see you've inherited the Visigoths' blonde hair and languorous eyes that give me the impression you are an admirer of the Arabian Nights."

Robert laughed and had no idea where this was going.

"We're both men from the great State of Texas, Senator," the president continued. "How ironic, in some ways. The Spanish controlled Texas for many years and here you are the pride of such a splendid race. You certainly have the looks of a JFK and the charisma of, well, me. I like the way you carry yourself. Dick, here, thinks your polish and mystique comes from the caballero. Must be the courage the Spaniards have from all the bullfighting."

"Thank you, Mr. President. I haven't experienced such an event myself but it is true that the bullfighter must master his own fear before he can master the bull."

"It is sad that death is the penalty for clumsiness or cowardice," the president implied, as if this went beyond bullfighting.

"Maybe it is why the toreros use a red cape, to hide their own blood from himself and the crowd. Yellow would be the better color because it inflames the bull more than red, Mr. President."

"I didn't know that," the president replied. "Anyway, Senator, the party needs a progressive, well-educated leader, one who will take my policies beyond my tenure and lead this great nation to the future. We think you're our man. What do you say about this?"

Robert was stunned. For a second he was speechless, which was unusual in and by itself. "I'd be most flattered, Mr. President."

"Dick tells me that if you push a Spaniard he'll come to a dead halt." The president laughed at his own joke. "I'd like you to consider my offer for a few days, then we'll meet again for your decision. Hell! They just might as well allow Texas to supply the Presidents of the United States every four years; it will save the nation a lot of time and money."

Robert thought this was hilarious. "Mr. President, I heard a joke about Texas lawyers the other day. Guess there are a few of them who haven't been too successful so they decided to measure each other, and anyone over five inches could join their new organization called The Hung Jury." Robert smiled and waited. The president had to think about it for a second, then chuckled.

"I'll have to inspect my own, Senator. I'm not sure I've even seen the thing in the past few years to qualify."

Robert laughed and figured this pompous, narcissistic fool was probably telling the truth for the first time.

Chapter Thirty

Leif and Sveinn had stayed the night in Washington, then visited McPherson Square two blocks west of their destination. Leif admitted the tall statue of Union General James B. McPherson, who died in the capture of Atlanta during the Civil War, was his favorite. The cherry blossoms were in full bloom and the weather was just perfect to stroll amidst the splendid architecture of chaotic symmetry that befits our nation's capital.

The timing was to the minute when Sveinn rang a buzzer that would allow them access to an apartment on the third floor. A young lady answered and Leif thought the girl was closer to his age than Sveinn's. She introduced herself as Lauri Fisher and invited her two guests in.

Leif eyed the girl's flightiness, reminding him of a few girls at his school. She appeared more than nervous and kept questioning Sveinn if this magazine would really pay her for tips leading to more information concerning the disappearance of these great people in Europe.

"You can trust them," Sveinn reinforced. He and Leif knew by now that they were being followed and had kept all their important papers on themselves. After returning from breakfast their hotel room had been ransacked by someone looking for the same documents they'd offered for money.

Leif wanted to sound important so he said, "The men you speak of in Europe, it's not their bodies that were stolen but only tiny fragments, hair samples, nails, bones—things like that. We were told you had information that may help us."

Lauri showed the two handsome males the tabloid which they'd long seen before. Leif wondered if this was the vital tidbits. The girl began to imply she was an intern for a senator from Texas, and, occasionally, the man showed up at her apartment for coffee. Leif believed everything and didn't read through the words like Sveinn was doing.   Apparently the senator had arrived the day before and noticed this magazine sitting on her coffee table.   The man had become livid, jumped up and said, "This is it!   I've got you bastards!" He then stormed from her apartment without so much as a goodbye.

"So what did that mean to you?" Sveinn asked ever so cautiously because he sure didn't get it.

"Robert's been talking…oh, that's his first name; anyway, he is always talking about this academy in Texas that he despises because this school stomped his kids' school in something.   Who cares, right?   He tried to screw them over quite recently but came back to Washington with his tail between his legs.   It was like, he saw this story and went berserk.   I'm thinking this academy has something to do with the article."

"Do you know the name of the academy, Lauri?" Sveinn asked.

Lauri was again concentrating on Leif.   "You're a cute kid."

"Real cute," Sveinn intervened.   "He's not available.   Anyway, Lauri, as I…"

"Yes, the name," she thought and searched her purse.   "I will get credit, right?"

"I'll make sure of that," Sveinn said and took the scrap of paper she'd written the name and handed it to Sveinn.   KINGS ACADEMY

****

It had been a long night and Robert had gone to his own condo in Georgetown.   After the president's endorsement that Senator Franco was throwing his hat in the ring of running mates, Robert wanted to be careful and not be seen going to a woman's apartment in the middle of the night.

He was still flying high when he woke up and decided to visit Lauri as soon as possible.   There wouldn't be as much scrutiny during daytime hours if he was seen entering her residence.   Interns were important for Congressmen and Senator.   Then there were the pages for those who liked younger skin, especially the boys.   Robert knew the government code of protecting your colleague's ass, knowing this colleague was into a boy's bottom.

Robert kept repeating to himself, President Robert Franco! Yesssirreee! As quickly as he had thought of his future supremacy he remembered what the Attorney General had said about running a very thorough background search again. The government lawyer had asked Robert if could remember anything at all that someone might use against him. Robert had given it a moment's consideration with a blank mind before shaking his head. Shit! Thoughts of his wife and children flooded his memory. Lauri's young body also came to his senses right after them. All he needed was to be labeled as a womanizer, an adulterer, and his quick career would come to an end, even as a state senator.

He regretted being so excited as to ask the president if he was aware of someone cloning Europe's historical figures. The president didn't laugh as Robert expected, only to give the senator an inside concern that several world leaders, very important to the United States, were keeping abreast of the developments. Robert felt obligated to inform the Head of State that he might have a few leads on this, but, for the time being, they were only speculations. It was then the president offered an open invitation to use particular government agencies, if they were needed, to assist in Robert's investigation. This truly intrigued Senator Franco.

Robert knew any such candidacy depended on a healthy and happy family. He and Debbie were in a position that, even though they weren't talking to each other, they were still communicating. They each knew they didn't want to hear what the other had to say. Then there were his boys. Whatever wedge had been driven between them and their father needed to be remedied pronto. Travis was hopeless. Robert just wanted to get the boy off to college and have him vanish like so many other presidents' kids—out of sight, out of mind. Lauri came to mind. Yes, he would bring Niqui and Jay to Washington; let them meet the big wigs and enjoy being the senator's kids. An emergency appointment could be arranged while he and the boys were at one of his campaign heads, which would just happen to be Lauri's apartment, then she could flash a little skin and seduce his sons. This plan would be perfect. The boys would be happy and look forward to being with their old man so they could screw this willing supporter of their father's. It wasn't like they were virgins anymore; hell! they'd done it with their own sex and with a female. What was with kids these days? This was a sure fire solution. Now all he had to do was put it in action.

No more had he arrived at Lauri's apartment building then two young males passed him in the lobby. They were smiling like they'd just had sex, and Robert knew he was minutes away from his own smile.

Chapter Thirty-One

Dr. Kerho sat at his desk figuring out the total expenses of this debacle, before sending the estimate to Judge Trayson for reimbursement. He reclined in his chair and stared out at the beauty of the academy grounds. In the distance he could see a dozen or so of his boys, their nude wet bodies having fun on the water slides to using the wave pool to surf in. They were no longer bouncy, flighty twelve year olds discovering puberty. Practically young men, many were now taller, if not bigger than their fathers. Time worked against fathers who wanted young sons to mold and smile up at their dad. He almost wanted to join them.

The academy had survived its first real test of someone feeling uncomfortable with a program that was designed to be different from the very beginning. Toy was glad that Mr. Howard had asked the judge to seal all of the court proceedings. The boys didn't need to be known as 'That gay academy' in the State of Texas.

Society was perfectly happy to allow status quo, as long as it didn't make them feel uneasy. It was when the message became too clear. The sexualizing and non-gender-defined influence of adolescents, gay directors, and mature sexual interactions with boys their own age was considered very threatening to a nation that relished violence more than sex. Such a repressed social majority, viewing homosexuality, teenage acceptance of sexuality, and everything associated with a proud human body as unrestrained pleasure, were ready to call it all a sin, a crime, or a mental disorder.

The summer was taking shape quickly as it was approaching in a matter of weeks. Toy had invited Prince William again to the academy. His first visit three years ago was private and not publicized. The prince had loved the boys and Wendy so much he'd invited the academy to London. The prince felt young again in the presence of the boys, and wasn't the least shy about being naked in their dorm. Next year, Toy considered, they'd make this trip to London. He had so much wanted to tell the prince who these kids were; yet he was afraid of the response and losing such a close friend.

Another visitor in a few weeks was Trevor's friend from St. Petersburg. The boys had remained in close contact and now Sergei was given permission by the academy for a month's stay. This would be an exciting summer before they even headed for China for competition with their junior teams in various sports.

Toy and Kami had spent the previous evening at the boys' dorm before calling a quick meeting. The two fathers rarely had a reason to address their sons with a problem. This time it was their concern of their sons' lack of participation in having night stays at their fathers' house. Thirty-three naked adolescents sat on the white sand of this artificial beach.

"Over the last year we've noticed quite a decline in any of you wanting to sleep over. I realize our sons have grown up, but it seems odd that none of you want to talk to us or spend time with your fathers," Kami expressed in all seriousness.

"We thought you were tired of us," Tomas spoke up.

"Why would we get tired of having our own sons sleep over?" Toy questioned.

All the boys glanced at Alex, who appeared resolute and embarrassed. "I suppose it's my fault, fathers. I overheard you two talking last year that you needed to get together to have some fun before the evening came. I just assumed you weren't having much fun when one of us came up."

"You could have expressed your concern with us, Alex," Toy admitted. "The word fun to us, in that case, meant a sexual romp in the hay, so we could focus on you guys at night. It certainly didn't mean we don't have fun when you come over."

Several boys pushed their brother around in a playful tease. "Way to go, Alex!" Zach spoke for the group.

Trent waved his hand. "Dad, it's not that I don't love to sleep with my fathers. We're all kind of used to having sex at night and maybe that's a concern for you guys."

Kami stepped forward and wanted to nail this thought out of their minds. "We've had you boys sleep with us since you were knee-high to a grasshopper. What boys need from their father, in this case, fathers, is affection, not masculinity. We've always stressed that you show us emotions and vulnerability. How many of you through the years have climbed upon one of us and gone to sleep body to body?" Kami watched every hand go up, including Wendy's.

"Exactly! Few boys ever get that opportunity because most fathers aren't that willing to allow their children that type of contact. We are there for you, to be intimate and caring. Yes, over the past few years you've become sexual beings and it's not easy sleeping naked with other males in the same bed without responding with an arousal. But love isn't always about sex, and the love we share with you is just being able to hold you and talk about things you might not talk about with your brothers. Do we care that you get an erection or get all kissie and snuggle? Ask a few of your peers and you'll get your answer."

The boys chuckled but they knew what their father meant. All the boys had bonded in one way or another by learning how to be a man and, especially, a gay man from their two dads.

"Sometimes a snuggle leads to pillow purring," Tim humored the conversation. His brothers laughed and readily agreed.

Alan sprang to his knees from his sitting position. "So we still get our hair washed and get back rubs?"

"Don't forget the hot chocolate," Will mentioned and got two pats on the back for this reminder.

"If you don't know, all you boys have one weakness, it's getting your back rubbed and your spine tickled, all the way to your balls. Wendy doesn't need to know how to get to a boy's heart, so forget what I just said, young lady. By the way, she loves to be tickled too."

"I want on the schedule!" Cole shouted.

"Now that I know why we haven't seen our boys lately, we'll make ourselves available," Kami admitted. "Guys, I'm not sure when a son no longer needs the comfort and security of his dads. I'm hoping it's not until you go off to college. You've grown up in an environment of safety and love, a home that you will all depart in a couple of years to be replaced by a new generation of boys. Your fathers want you to explore the world, challenge its hypocrisy and decide for yourself how you best can contribute to helping mankind survive and improve. Many of you will decide to return, to build a home upon the hill and help your sons and your brothers' sons develop and find out who they are.

"Within your arms they will also find love and acceptance, the affection a boy needs to believe in himself and obtain the talents and successes that thrive in a gay community, such as ours. Within our arms is the love that will always be here for you, not these rooms or the toys you have enjoyed. Many of you might choose the love of your brothers as your companion for life; others will find new loves and challenges in your college days or vocation."

Jon had tears in his eyes when he honestly said, "I don't want to leave."

This might have been a comical gesture but for the seriousness of Dr. Marion's speech. It was the first time the boys had been impacted that their childhood and freedom of living as open gay teenagers were coming to a close very quickly.

Dr. Kerho moved to his lover's shoulder. "Boys, we will never abandon you. Our home is your home, but society outside this academy will offer you another education. Whether you accept this is your choice, but we will always welcome you home."

There was immense silence and more excited conversation of spending time with their fathers.

"Can we still come in pairs, Dad?" Tad asked. "And can Michael and I have sex if we want to?"

Toy shrugged his shoulders. "We've always said you can come two at a time if you both agree. There's nothing wrong with anything you wish to do within reason. It's your night with us."

"Then you two can relax, as well," Alex said and every boy there knew what Alex meant.

"Yes, I suppose your right, Alex. Your fathers can relax with you in our bed and not have to worry about you out terrorizing the neighborhood in your ninja outfits. And Alex, you will likely be seeing more of our bed than anyone; remember how much fun we had together reading Winnie the Pooh and Harry Potter? And how you liked me to make Harry a gay swimmer who teases all the boys in the locker room?"

Alex blushed and took the humor from his brothers quite well. All of them loved to be read Winnie the Pooh as children, and they all believed that Christopher was a gay boy, according to the author of the book who was into boys himself.

Toy glanced at his significant other. "We've never hid our sexuality from you boys, as you've never been embarrassed at your own expression. By the same token, it's not that difficult to refrain from sex for one night."

"Boo!" several boys shouted.

Toy and Kami departed with a full list of stay overs for the next month and whatever insecurities they felt about being neglected were solved with this impromptu meeting.

"Whatever happened to quiet evenings reading a story to the boys and rubbing their backs?" Kami asked with a smile.

"They are used to at least three orgasms and having other things rubbed. We'll probably wish they were ten again when they relished our words and not our actions."

****

The academy was off to Austin in two weeks to participate in the state's track and field finals. Kings had qualified fifteen boys and Wendy with state qualifying marks in eight events. For a small school like Kings Academy this was an amazing feat. Regardless, the entire academy went as a group and rooted for each other.

Though the Kings Academy was categorized in the smaller division of schools in Texas, 1A, they enjoyed more competitive contests against the large 5A schools, though they lacked the depth at times in certain areas. It was one thing to be a swimmer and a volleyball player; quite another to be all those and train for four events in a track meet.

The year previous Dr. Marion and Oshi had debated whether to send their 14-yr. old group to various state competitions. Kami decided that the boys were just blossoming from children into young adults and another year of growth would assist them in physically handling the strenuous sports against juniors and seniors in the state.

One year had a made a big difference. Gone were many of the

boyish faces and bodies that lacked hardness. Fifteen-year olds had sustained tremendous growths in height and weight. The boys were more mature and ready to mentally tolerate the interaction of peers, usually older than they were.

A more pressing problem was the style of clothes the athletes would wear outside of the academy grounds. Dr. Kerho decided to allow the boys to vote on whether they would wear their skin suits or something more compatible with other athletes throughout the state. The vote was 100% in favor of the skin suits.

Tommy Wilde defended the team's decision. "Dad, those body suits those guys in Dallas wear show as much, if not more, than our suits. You can see their penises flop up and down when they run. Ours might show our complete rear, but our penis is kept against our body. Anyway, our suit is more than just our sexuality. It's a symbol of sexual freedom and a discovery of the body."

Dr. Kerho wasn't about to change their mind.

****

There was a knock on the office door and Toy activated a button to allow three boys to enter. "My lucky trio!" Dr. Kerho announced. "Just picked three names out of the hat and you were the fortunate ones. See the acreage, gentlemen? Magnificent, isn't it. Oh, did I pull you away from the lake? My bad. I was thinking that, on your spare time, and starting tonight with flashlights, I want every paper, twig, branch, or foreign debris picked up from every square foot on this property. I know how the three of you love nights and parks. Hey! You can even wear your ninja outfits. Plan on a couple of weeks and, if I go out and find anything, I'll drag you three from your comfortable blankets and you'll be running while picking."

"Lucky us," Alex said.

"Glad you reminded me, Alex. I want a ten-thousand word essay on this subject: Be aware of a man who brings a gun to a poker party because he doesn't believe in bad luck."

"Huh?" Alex seemed dumbstruck.

"I thought that would be your response. Think about it. Your martial arts skills are a means to protect yourself in cases of emergency, not a method to play law enforcer for everything you think is an injustice. We haven't created super heroes here at the academy. Leave that for the comic books. Dismissed!"

****

A few hundred miles from Kings Academy, Maria had completely enjoyed her workout with her mother on this Saturday morning. The women had risen early and combined their new exercise plan with assorted Swiss balls, foam rollers, balance trainers, cardiobells, body bars, and a slide board. In blue leotards they were devoted to keeping their bodies slim and attractive.

Debbie had never been as open in conversation with her daughter as she was now. They discussed everything and Mrs. Franco had made sure her daughter was on birth control now. She even left a package of condoms in the boys' toy chest in the garage. Not something they'd use now but it was the perfect hiding place from their father.

The family had talked about going to Austin for the boys' high school state track finals. Kings Academy had quite a few participants, as did St. Bonaventure. Usually Dallas Carter and Dallas Central were the powerhouses of black schools that controlled track and field in the State of Texas, but St. Bonaventure often had a few of their athletes place in the top five in one or two events.

After their strenuous workout Maria scampered to the bathroom and walked right in. Not something she would have thought of months before, she sat on the toilet while Jay was showering. Her humor got the best of her and she flushed the toilet.

The frosted sliding door flew open. "Hey, dick head!" Jay saw his sister and his first instinct was to cover up. "I thought you were Niqui."

He watched Maria stand up, step out of her leotard and all but push him aside to join him in the shower.

While she was nice enough to wash his back, Maria offered the latest word from their father. "Guess what? You and Niqui are going to Washington. Yup, Dad just called and said the president wants him to run for president.

"You're kidding?" Jay said.

"He wants our family to look all normal and happy. We are a lot closer than we used to be." Maria laughed. "Mom thinks this might turn Dad into a better family man and change his outlook on Travis."

"I really despise him when he calls Travis a mariposa."

"Why? It's only a name for a butterfly."

"In Spanish it's a name they use for gays."

"He won't do that now if he wants to be elected president."

"I wish you were Wendy in the shower with me. Me and Mr. Happy would be a whole lot happier."

Jay twisted his head with a wicked grin. He received a smack on his wet ass.

"Your turn. Wash my back," Maria ordered and turned.

"What if….."

"As if I haven't seen it before. You have nothing on Lance."

"I'm a bunny, just ask Wendy."

"Well, Mr. Bunny, you've met my pony."

"Yeah, but we both find the barn door with ease."

"Boys! They're so crass. What happened to my little baby brother?"

"I'm still him."

Marie twisted around and kissed her brother. "I do love you, even if you've grown balls. How about we shave each other?"

"Now you're talking."

As hard as Jay and Niqui protested this trip Mrs. Franco reminded them that they lived very comfortably compared to other children. The senator was still their father and the trip would be an excellent education for them both. The boys were pissed because Travis wasn't invited to go on this excursion to Washington. Even Mrs. Franco was wondering what this snub was all about.

Chapter Thirty-Two

Leif and Sveinn flew into Dallas-Fort Worth Airport about the same time Senator Franco made the front page of the Washington Post as a new candidate for the next Presidential election. They took a taxi to the Hyatt-Regency. Sveinn had made a crucial decision before the two left Washington. Being followed, having their room ransacked, and with the information they possessed, they couldn't afford to lose it all without compensation. Consequently they used their own money to buy tickets to Dallas and for the hotel. Their journey was no longer the business of a tabloid magazine in Tampa. To them the highest bidder would be the winner. What they had was now worth a million easy, according to Sveinn and Leif.

Their suite was far more elaborate than what they were used to. The focus had changed to the location of Kings Academy, though it wasn't to be found on maps or by anyone's knowledge around this hotel. The talk at the dinner table was now on what approach to use when they got to the academy and walked into their principal's office.

Leif had his own ideas. "What if we walk right in and say, "Look, we know you have Abe Lincoln." He laughed at his own gesture, though he considered it not a bad plot.

"They'd laugh in our faces, Leif!" Sveinn said. "No, we'll have to be diplomatic, like we're reporters and have inside information. They know that newspapers write what they want, whether it is true or not, just to sell papers, so they think about this and give us the full scoop. Then we're millionaires and we can settle down in California. Wait! Can't do that. Their age of consent is eighteen. We'll have to go to Hawaii or Pennsylvania."

"I think Hawaii would be warmer," Leif agreed.

"Yeah! We'll get married and adopt a couple of kids and live happily the rest of our lives. Course, we have to invite your mother to live nearby, just not too near."

344

Leif cracked up. The marriage thing scared him; his thoughts went back to Fire Island where there were all those men in tight Speedos and adoring eyes. He wanted to explore the gay world more, have relationships before he decided on just one man. "Why don't we just tell them the truth?" Leif changed the topic back to the academy.

"They'll just deny everything, Leif. That's what I'd do and leave it up to us to prove them different. An old man in Florida isn't enough evidence. I dare say we're dealing with men who might even try to murder us for what we know, just to keep their secret safe. I'm not so sure we're the people for this because it could get real dangerous."

"I'm not giving up my share," Leif said. "Look, we can't even find this place because no one has heard of it. Maybe that girl was giving us a wild goose chase."

"I think it's real. We'll try the Chamber of Commerce, they know everybody. This is how I see it. Tomorrow, first thing, we'll rent us a car and find this academy. Let's say they're all gay. What are they going to do when a pretty thing like you walks in their dormitory? You can pretend you're new, a visitor who's lost his way. They'll want your body and you can be like a spy; get all the information in exchange for sex."

Leif found that hysterical. "I might like that," Leif admitted.

"Leif, I was just kidding."

Adjourning to their room they showered together and prepared for a great round of sex in bed. An hour later Leif snuggled under the covers and felt the warm caress of Sveinn's arm over his chest. He had his eyes open and stared upward into the darkness.

"I think me going to the dormitory was a good idea. Goodnight." He was so sweaty after Sveinn tickled him, they had the energy for one more orgasm each.

****

The two Icelanders had an early breakfast, if ten-o'clock could be considered early. They took a taxi to the city's Chamber of Commerce office. These people had heard of such an academy, but it was north of the city and they recommended that the State Board of Education might have an address. With this lead the two found their destination.

345

Instead of renting a car they decided to simply pay for a taxi. This stay might be rather prolonged, they figured, so they packed their suitcases for an overnight visit because they were pretty sure that the academy had dormitories.

It took an hour-and-a-half for the taxi driver to find the Kings Academy. He wasn't extremely happy, except for the amount he was receiving on his meter. The location just seemed fitting for a school with so much to hide. The two exited the taxi and were immediately approached by a Japanese security guard who spoke in perfect English.

"Yes, gentlemen. Are you lost?"

Sveinn had the taxi driver wait a minute and moved ahead of his protégé. "We'd like to speak to the man who runs this place."

The security man smiled. "Do you have an appointment, sir?"

"I don't think that will be necessary. He will no doubt want to talk to us."

With the names of the two visitors this inquiry was reported to the main office and the head of security. The answer was immediate with a negative reply.

Sveinn didn't take the refusal too kindly. "Look, we know what goes on here. If I have to go back to my people they will print enough smut on this place to call for an investigation that will have this academy on the auction block and everyone working here in prison. Now why don't you relate that to your boss."

There was an expression of disinterest from the security officer. "I think you two need to run along."

Sveinn got right up in the guy's face. "Look, buddy, try one more call and say the word cloning."

There was a brief hesitation whether to call in assistance, but the officer made one more contact that there were two demonstrators who wouldn't leave and threatened some form of retaliation if someone didn't see them. The head security man ordered the gatekeeper to get identification, whereupon fingerprints would be copied and recorded for future reference. If need be, the Highway Patrol would be contacted to escort these people off of private property.

Sveinn thought he was getting attention when he slipped out his driver's license and presented it to the guard.   Leif said he didn't have a license but brought his student ID from school out of his pocket and handed it to the guard.   In both cases the guard handled the cards on the edge to protect the fingerprints he needed.   In addition the officer ran a Bioscan over both males, then typed in the two names.   He was handing the two cards back and ready to point them to return to the taxi when a beep was heard in the guard shack with a red light flickering on the console's dash.   Figuring this was more important than chasing these guys off the property the officer reentered his station.   The console had bright red letters:

EMERGENCY—IDENTITY VERIFICATION—DO NOT RELEASE!

"There appears to be a problem," the guard told Leif and Sveinn. "It will be just a moment."   Within a minute there was a security car, lights flashing, which pulled up right behind the taxi to make sure it couldn't back up.   The tram had arrived and five security men ran out and surrounded the two visitors.

Sveinn held up his hands.   "Whoa there!   We're not terrorists! All we wanted to do was see the director of this place.   There are people who know we are here."

The academy's security chief came up last with a printout of a student ID and a fingerprint file.   He looked at Leif.   "I assume you are Leif Bjornsson."

"So what?   Don't tell me I'm on America's Most Wanted list," Leif humored.

The chief grabbed the boy's arm.   "You need to come with us."

"Hey!   What are you doing to the boy?!"   Sveinn protested.

"We're not harming him, sir, if that's your concern," the chief replied but then stopped when a second tram arrived.   Two men in casual wear stepped out.   Their smiles were somewhat of a relief to the two visitors who considered themselves hostage by now.   Sveinn turned his head around and saw that his taxi had swung around and departed. He began to shout in protest as the driver sped away in the distance after leaving the luggage sitting on the road.

"Good morning," Toy greeted with warmth and came right up and

347

introduced himself and Dr. Marion as the directors of Kings Academy.

"Now we're getting someplace!" Sveinn said with satisfaction. "Sir, we're reporters and what we know you'll be very interested in."

"I believe you're right," Toy said with a nod. "Reporters, huh? The boy here doesn't look much older than the kids we have here at the academy."

"Well…ah…he's my apprentice. He wanted to see the place, as well."

"Excellent. We want him to see the place." Dr. Kerho reached out and ran his hand through the boy's platinum blood hair. Leif jerked back.

"Cool it, man! I'm not that easy."

Toy apologized and removed the three blond hairs between his fingers and gave them to his security head. "First we need to put a band around your ankles—purely a security measure, so don't panic."

"This isn't some kind of virus that will make us sick, is it?" Sveinn challenged.

"Trust me. This is only to keep you safe while you're on the grounds." Toy assisted the two into the tram and they sped away. The view of the lake and surrounding athletic facilities had Leif glued to the window. "I suppose you two would like to know what's going on?"

"I'd pictured that would be your remark," Sveinn said.

"Yes, I'm interested in how you found us, but I'm more than happy that this miracle has happened."

"Miracle?" Leif asked.

"Let me put it this way, young man. You might be who you say you are but your picture resembles a painting of a young man who became King of Sweden many years ago; plus, your birth date is around the same day as all of our boys here. That's not to mention that a Dr. Jolson took fingerprints off of documents that men in history had signed or handled in one way or another. Yours is a close match to one. In a few minutes we'll know for sure when we get your DNA sample analyzed. You also appear not to have any pubic hair and your genitals are unrivalled."

Sveinn decided this was all a trick. "This is a ploy to get us in your control, right? I've known Leif most of his life and he's not one of these clones you have here, Doctor, I assure you. And how to you know the boy has an exceptional penis?"

Toy's and Kami's eyes opened wide with surprise. "I'm not in the business of lying, as you tried to deceive us. This academy wasn't responsible for another man's vision or aberration, as it may be. I am only accountable for thirty-plus children and your well-being. I owe Leif an explanation for his existence, if that ends up the case. I'm quite willing to listen to what you know or don't know, but I assure you this is not a ploy to do harm to either one of you."

The tram stopped at the home of Toy and Kami. They led this party to the second level and a huge office overlooking the academy. Toy's secretary came in with a full report of the hair sample and physiological report of the visitor's health. Toy raised his eyes and was straight forward with both Leif and Sveinn.

"You both have sexually transmitted diseases," Toy informed and let it sink in.

"No we don't," Leif shot back.

"You're not much for the truth are you, Mr. Bjornsson? Here's the report taken from the sensor which analyzes your body fluids and every other component. Care to read?"

Leif glanced at Sveinn. "Did you cheat on me?"

Sveinn wanted to lie but this whole scene was already chaotic enough. "It was a quickie, I swear, Leif. The guy came on to me in Florida. You weren't exactly cordial at the time."

Leif moved away from Sveinn's arm. "I should have figured that out. You told me not to stare at or tease other men because you think you own me, but then you go out and play around. You're nothing but a hypocrite, Sveinn. Screw your marriage proposal!"

Sveinn fought for words to defend himself and gain Leif's favor back until Kami removed this thought.

"Gentlemen, whether you believe this or not I think it's best that you both spend the day with us. Let's get you both on antibiotics and allow you time to think. As for Leif, your DNA matches our missing boy. I know you don't believe this but your progenitor or, to put it another way, your father, was Charles the Twelfth of Sweden."

Leif was stunned. He knew of this king's tomb being tampered with but he never would have imagined that this is how he arrived in the world. His thoughts switched to Rodney Miller. The boy had even said he looked like Charles the Twelfth. "I want to call my mother before I believe any of this."

Toy slid his cell phone over in front of the boy and Leif punched in his home number in Iceland. "Hi, Mom, it's Leif……….Yes, I'm okay……..Sorry I haven't called earlier. Sveinn is here with me, Mom, I……….Mom,……Mom, listen a second, will ya? Aside from you not liking my sexuality, were you ever like a surrogate for a doctor? ……...Mom, don't cry, please …….Mom?.......Look, I'll call you back, I promise……You don't have to explain to me now. Don't worry, I know everything." Leif handed the phone back to the doctor. "I don't feel very good. Do you have a place where I can stay?"

Kami began to escort the boy out when Leif informed Sveinn that he wanted to be by himself. With a swift tram ride Leif found himself in a waiting room with lockers, overlooking the academy below. He was told to remove his clothes, before Kami gave the boy a quick physical. Throughout most of this Leif didn't mind the examination, but his tears increased gradually from the hurt he felt and being scared that he truly was this Charles the Twelfth and not Leif Bjornsson. The shot in the butt made Leif woozy and he never knew he was carried by Dr. Marion to a room, numbered thirty. Wendy had deliberately moved to room thirty-four to allow Leif his true entitlement.

* * * * * * * * * * * * * * * *

It was in the early evening when Toy and Kami met with Sveinn again in their guest room, where Sveinn was allowed to stay and freshen up after a most hectic and surprising day. Sveinn was truthful in how he and Leif had investigated the grave robberies of so many American figures of importance. Their relationship with a tabloid newspaper in Florida was also revealed.

"Toy, I'll get our secretary to order the past few issues of this magazine. I've called Oshi; he's returning tomorrow a day early from his vacation. Leif has woken but refuses to eat anything or meet with the other boys.   Obviously he's extremely embarrassed and confused."

"Of course he's confused, you've told him he's the King of Sweden," Sveinn said in defense of the one person he loved in this world.

"Mr. Eldjarn, Leif is not the King of Sweden.   He's a boy who was cloned from a past King of Sweden.   Leif has his own personality, habits, hobbies, feelings and thoughts that Charles the Twelfth didn't possess."

"But are you saying that Charles the Twelfth was gay?"

"Yes, that appears to be the case," Toy agreed.   "You're free to walk the grounds, sir, but don't try to contact Leif."

The two doctors left the guest room and had their own conversation.   "Kami, if you don't mind, find out everything you can about Leif from Sveinn: his favorite meal, his hobbies, likes and dislikes, what ticks him off—besides his best friend cheating on him—and his soft spots.   You might get out one of Oshi's samurai outfits; I think he's going to need it again.   This kid could be a tough case."

"How about if Sveinn asks to see him?"   Kami asked.

"It's best we keep this teacher away from Leif until the boy gets a clear head.   They love each other but the boy sees the possessiveness of the adult.   Not that I blame the guy for desiring a teenager that pretty, but the boy has the right to explore the field, experiment and discover boys his own age and desires.   We have to be ready for a backlash if any information has already escaped from these two.   This was bound to happen with Senator Franco snooping around.   I think we're in for a battle to keep our foundation and the boys' identity a secret."

****

Leif woke up the next morning and refused breakfast and told any boy who stopped by that he wanted to be left alone.   He didn't like not having any clothes to wear, but he also didn't mind seeing boys his age and size moving around naked, as well.   He had pretended not to look at the boys' faces, yet his quick scans had excited something far deeper within him than he'd known before.   He was tempted to ask them which one was Abraham Lincoln, but who was he to speak?   The one positive feature, they were his age, handsome, and athletic bodies with penises like his.

"Hi!" Erik greeted when he swung his head around the open glass door.

Leif was ready to wave off another boy who had intruded on his space. Instead he saw a kid his size, not with anyone else, and appeared as forlorn as he felt.

"Hi!" Leif replied back with less enthusiasm. "I'm not coming out, if that's what you came for."

"That's okay. I stopped by to see if you had any questions about you-know-what. I have a brother who has a friend who had the same thing, but he's all better now."

"You guys get these diseases here?" Leif asked.

"No. He's a friend we know in San Antonio. The shots are a pain in the ass, but you'll be 100% in no time."

"Really? Thanks. I'm pissed at the guy who gave me this; he's my friend and teacher. You a student here?"

"Sure am. I'm Erik and you are our missing brother. You're going to find that your brothers here are really cool and can't wait to speak with you. I've never seen them this excited about anything. It figures you'd be one cute boy."

Leif chuckled for the first time in a day. He invited the boy in and desperately wanted to ask a million questions. "You mean I can just get up and leave?"

"Sure! Why not? If you decide to leave I'm sure my dads will allow you. Nobody's keeping you hostage. Not sure why you'd want to leave, though. This is the coolest place for any boy on the planet."

"You have more than one dad? I don't have any," Leif said with some embarrassment.

"Yes, you do. Our fathers are your fathers. You're also our brother because we share many of the same genetics. Nothing like knowing who your brother is by counting their pubic hairs and seeing a long penis."

"And here I thought I was an exception," Leif humored. The more Erik talked, the place sounded like a sports academy with coaches busting your butt. Leif downplayed the praise of this boys' Disneyland

352

and wondered if the place was right for him.

"I know this is all hard to believe, Leif. Here we learn to express our feelings, our love for each other. Dad says when someone enters your life unexpectedly, look for the gift that person has come to receive from you. I knew immediately your fears about being infected because I've seen those fears. They teach respect here for other people's feelings and paths."

"Sounds like a monastery," Leif said.

"Maybe a little. There are three laws: love God, love others, love yourself. Sexuality here is seen and treated as totally natural, totally wonderful, and totally okay."

"You mean adults don't trip on being gay?"

"You won't believe it, Leif. Want to play chess?"

The boys played two games and Erik was very impressed with Leif's knowledge of the game. He even let the boy win one. "I now know why they call that place Iceland. You stay in all the time and play chess."

"That's Greenland, Erik. They named the countries like that to keep people away from the most beautiful country, and that's Iceland. Greenland is an iceberg. I even surf. Do any of the guys here surf?"

Erik chuckled and said he knew a boy who was really good at surfing. He ran to get Dow Whitman. When Dow introduced himself Leif asked the boy what his last name was, then grinned.

"I've seen your father's grave," Leif admitted and hoped he hadn't said the wrong thing.

"I've seen it in pictures. I wish I had his talent," Dow said.

"He does. Dow is only being modest," Erik replied.

Dow pointed out toward the lake and explained where the wave pool was for surfing. It was a little too perfect compared to the ocean but it was great to keep a surfer's technique up and to experiment with new skills.

When the boys left Leif knew he'd given up more about himself than what he had wanted to. He was still too scared to admit he wasn't equal to these boys; if they knew his past they wouldn't accept him, he was sure. His mood turned dour again.

****

Dr. Oshi Makawa didn't mind scurrying back from Hawaii to the academy, what with the new addition of their one and only missing boy. To say this was a blessing from God was an understatement. Psychiatry was one of Oshi's specialties, a major challenge when dealing with a group of boys who had a variety of skills and mental aptitudes.

Oshi met with Toy and Kami; the three of them decided amongst themselves what the best approach was for Leif to make his transition as smooth as possible.

When Oshi entered the boy's room Leif was experimenting with the computer panel and the various games available. So far the boy had only trusted Erik and Dow. Dow had brought his newest brother a protein shake that was loaded up with nutrients as his only source of nourishment, since Leif had refused to attend any meals. The boy drank it down only after Dow left.

"Konnichi wa!" Oshi yelled and nearly scared the boy to death. Oshi was all dressed up in samurai regatta, a dai-katana hung in his hand, with other weapons in his obi. As Oshi began to swing the sword with graceful ease Leif was scooting backwards on his butt, mouth wide open, and eyes nearly as agape.

Leif thought his life was over, as young as it was. He was both mesmerized and in shock with the whirling blade, then followed two sticks with a cord between them which whizzed in rapid movements near his body. The whips of the sticks over the man's shoulders and around his sides weren't the sounds of rubber playthings. Then came the shorter blades from the man's belt, a mere slip of a finger would have one of these sharp objects come flying at him. In a few minutes it was all over and this crazed creature knelt on his knees and appeared to go in to meditation. Leif knew this was his chance to break for it if he was to survive. No more had this thought come, and he had begun to leap up, when a blade flashed up to the side of his neck.

"Sit!" the voice came from the rear of the sword.

Leif decided not to escape; no use in riling this man any further.

354

Finally this warrior from Japanese feudal times lifted his head, looked calmly at Leif, and spoke, "How do you feel?"

"Who are you?" Leif asked.

"That is not a feeling."

Leif searched his body for an answer. It was quite obvious. "Frightened. Terrified."

"Good feelings. Terrified of what?"

"You"

"Me?! I know not how to defeat others; I only know how to win over myself. I am your protector."

"If you're my protector then I can leave, right?" Leif felt a new sense of confidence and began to rise, only to have a smaller blade appear underneath his chin, pushing him upright and backward.

"One who is careless in small things is careless in all and will never accomplish great things. Sit up straight. One cannot leave without knowledge. One who knows, does not speak. One who speaks, does not know."

"I don't know anything, really!" Leif said with certainty.

"Then I must teach you. Do you have a few days?"

"If I say no, will you kill me?"

"I am your doctor, why would I hurt you? Any other feelings?"

"Who are you, really?"

"That is a question, not a feeling."

"Okay, I'm confused. I apologize if I've offended you."

"You haven't offended me. You are who you are. I am who I am. Ask me a question?"

"Where is Sveinn?"

"Your friend is having a great time in the company of our directors. They are currently eating steak and lobster, pie and ice cream—cherry pie, I believe. Afterwards, a movie to help their digestion. Would you like to call him?" Oshi began to place all his weapons on the tatami: nunchuks, throwing stars, chains, swords, blades, knives, all in search of his cell phone. Finally this was pulled out and had Leif laughing.

"Nuh! I'm still mad at him. The butt head! He's up there eating while I'm starving. He doesn't care about me."

"You care about him or you wouldn't care about what he thinks of you. You have many brothers here who care about you, other boys and a girl who would love to meet you."

Leif put his finger in his throat with the mention of a girl. Oshi couldn't help but break a smile. "I thought you tough samurai never smiled," Leif said with a grin.

"Tough? I just learned all this not too long ago. Even samurai have sense of humor and agree with finger in throat."

Leif chuckled and began to relax, but then found that little blade under his throat. "You make fun of ancient samurai?" Oshi asked.

"No sir. I….I would like you to teach me. Do you teach martial arts here?"

"We have several sensei. All the boys are skilled and think they are Shaolin priests, but more like cartoon characters."

Leif would've laughed but the blade tickled his Adam's apple. "What makes these boys my brothers?"

"What you share gives each of you a bond; yet, the most important thing for a human being is not what is between his ears; it is what is in his heart. If the spirit is strong, one can accomplish anything. You have proven this out to be so. Hai?"

"Hai!" Leif said back and hoped it meant yes.

"Am I really Charles the Twelfth?"

"You are Leif Bjornsson. There were great men who did not sire an offspring for certain reasons, one of these was because of their sexual orientation. With exceptions, others, for political gain, social expectation, shame or denial, still managed to marry and have children."

"Like Abraham Lincoln," Leif surmised.

"Yes. So you are aware of Jon?"

"No. Is he Abraham Lincoln?"

"No more than you are Charles. Just don't call him Abe. His technique with the sword is not as polished as his sensei; thus, your neck will remain nicely balanced. Our boys don't always choose the names of their forbearers."

"Good. I don't think Charles is better than Leif."

"I'm sure Charles would say the same thing for you and himself.

Unfortunately, in days of old, there cannot be a twelfth without an eleventh or even a thirteenth without a twelfth. This can confuse even a Japanese man."

"If I decide to stay here, will you be my friend?"

"I cannot teach one who is terrified, undernourished, confused or uncomfortable. The best attitude toward life determines its outcome. You must foster the life force by being healthy, courageous, decisive, resolute, and vigorous. Real budo is a function of love."

"I can love. I'm just not sure if anyone loves me," Leif admitted. "You wouldn't like me if you really knew about me."

"Hai! We come to the truth. A boy's burden: no one could possibly understand who I am. Love will consume us only in the measure of our self-surrender."

"But it's true. I don't even like myself at times."

"It's because you question yourself. What if I told you there are no rights or wrongs. The things you do are all done to find out who you are in relation to God. I know much about you, Leif Bjornsson. You are gay, which would confuse most every boy. You're also very intelligent, but school bores you. Of course the subject of education has improved with your love for your teacher."

"How did you know all that?"

"I am one with you. Is it not better to embrace love and crave the touch of another? Why not experience the joy that's part of honoring life, rather than harboring guilt and shame for something you shouldn't feel shame or guilt for to begin with?"

"This is too weird—how I got here and all."

Oshi handed the blade to Leif to feel and hold. "There are a lot of things in this world that just don't happen by coincidence. A creator, far superior to our limited knowledge, has a great sense of humor. As short as we live each of our lives, wouldn't it make more sense to spend it in love, serving others?"

"I don't deserve to be here. These boys are probably far more advanced than I am."

"They will accept you as you are, and you will blend in with them in no time. This dark side you think you have, did you ever consider that it's simply waiting for you to accept it, turn on the light and say, 'I'm here, this is me?! Can you handle the blunt truth, my young samurai? Have an adult lay it out for you, piece by piece?"

"That would be a first, an adult not treating me like a kid. Sveinn is different. I was fourteen, though. Maybe that's not right. Oops, I forgot, there are no wrongs."

"Ah, my teaching is progressing. What was he to you?"

"He was kind of a....I guess, a friend. It wasn't his fault. I wanted it."

"Did you struggle with your masculinity? Your sexual identity? The shame of feeling complicit?"

"Yes. He made me feel that I was right."

"You are right, son. We teach boys here to fall in love and to take this love to formulate boundaries for future relationships. An adult should have boundaries from letting children into their lives, sexual lives, that is."

"I'm not a child."

"If you were twelve I'd say you lacked the mental capability to partake with an adult. The power differential is too much. At your age it's a means to discover sex, an expression of presence and consciousness. The opposite is someone who is exploited or forced into something before they're ready. Sex is a divine communion of pleasure, a freedom and responsibility that is made with your own mind. If you second guess yourself or allow your mind to feel guilt over something totally enjoyable, we have a problem."

"I feel better already telling you all this, but I really enjoyed it. I'm thinking you want me to feel all guilty about having sex."

"That's not the case, young Icelandic boy. As in any relationship I don't want the sexual aspect to be the nucleus, the one and only thing you have in interest with the person you're with. No one really discusses their sexual truths or we'd all be laughing ourselves silly over sex. There's a difference between what's private to you and what you keep secret. Revealing a secret takes away its power. If you don't it eats you from the inside out and you spend years searching for a way to tell someone through compulsive behavior that's often destructive.

Secrecy and guilt can destroy what's natural."

"Dumb guilt!" Leif's smile was genuine, but an attraction was being built with this new father figure in Leif's life.

"Hai. Sex isn't a code or something so mysterious that answers will come as you get older. It's an orgasm, a foresight that the ultimate aim is going to feel so great, it is the ultimate high. Afterward your mind should be asking, "When can we do it again? Forget the bad, the poor me, the 'He made me do it.'"

"You're pretty smart," Leif admitted.

"So are you, young man. You have street smarts that your brothers here aren't privy to. Be aware of your ability to manipulate people, Leif. These attempts make you shallow and distant; plus, when the person finds out they've been duped, you've lost another friend. Get my drift?"

Leif hung his head because he knew this man saw right through his heart. "I really want people to like me."

"Leif, you're as handsome as any boy we have at the academy. Just be yourself and let nature take over. Don't be someone that you think the other boys will like; it won't work. These kids are as sharp as a tack around here. If you're genuine, they'll fall in love with you in minutes. I don't have all the answers, young man, and sometimes my weapons knock me in the head and convince me otherwise. What happened to you instinctively taught you to find a way to meet your needs. You are to be complimented."

"It led me here," Leif said. "I'd never found this place if it wasn't for wanting someone to love me."

"The fact is, this is who we are, what is. Don't blame God, because you are God. There is no right or wrong in God's eyes, only what is derived from the action do we make it what it is."

"Are you God, as well?"

"Oh, yes, we are one in the same, living God's life for Him and letting Him experience life through us. Possibly we both pursued a few experiences before we were ready, but we're here now, aren't we, kiddo?"

Oshi clicked his fingers and Wendy came in with a tea service.

Leif scanned the girl's nudity, and even he thought she was beautiful. "We have a tea ceremony together. I've brought you a scroll of calligraphy and a book of haiku. They are gifts from Japan; my country desires to be friends with our thirtieth king."

"I'm not a king," Leif said with modesty.

"Yes, I have forgotten your modest appraisal. What I mean is we'd like to be friends with a boy from Iceland, who we are honored to have as a guest and hope he accepts our invitation to stay a few weeks so he will know us as we discover his beauty."

Leif accepted the gifts and eyed the man's sincerity. From a scared teenager Leif felt honored and appreciated. He made sure to do everything exactly like this humble gentleman in front of him did.

By evening Leif was hungry, starved for attention and desiring to be accepted by everyone. He went in search of Dow and was directed to the theatre by another boy. Leif walked in and saw a multitude of boys his age watching, IT'S A MAD, MAD, MAD, WORLD, on a large screen. There was so much laughter and it was so dark it took a few seconds for Leif to notice the many naked bodies.

Dow grabbed Leif's hand and they went to sit on a stuffed zebra. In seconds Leif had laid back into Dow's arms and they were soon in hysterics together.

Afterwards Dow showed Leif the entire dormitory and the boy's consciousness had trouble absorbing all the sensory delights and possibilities. Every boy that passed called his name and smiled. Trent Marion approached Leif and hugged him.

"I've been praying for you since we found we were minus one," Trent said with tears coming from his eyes.

"Thanks," Leif said and felt bad like he'd disappointed these boys, yet it wasn't his fault his mother had kept him.

Leif was quickly absorbed in fun with more brain challenges and physical dexterity than anything he'd ever played. Nudity was long forgotten with the laughter and trash talking to each other, yet it was never cruel or demeaning like the boys at school. His classmates in Iceland treated boys who were different like pariahs. Leif found the time to ask a few boys about the crazy guy who dressed up like a samurai.

"Dr. Makawa. A man of honor and skill. Always bow to him when you cross paths and never tell him you've done something wrong. Tell Dad instead so you'll live to see tomorrow," Tomas Steuben spoke without a smile. "It's not that we do a whole lot wrong here, but we act like crazy kids, at times, and he acts like crazy samurai. Dr. Makawa respects everyone, but you better respect him or he'll make you into a pretzel. He is a man who expects one-hundred percent loyalty and dedication to the art; a true samurai, like Dr. Marion. When we were young one of my brothers ran from him and, when Will began to open his stride up, he ran right into Dr. Makawa. To this day Will can't explain how he dashed thirty yards away from a sitting man to where this same man was standing upright with his hands on his waist waiting for the boy to run into his arms. It's a samurai thing, we all know it, but once he touches your soul you're forever in love with him."

Leif was awestruck. "We drank tea and the man touched my forehead for several seconds. I felt almost paralyzed."

"You've been Bushidoed, Leif," Tim Melville said with a smile. "No hope for you now to ever like girls again. Every time you see one you'll start acting like Bruce Lee and doing kicks and swirling all over the place."

'I don't like girls anyway, though the one who brought us tea was cute. Too bad she doesn't have a penis."

The boys laughed and pointed to where she was playing a video game. "When we were younger Wendy asked her dad for a penis every Christmas. She might as well be both sexes for hanging around us all her life," Tim said.

Wendy soon strolled over and Leif went into his best Bruce Lee antics. He had the boys laughing. "How quickly my brothers have ruined you," she teased.

Dow imitated Wendy's walk as she moved away for even more laughs. He turned back to Leif. "Come on, brother, I'll teach you to pole vault on our mechanical vault. Sort of like riding a bull but less dangerous." Twenty minutes later Leif was sweating and had the hiccups from laughing. He didn't jerk away from Dow's kiss on the lips.

Chapter Thirty-Three

Senator Franco spent several days doing his research and reading everything ever printed by Stern Magazine and all the tabloids. So much of it was sensationalism; yet, a great deal of it was fact. He was determined to destroy this school as quickly as possible to alleviate any future problems with his family. The key was devising a plan that would allow him to use the various federal departments and personnel available at his disposal.

With the assistance of the president, Judge Trayson was put up for a district federal bench appointment. Robert had also remembered a DNA sample turned into the judge when a Kings' boy had been accused wrongly of impregnating a girl at St. Bonaventure. He had Tad Barrie's DNA mailed to Washington where it could be compared to samples found at the James Barrie residence and hair found among his historical records. The sample was sent to Scotland Yard for further review and comparison.

Robert was enjoying his new found fame and independence as a viable candidate for president. Though the election was well over a year off he'd come in at the right time after several candidates had beaten themselves in the form of debates and through smut dug up by their rivals. The Iowa caucus was advancing his potential as a valid nominee to continue the family rights' agenda of the current president.

His self-instituted bachelorship meant a nightly jaunt to Lauri's apartment for frisky sex. He made her an assistant campaign manager, which gave him the excuse for being there constantly. Robert was also making promises of a position much higher than the mistress of a president, a traveling secretary like the one President Kennedy enjoyed for so many years.

Lauri was extremely pleased and content; yet disappointed that she wouldn't be the First Lady, but this was no time for a divorce to foul up the senator's chances or hers. She took her role with exuberance, furnishing her body for the good of American politics and wearing the black leather boots, with whip, to keep the senator smiling. It was really weird, but fascinating, to spank a U.S. Senator.

In the meantime, Robert began to work with the FBI in formulating a plot to unravel the true nature of this academy in Texas and protect the United States government at the same time. No one wanted the repercussions from other countries that the government of the United States knew and condoned this cloning of foreign citizens, especially of heroes and icons.

There could be no mistakes this time around, no wiggle room for the academy to escape from in a court of law, if it ever reached such a level. Robert was trying his best to alleviate this nuisance quickly. This had to be done cleanly with simplicity without publicity or public knowledge.

The president was in support of everything Robert was doing and glad to stay in the back while others did the dirty work. With the president's recommendation a tiny microrobotic fly used for military reconnaissance was to be implemented to survey the academy grounds for suspicious activity.

Within days the pictures were on the senator's desk to analyze. The senator watched his monitor as the robotic fly flew around this majestic campus, far more beautiful in the daylight than Robert remembered. It was truly an architectural work of art.

There were few human forms moving on the campus ground. The fly swept over a grass knoll and positioned itself over a recreational area, and the senator saw a group of teenagers skinny dipping in the lake, sliding down water slides, windsurfing, and playing with dolphins. The size of the boys' genitals had the senator's eyes glued to the zoomed image on the beach where three boys prepared their boards for surfing.

"What the hell?!" he exclaimed in his office chair and did a double take at the length of the boys' penises, minus hair. As the fly flew further over the lake, Robert began to rub his erection, which had begun to swell with the view of genitals that he'd rarely seen, even at college. When the camera focused on the young lady of the group, her nubile breast bounced above the water line as she played with a beach ball with two other boys.

The senator watched closely as she swam to a raft in the middle of the lake, before she flopped up on the rubber flotation device and spread out on her back. The fly, thanks to an eager military officer, a mile-or-so from the academy, hovered over her and then went stationary above her shaven sex before zooming in between her legs.

It appeared this bug was going to land and crawl into this wet opening, it was that close. The senator recognized the young girl as the one who befriended his son, Jay. Of course, he knew his youngest probably hadn't even garnered a kiss from this beauty, but he considered that the boy might be real pleased with his father if he had a chance to see this girl in this pose. He decided this would be an extra treat for his sons. It might even give the youngster a sex education course to the anatomy of a female.

These military guys sure knew how to do reconnaissance work and manipulate a camera lens, Robert ruminated. Through the tiny audio transmitter Robert heard the splashing of the boys in the water, before one of them spotted this insect a few feet above this girl's sex and swamped this mechanic marvel with a handful of water. The pictures ceased at that moment, but not the sound.

"Hey, Creeps!   Cut it out!" the girl had shouted.

It was apparent the boys had picked the little device up and showed it to the young lady. There were moments of silence and then the picture resumed with the swaying of the bug in someone's hand as they walked up the knoll. The lens had a bird's eye view as it swung forward of the lad's penis, a true marvel of masculinity and an envy of almost any male. Inside of an office the camera lens still operated and picked up nude statues and paintings on the wall as it was passed to another person.

"We found this flying over the lake, Dad," the boy said.

"Looks like someone wants to spy on us," the adult replied.

"Can I have it?" a boy asked.

"Son, it's not a pet."

"I realize that, but maybe Bobby can find a use for it."

"Sure, take it. I suppose our security people can use it in the future."

The total information and pictures were hardly useful for the senator or intelligence. It did seem to verify that this academy was very liberal in their dress code, but that didn't mean enough to prosecute.

This bit of military hardware and useless video was passed around to the different agency heads, only to be viewed in private and stored as a sexual aide for later use. This wasn't something the president or the FBI chief would make known publicly.

The president spent a few minutes in his own observance and was glad his erection was hidden behind the desk when he perused this fine young lady on the rubber raft. He was sure that the academy had to be doing some sort of mysterious medical science to give boys penises that would have women storming the gates to get in.

The president didn't dare say anything because of the constant audio taping in his office. He knew the less he had to explain to the public and other countries' Head of States, the better.

The president slid this surveillance video back in his drawer and pulled out a wooden box given to him by the senator. While he contemplated all that he had to do during the day he fingered Senator Franco's gift to him: a twin set of Remington .44 magnum revolvers with pearl grips and silver barrels. Having a flare for the cowboy image the president grabbed one of the guns, stood up and unzipped his trousers to allow his half-erect penis to flop out and point straight ahead.

"Take your pick, cowgirl. Which will it be?" He checked their loads, then played with them briefly, spinning them, doing a couple of border shifts, and then dropping them into their holsters. A clumsy quick-draw almost had the pistol fly from his hand, but he found control and twirled one the revolvers around his finger in pretending to stare down the likes of Annie Oakley. He was just hoping his secretary didn't come in at that moment and figure out that he had the mind of a twelve-year old with an erection that would be tough to explain.

Chapter Thirty-Four

Sveinn was enjoying his stay at the academy. He had made quick friends with Toy and Kami. The lodging was free and the food great; yet he had other plans to gain the information to attain the wealth he desired. All he needed was a collection of evidence, and this he hoped Leif was gathering by the day.

The directors had been open and honest about the academy's program—how they inherited twenty-nine babies and discovered through the memoirs of Dr. Jolson how he'd pulled off this great heist in front of a blind audience. Sveinn knew he was sitting on a gold mine that would make Leif and him rich beyond all their imagination. He was able to tour the campus and take in the sights. To see the bodies of so many gorgeous teens had him horny with anticipation for the next time he'd see Leif. Hopefully their diseases would be healed.

The science exhibits were exceptional and academically superior to anything Sveinn had seen in his high school. The amount of research here and the enormous amount of information was enough to distract him until he found the opportunity to grab Leif and get out of this place.

Sadly Dr. Marion had informed Sveinn that he wouldn't be permitted to have sex on campus because of the age difference and Texas laws. Sveinn would have loved to have intimacy with any of these dynamos, but he was at least responsible enough to know he had to be celibate while he was under medication and healed completely.

Not wanting to cause suspicion or create an investigation that could edge him out of the way, Sveinn decided not to contact the tabloid while he was here. They were no doubt wondering where he and Leif had gone, but this hands-on experience might make them offer more money in the long run. He did feel guilty for giving the boy a STD, though he didn't want to take all the responsibility, since Leif hadn't exactly been loyal himself, teasing men in the pool and wearing provocative clothing. Sure, he'd bought the clothes for Leif, but the boy didn't have to wear them like he was a male Lolita.

So, for the time being, this teacher from Iceland was treating this as an educational sabbatical, though his career as a pedagogue would soon come to an end with the millions of dollars he expected to receive with this discovery. Leif would come around to see a future with a man who had looks and money; one who would show the boy the world with every gay beach, nightclub, resort and romantic island available for two adventurous males like themselves.

****

Leif was in his second week at the academy but he already considered himself as one of the boys. He loved every new day, the challenges and inspirations that were his for the asking. His brothers had persuaded the academy's sculptor to do a thirty-fourth statue of their newest brother, and Leif had a smile a mile long when he posed for pictures and physical dimensions.

Already one of his favorite spots to think, Leif would often walk down to the track and sit at the base of Dow's statue and ponder his studies. It reminded him of Iceland and the statue he used to admire there. In this case his physical presence would soon be part of history.

Dow often stayed back with Leif through the morning runs and swims to encourage his fitness and push the boy when the going got tough. After five days of aching muscles and burning lungs Leif began to stay in eye range of the bare butts ahead of him. His strength was in his swimming and cycling, two skills the Kings' boys were only slightly better at than he was. He enjoyed impressing the boys with his surfing skills, his aerials and turns. It was even more fun to do all this naked in water that was twenty degrees warmer than Iceland's.

With the athletic venues Leif discovered through digital tape how to improve his technique and form. He was shown his weaknesses and how to improve his style and performance. Not too happy that school was year around here, the classes were interesting and fun to go to; something he really began to look forward to every day.

Weight training was a regular daily event along with martial arts training. He felt inadequate but the other boys encouraged him and didn't give the impression they were better than he was. Being the same age and feeling inferior to his peers forced Leif to exert twice the effort.

Dow had explained the lighting system in his room and by the third night they were sleeping together without sex. It wasn't like Leif wanted to wait for the doctor to give him the okay, so he pleaded for the warm body next to him and Dow was only more than happy to comply. Masturbation was far more fun with someone who required two hands around their erection, and Leif was glad to show his skill at fellatio.

The adjustment to the daily regiment fit Leif's character. He liked and adjusted to structure, organization and a purpose to life. When his VO2 was measured he was nearly as fit as the Kings' runners, so his swim coach asked him if he'd ever tried a triathlon, and Leif said he hadn't. To find a place for this missing youngster on the track team was very difficult. Leif loved the idea of this individual test of courage, fitness, and stamina. Daily the triathlon became his obsession.

Dr. Makawa met with Leif every other day. Leif was always ready for each session, usually held in his room, and his love for older men devoured this Japanese martial artist when he entered Leif's room without clothes on. Leif began to learn more about his feelings and thoughts with every session, a sure aspect of a boy who was becoming self-centered and relaxed with the surroundings.

He also was falling in love with his shrink, a fact Oshi knew only too well from years of experience with 29 other boys. Leif knew he could talk about anything and everything with the doctor. His new friendships with Dow and the emotions that came with it; plus, the arousal that had to be postponed was admitted with a blush, then laughed about when this Japanese sensei, who exemplified what the word PROUD defines, smiled from ear to ear. Leif couldn't hold his attraction for this man and sprang upward at the groin.

"I can admire your response, young man, only because I was in love with your father at your age," Oshi admitted and didn't dare embarrass the boy in front of him with a full erection.

"My father?"

"Yes, excuse me. I consider Dr. Marion and Dr. Kerho the fathers for you all. You aren't at that stage, yet. I've always loved Dr. Marion and we had much the same relationship that you and Dow have."

"What if I love you, too, Dr. Makawa?"

"I accept your love, Leif, as I do any of the boys; yet, I uphold the laws of the society within which we live. My integrity comes before my desires. I understand you have been deprived of a father figure for most of your life. My goal is to give you that authority parent with boundaries. I respect you; you respect me. Sex would change all that because it involves an intimacy of friendship and equal power."

"Why haven't you enlarged your penis?" Leif asked with his full view of a man's nakedness and not wanting to accept the rejection he felt.

"I could have received this dystrophin gene but it might have interfered with a few of my judo throws. I'm not so sure it wouldn't show up in my system during drug tests. See, this gene produces an insulin-like growth factor known as IGF-I, a powerful hormone that drives muscle growth, cellular expansion and can repair damaged tissue. Born with it is one thing; receiving this as an adult might be something else. I might end up with a foot-long one."

"That would be so cool!" Leif said and had them both in stitches. Leif had never accepted no for an answer and split his legs to allow full view of his anatomy.

"Leif, there is a concern because you're underage and lack the ability to consent under state law." Oshi didn't dare look down because he knew what would happen to his own libido.

"Sveinn is my guardian; at least that's what my mom told Sveinn if he found me. He can give permission for you to keep me. I'm quite capable to giving my consent for what I enjoy and desire. Do they think teenagers are dumb? If I commit a crime they will charge me as an adult and say I knew what I was doing; yet, if I have something as natural as sex, they'll make me out to be a child without the ability to think. That's not right. I knew what I was doing with several men who wanted me, like I wanted them."

"As tacit as that might be we view this relationship with Mr. Eldjarn as one that has serious violations. An adult given the responsibility for a minor carries a great deal of trust. The man could have given you HIV. His actions were selfish and not in your best interests. How do you feel about this?"

"Why do you always make me want to feel emotions? You are sounding way too much like my mother. Okay, I'm sort of ....mad. I expected more honesty. Is it not important for the adult to protect me, as I protect the adult from harm? I read Dr. Marion's book, THE ART OF LOYALTY. Erik gave it to me, and I'm not that much of a reader, but this was a great book. I can't expect to have you think of me as a mature person who loves you and desires your love, than betray you when acting like a child serves my interests. I'm loyal to you, as I am to myself," Leif admitted and reached out to hold the doctor's penis in his hand. "I promise you my loyalty."

"Ah! You've learned this biblical oath," Oshi chuckled at this display of integrity. "Leif, you make good points, and loyalty is something few boys possess. I make you feel because it puts you in touch with your feelings and people then know the real Leif. You've told me how the boys treat you so well and share what they know and have. It's a far cry from boys in school, isn't it?"

"Yes. I can't even swear. They give me this look like I've committed a major crime. I can't help but be good; worse yet, all my brothers are tougher than I am."

"Good, though I look forward to teaching you how to defend yourself and discover the many gifts your body possesses, but hasn't been trained properly. Here we never use violence as a means to control another, though boys have used violence to impose their own sense of justice. This we abhor because I believe there are better ways. Love is the core of life and sharing your experiences with another. Leif, each emotion has a relationship to time. Regret is an emotion of the past; you regret something you've previously said or done. Love, in its purest form, is an emotion of the present. If you love in the past, it becomes something you're missing and turns into regret. If you put love into the future, it turns into longing and pining. The emotion of love can really only exist in the present. This is also true for the emotions of peace, contentment, reverence, and joy. Don't fantasize or daydream about the past. Take responsibility and change yourself today—how you want to feel the joy and happiness of knowing Dow."

371

"And you, Dr. Makawa. I really like you."

"Thank you, but you're still not getting me in bed."

Leif chuckled. "We could do it on the tatami." Leif leaped at the doctor and they wrestled on the mat. Oshi had no trouble pinning this spry boy but it was the temerity and boyishness of this young lad that impressed this Japanese man of great distinction. With just this quick horseplay Oshi discovered his own arousal had been too quick to resist with Leif in his arms.

When he made Leif giggle and give up, Oshi turned Leif toward the phone to cover his own erection and had the boy dial his mother in Iceland to make a few decisions that would affect the rest of his life. The speaker phone was turned on so both of them could hear the conversation. Even then Leif twisted his head and caught a glimpse of his accomplishment.

"I knew you liked me!" Leif said and got a knuckle in the rib cage again.

"Dial or you're going to get the spanking of your life!" Oshi warned and found his student way too willing when he spun his butt around. Oshi gave it a whack anyway. To stop this insanity Oshi sat on the boy and made sure Leif punched in the correct numbers.

"Hi, Mom, it's me again. I wanted to ask you a few questions, if it's okay."

"Oh, Honey, of course it's okay. I'm not upset; actually, I have a confession to make. Remember when I left one of my tabloids out on the table? I did it on purpose so you would find it. When I read the issue I put two and two together and knew this is what happened sixteen years ago when I was asked to do this. I accepted the doctor's money and didn't fulfill my end of the bargain. Working at the hospital I managed to hide paperwork and your birth to keep anyone from finding you. I loved you too much to give you up. Now I really want to make things right. It was when you turned five, and again when you reached ten, that I tried to reach this doctor but the number no longer worked. I've always felt you were destined for something greater than I could offer you here in Iceland. I hoped that you would read the article and become interested or ask questions. Something beyond my

372

understanding guided you to do that research and find this academy you talked about."

"Does that mean you want me to stay?"

"It means I love you enough to want the best for you. It is still your choice but your relationship with Sveinn needs to be considered. I feel terrible that I made you feel shame for your sexuality. You're such a beautiful boy with so much to give to another person. Seeing you with an adult made me jealous—a mother's instinct, I'm afraid."

"Sveinn isn't the man I thought he was, Mom. He gave me a venereal disease and all he thinks about is money. Oops! My shrink tells me not to do this blaming game, so I don't entirely mean what I just said. His name is Dr. Makawa, and he's real cool, though he's sitting right here getting embarrassed. I didn't know the Japanese could get red in the face. You should meet him, Mom, he's so strong and you should see his..ouch! Never mind. He has this thing for spanking. I was going to say that you should see his samurai outfit, though he's even more gorgeous when he doesn't have it on. Ouch! Anyway, he has all these quotes of wisdom like the fortune cookies we always get at that Chinese restaurant downtown, but my sensei uses all these weapons to make us believe them."

Leif got popped on the head for that remark, but he was giggling and his mother thought it was funny. "Leif, Honey, I'm real proud of you. Dr. Kerho has called me to keep me informed of your safety and health. He's such a gracious man. I've been dating since you've been gone. Seeing you that night….I meant, morning, convinced me that I'm way too young to sit home and do nothing. Needless to say, my sex life has improved and I think you'd like the man; he's even considering moving to Arizona. Of course he wants me to go with him."

"Mom! That's so close to here. You need a stud in your life since I'm not there to humor you with in my birthday suit while fixing breakfast." Leif sensed his mother's laughter meant she was happy and not totally depressed because he wasn't there. She did sound really upbeat, so she must be falling in love with the guy. "Mom, I've been thinking, can I stay here?"

"I'll call Dr. Kerho in a few minutes and we'll get the paperwork done. You will behave yourself, right?"

"Mom! You know I'm practically an angel. I'm in love with a boy here—actually, several."

"Can you blame them for loving my boy? Your penis is an attraction by itself."

"At this academy I'm no exception, Mom. And I know now why I'm hairless; though it's really cool having twenty-nine other brothers the same way."

When Leif hung up he had a few tears in his eye. Oshi hugged him and his heart was quite fond of this thirtieth boy in his life. "I believe we need to put some of that energy to work before we do something we'll both regret," Oshi informed the young man. "How's your cycling coming?"

"Awesome! It's the same cycle that Lance Armstrong used; all carbon and titanium, handmade tubes and it's so smooth. Coach Larsen is having me do interval training. I've already improved due to physiological changes in my muscle cells that make them better able to use oxygen and tolerate the build-up of metabolic waste products. I'm thinking of majoring in molecular genetics and biochemistry. And I promise I won't regret anything I do with you."

"You are incorrigible and way too cute for your age."

"You are, too."

"I have two more boys to talk to. Get your butt out there on that bike and make that erection disappear. You're way too happy to see me. I believe we can eliminate our daily sessions, but I'll stop by on a weekly basis," Dr. Makawa suggested.

Leif appeared very disappointed. "I don't know. Maybe we should talk more often so I don't get confused."

Oshi smiled and knew the transference that had developed. It was a love he would have to monitor closely and not be weakened by the beauty and temerity of one so young and tempting.

****

Sveinn Eldjarn had committed himself to his goal of making money and undercover investigation to disclose everyone in this academy. His attempts at night to sneak into the dormitory had been hampered by an elaborate security system. The consequent attempts to get into the athletic venues to search out Leif were also restricted by security personnel.

With frustration and built-up stress Sveinn was plotting another way to see the love of his life. Before disclosing the results of their investigation, Sveinn wanted to discuss with Leif how they should formulate their attack and present all this information to go to the highest bidder for an outrageous sum.

With daily observations Sveinn began to calculate Leif's appearance on a two-lane, paved course around the academy grounds. The course was five miles in length and swung up and over the hillsides with gradual elevations and some very steep climbs, but a true test for a cyclist in training. Leif would not always ride by himself and, on this day, Sveinn saw two boys pushing each other in a time trial, side by side, as they approached his position behind a tree.

Sveinn jumped out in the pathway, waved his arms, which had the intention to have the boys slow down and stop. He could not exactly send the other boy on while he conversed to Leif. It was all too quick and left no room for the boys to detour or recover from their speed. Michael had no choice but to lean off to the right to avoid hitting Sveinn. His bike went straight into a tree, which sent the youth tumbling over a bush and crashing to the earth. Leif barely survived when Sveinn caught his handlebars and made Leif's fall with less severe trauma.

"We need to talk, Leif. We are close to having our lives gifted with gold." He found his hand quickly disengaged from the handlebars with a wrist hold from Leif.

"Hey, where did you learn that?"

Leif gasped for oxygen with this sudden interruption. "From a teacher I trust!" Leif spoke harshly. "Don't you even care that you knocked someone down?"

Leif leaped over to where Michael lay, his moans were from several injuries. The boy was not only the school's best cyclist, but one of the finest hurdlers in the nation. It was only six days away from the Texas State Track and Field Championships. Leif wanted to make things all better by touching his friend, but the wounds were far more severe than he could deal with. He had to get help, but Sveinn was more intent on keeping him there and talking.

"The boy will be all right; probably just a bruise. Have you missed me?"

Leif didn't answer and began to run back in the direction they had come. His yells through the woods were heard by many. In less than a minute an emergency cart was zipping up the path, just as Leif was sprinting downward. Michael was quickly surrounded by trainers and helped upon the cart.

"What happened?" one of the trainers asked.

Leif glanced at Sveinn, then answered the question himself. "Just an accident, sir. We collided and Sveinn showed up." Leif didn't like lying and felt doubly worse to see Michael on the stretcher. Both bikes were contorted with the sudden spill. One of the security men arrived and escorted Sveinn down the path while the other stayed with Leif.

"Son, I can appreciate you sticking up for the man. It's honorable to a point but, here at the academy, to betray yourself is still a betrayal, nonetheless. Care to tell me the truth now?"

Leif thought what Dr. Makawa would say, then confessed exactly what had happened.

****

Mr. Eldjarn protested vehemently as Dr. Marion and a security man escorted him off the premises. "I'm responsible for Leif's safety and security! He knows I'm his designated guardian and this academy is holding the boy hostage!"

Kami showed the man a signed, notarized paper that authorized Kings Academy to take guardianship of one Leif Bjornsson. Sveinn examined the document closely and realized he had lost the biggest opportunity of his life to possess something so young and beautiful. There were still the possibilities of financial gain from the information obtained during his stay, and Sveinn was already putting together a story that would shock America and the world.

Kami put $10,000 in cash in the man's hands. "Return to Iceland, Mr. Eldjarn, and forget about the boy. Leif says you're an excellent

376

teacher and this will give you a fresh start. Allow the boy to grow up and experience love with his own peers, because anything you do now to embarrass the academy will embarrass Leif. You are young and handsome; let that be your future to find someone to love and share life with."

Sveinn gripped the stack of hundred dollar bills with his eyes glued on his fingers squeezing the money with despair. His conscience wasn't on this payment to remove him from the academy grounds, but it was an attempt to hide his tears as they began to flow down his cheeks. Sveinn kept his head down in surrender and loss, before solemnly stepping into the vehicle that would return him to a mundane void and end all his dreams of wealth and love.

****

The MRI showed no fracture in Michael's arm, only a strain and several contusions to go with both boys' skinned knees and elbows. The clamps on the toes of the bikes didn't make for easy dismounts on these $6,000 racing bikes. Leif must have apologized ten times for his former teacher's stupidity. Michael wasn't upset, only more determined now to use every therapeutic remedy available to be ready for the state track championships.

Leif ran into Bobby and Dow on his return to the dorm in preparation for swimming practice before a biology class. Dow was practicing his saxophone and Bobby was accompanying him on the piano. Leif stood back and admired the jazz composition of two superb artists. It appeared everyone played a musical instrument but him, though Jacob was giving him daily lessons on the vibraphone.

When they were finished Bobby took the time to review a further understanding of the physiological transformations of gene therapy, or as Dow expressed it--gene expression. Bobby was amazed with Leif's interest in biology and chemistry, two of his favorite subjects.

"Leif, gene therapy is the opposite of the research to control cancers and other diseases," Bobby implied. "There are some genes that are so subversive that they should never be given freedom of expression, though we doubt if our penises are one of these."

"You mean, like, we were guinea pigs for this experiment?" Leif realized.

"Sort of," Bobby didn't lie. "We've worked with animals and only one required a censor; that's a RNA interference, or RNAi, which silences the gene by intercepting and destroying only the offender's messengers' RNA, without disturbing the messages of other genes."

"Did the animal's penis shrink?"

"Not exactly," Isaac replied. "Fifteen inches and holding, I'm afraid."

"A horse?" Leif asked.

"Gee, Leif, a horse is longer than fifteen inches without us messing with its sex. Actually it was rabbit, a white, fluffy rabbit who has no female friends. Leif, if you have a penis fifteen inches long I'd be your friend for life, which I am already. I might wear a rear chastity belt though."

Leif punched Bobby in the arm. "I think we ought to give Dr. Makawa a penis like we have. He only has, like, three inches hanging, probably six hard. Not like I've seen it, mind you." Leif smiled which gave that one up.

"If he was five-years old, no problem, but I'm not sure of the results with an adult. Even if it goes awry we have a mechanism that blocks translation of the messenger RNAs into productive proteins, and we know all 200 and 255 microRNA genes, which, on the bright side, is only one percent of the total number of human genes."

"You're way too smart for a kid, Bobby. If the doctor ended up like the rabbit our asses would be grass."

"Dr. Makawa is a shrink, right? We can shrink a mistake, is all I'm saying. Actually, I like penises. My forefather will vouch for me, if you can somehow ask him."

"Who's your forefather anyway?"

"Isaac Newton, silly. We inherited the same scruples for experimentation. The rabbit was fun, but you I love. No sweat, I'll have Dr. Makawa bouncing a nine-incher in no time with twice the nerve endings, if you want."

"I'm at the mercy of a boy who gets acne like I do and wants to jack off half-a-dozen times each day. This is way too much like Frankenstein."

"You lack the confidence in your peer who yearns for your loins. Maybe you'd prefer an old man with no sexual interest and despises the young for having them."

Leif thought about this. "I'm putting my doctor's balls in your hand. If he doesn't appreciate this gift, my head is off with his sword."

Bobby, Dow and Leif laughed their way through swimming workouts with ideas on how to proceed with Dr. Makawa. They knew they just couldn't walk up to him and give him a shot in the butt. Leif suggested a tea ceremony with a mild sedative in the doctor's tea cup, then an injection. He'd never know the difference. Not surprisingly, Oshi loved the idea that the boys wanted to learn more about the tea ceremony, an integral part of a samurai's daily life.

Oshi left the dormitory a whole lot more tired than when he entered. He couldn't believe he'd dozed off in the middle of demonstrating all the nuances and intricacies of a tea ceremony for the boys. He sensed a slight soreness in his rear, but didn't give it a second thought.

A day later he noticed that, when he urinated, his penis hung loosely in his hand, as if he'd just recovered from an erection. For the time being he didn't pay this any mind, nor the constant tingling in his groin and the yearning to have sex. He stayed away from Leif because the boy had affected him way too emotionally to tempt this relationship.

Oshi felt so randy that evening and was hyper enough in bed that Brittany was excited to see her husband in such a great mood—three times in one hour and she was spent before he was. Her orgasms had never been better. Oshi also felt rather strange, though he assumed, since he hadn't had intercourse in several weeks, his erection was at its peak.

The third day his penis was bothering him with soreness, much longer flaccid than he remembered ever being hard. He went to see his best friend Kami. Dr. Marion examined the swollen member and asked several questions. The two had been boyhood lovers and shared many a night in bed. In this case Kami began to manipulate the organ to an

379

erect state and discovered its vast increase in length.

"You didn't?" Kami questioned as if this was a joke played on him by Oshi.

"Didn't what?" Oshi asked.

"You didn't get a genetic transformation, did you?"

"Trust me, I've been in awe of yours, but it would give my opponents another object to grab a hold of, I'm afraid."

Kami checked Oshi's rear and discovered a needle mark. Several questions later they knew what had happened. "I'm going to ring their necks!" Kami shouted. "These boys have gone way too far with this one. They can't think, just because they can, that they can manipulate others without consent."

Oshi saw the humor in this ruse. "Look, Kami, Brittany thinks I'm this new sex god who possesses the ability to satisfy her beyond what she's known before. She thinks it's my new diet and exercise program."

"I'm not going to tell the boys that you're pleased with this. That will only give them permission to continue this type of behavior. We have to teach them a lesson."

Kami arranged from his science teacher to make a thirty-six inch penis out of rubber that went beyond Oshi's ankles. That evening he called the boys to his home, then invited Oshi in. The outline of the long penis was very evident through the man's trousers. Oshi had a stern look on his face with a katana sword in his hand.

"Care to explain?" Kami glanced at the three boys.

The boys bowed when they saw Oshi standing by the door. They all gulped and stared and stared and stared. "Holy shit!" Leif shouted.

"Leif, we don't use such descriptions. That's another restriction on top of the others you're about to receive," Kami said.

"Sorry, sir. I'm sure Bobby can reverse it. I just wanted to do something to show Dr. Makawa how much I loved him."

"You can do that by being his best student on the mat, not in bed. How can one possibly reverse a thirty-six inch penis? One thing about genetics, boys, you don't make a man six-feet nine, then expect to reduce him because he exceeded your expectations. You've ruined this man's sex life.   What do you have to say to that?"

Bobby spoke up as the genius that he was.   "Dad, I think I can reverse it.   It's all too experimental, though.   I know you've told me in the past not to do these experiments, but I was sure of my calculations. It was all my idea, Dad."

"No!   It was mine," Leif blurted.

"I'm pretty sure I thought of it first," Dow said.

Dr. Makawa expanded his chest and gave a look that he was hungry for three young adolescent bodies to destroy all at once.   Grim faces had the boys close to tears.   Oshi reached down his drawers and began pulling out the elongated rubber hose.   It took a few seconds for the boys to realize it was a prank.   Smiles creased their faces in relief.

"This doesn't excuse your behavior.   A week's restriction to your rooms when we get back from the state meet; also, I want two-page reports on ten different TV shows you truly love: The Brady Bunch and The Partridge Family."

"Ah, Dad, that's cruel," Dow complained.

"Yeah, that's torture," Bobby admitted.   "I would get sick writing two pages on watching these."

"Interesting.   You'll probably need medical assistance because I want two pages on each of the ten episodes.   They better be thorough and insightful or we'll do it again," Dr. Marion insisted.

"I feel sick already," Leif insisted.   He looked at the man he truly loved.   "Dr. Makawa, is your penis okay?"

Oshi pulled out his hanging seven-incher.   "Your father has been trying to convince me to do this for fifteen years.   A tea ceremony has apparently made this decision for me.   Must have been in the leaves."

The boys laughed until their sensei had the final word.   "We'll talk about it further in the dojo."

Kami approached Bobby, before swinging his son over his knee as they tumbled in a chair.   "Another thing, young man!   We found this mechanical spider in our bedroom.   You took that fly and did your own variation, didn't you?"

Bobby began to flap his arms like a boy who was in dire need to explain and avoid punishment.   "Dad, I swear it was just for

381

experimental purposes only.   We only watched a couple of times."
Leif had trouble not giggling, as did the other boys.   "You guys are kind
of kinky, doing it in front of the fish, and all.

"It's not that you haven't seen the sex act done every night.   I just
hope you didn't intrude on a night when one of your brothers slept over
here."

"I wouldn't do that to one of my brothers, Dad.   It's been a few
years since I've been in your bed, but I can respect privacy.   I really
can."

"And I can give you a reminder to help you out.   Next time you
can debate this with Oshi."   A few sound smacks on a bare ass were
tough to do, as Kami couldn't hold a laugh at his boys' antics.   He
dismissed his sons to their dorm, but Leif hung back for a second and
had a word with his favorite psychologist.

"Dr. Makawa, I know I should have asked you first, but I love you
so much I wanted this to be a surprise.   I've been doing some sensitivity
training about psychologists and I thought you'd be interested."

Oshi saw the seriousness in this boy, one who already had had his
trust violated by another adult figure.   He knew he had to handle this
openness close to the heart.

"I'm flattered, Leif.   I adore you, though you're one of my more
mischievous judokas.   I find it amazing that you're so angelic when you
sing in the boys' church choir, yet you are more tempting than
Ganymede was to Apollo."

"I think I understand, but Bobby told me that you and Dr. Marion
were boyhood lovers and that you probably married his sister to be close
to him."

Oshi found this information hilarious, if not overly insightful.
"Everyone is a comedian.   Bobby is too smart for his own good.
Remind me to spank the boy next time I see him."   Oshi turned his
favorite student over in his arm, lifted the yukata and lowered the
fundoshi, then spanked the boy a few good slaps.

"I have to get my share in tonight, as well.   That felt good!"   Oshi
said to himself.

Leif held his butt and was red in the face; the stings were something he'd never felt from a father he never had. "Yeah, I know, it hurt you more than it hurt me, but it's my butt that's sore." Leif lunged forward and kissed his sensei, then sprinted away.

"That boy is a handful," Kami admitted to his boyhood friend.

"When does a psychologist touch a heart and not feel he's accomplished something totally miraculous? Our boys give me one-hundred percent on the mat and on the fencing strip. Leif gives me one-hundred and ten-percent and then wants more when we're done. He scored on me in badminton the other day and paraded around the court like he'd won the Olympics. He says in a year he's going to beat me. I love his enthusiasm and spirit; plus, he's invigorated the other boys to stay a step ahead of him. I've never been this enamored with anyone since you, Kami. I can't explain it."

"I can," Kami answered. "Our sons have a long and devoted love for you, yet they know their boundaries and are in love with each other. Leif desires a father figure, but he associates age with experience, acceptance, and the power that he wants from a man to inspire to—all of this is exemplified in a sexual sense because it's what makes Leif come alive. Our boys have experienced this all their life, while Leif is just discovering this huge part of his being. You don't show the weaknesses of the men he has had in his life before, and you're caring and restraint excites him. My guess, the boy will break through to your heart and capture the love we found in each other. Is it wrong? Hardly. God will find it amusing, and it will make Him feel twenty years younger."

"I wouldn't want to upset God and His quest for youth and vitality," Oshi admitted with a wide grin.

Chapter Thirty-Five

Senator Franco was too busy with diplomatic relations than to be concerned that his family never seemed to be at home, nor was there any contact with his wife. He'd have time after he destroyed the Kings Academy and made his name a household word to make amends with his wife and children.

London had announced to the FBI that the sample from the United States was a match with J. M. Barrie's DNA. They had protested vehemently unearthing the remains of this writer of Peter Pan, but they were able to discover there were too many similarities in the DNA that was extracted. The truth was evident: James Barrie was alive and well somewhere in the United States, and England wanted the boy back. Why and how that was to be achieved was beyond anyone's scope.

It didn't take long for this leak to penetrate even the Parliament's tight gag order on this latest development. With the noted paparazzi of England's royal family, such a secret didn't remain hidden for long. Monetary rewards were often too exorbitant to keep this type of sensationalism private.

PETER PAN LIVES IN AMERICA!

Those were the headlines of the next London Times. The next day another issue hit the newsstands:

PETER THE GREAT LIVES IN AMERICA!

On and on it continued and papers sold faster than they could print them. Six other countries had their ambassadors on the doorstep of America's political figures, requesting updates and confirmations to their own missing icons. Under a blanket of secrecy, in spite of tabloid exploits, the president foresaw an international debacle if this thing exploded in his face. The accusations began to fly that the United States had intentionally stolen these great men as a means to further their scientific, economical, if not athletic superiority in the world. As bizarre as this assumption and exaggeration was, the president couldn't dismiss this accusatory finger.   How the dead had become such a thorn in his side had the Head of State in a corner protecting his reputation and his party's.   Certainly the ultra-right wing, which had assisted his reelection, would not approve of any government entity involving itself with God's sole right of procreation, at least that's how they saw it.

The entire situation, if not a crisis, was a godsend for Senator Franco and his political ambitions.   He was now the first name on the president's lips; a man who had knowledge of who and where these boys might be kept for whatever scientific reasons or madness some demented mind had conceived.   When he stepped into the White House Robert knew his importance had been magnified; he was taken directly to the Oval Office.

"Afternoon, Senator.   I don't have to tell you the importance of this meeting.   We have a multitude of foreign dignitaries breathing down our necks as if we've stolen their national treasures right out from underneath them.   We need to take the heat off our government immediately.   Any suggestions?"

"Yes, of course, Mr. President.   If I have your support and assistance from various military agencies I believe we can storm this academy and shut this factory out of existence."

"Hold on there, Senator.   This academy could be developing a dozen Sadam Hussein's, but we can't just break down the door.   We need hard evidence, sure proof and facts to garner a search warrant and do this by the law."

"I believe I can devise a plan to do just that, Mr. President.   Do I have your full support to use people at my disposal?"

"I'll make sure the directive so stipulates.   Senator.   In the meantime, we will attempt to keep these inquiries at length.   I am envisioning something far more devastating if other countries get a grasp of what we are doing or know.   What if this academy does accommodate these great men?   What protection do cloned humans, possibly having birth in this country, have to citizenship and our protection?   What right do we have to send these boys back to their countries?   Is Tad Barrie really J. M.   Barrie?"

The senator didn't really want to hear this polemic nuisance.

"Maybe it's best we bring in the Attorney General on this, Mr. President." Robert knew the A.G. was weak and couldn't make a decision like this if his life depended on it.

The senator left the White House feeling like he was the most important man in America. Diplomatic relations between the most powerful countries in the world depended on his resolution of this outrageous experiment. Maybe having his sons being friends with these Kings' boys would pay off. Once he had Jay and Niqui in Washington he would have all the information he needed at his fingertips. He also suspected he needed a backup plan, as well. For some reason the senator didn't trust his family; of course, once his sons realized they would soon have a father as president they'd likely see the error of their ways.

While Robert was being chauffeured back to the Capitol he reviewed a packet given to all government agencies and political nominees. The CIA had intercepted another transmission from a terrorist group in the Middle East. The pictures of the five suspicious Muslim operatives were seen in Canada, before they just as quickly vanished and were presumed to have crossed Lake Erie into Pennsylvania by private craft sometime in the past week. One of them was seen in Philadelphia but had escaped capture.

As if America didn't have enough problems Robert had bigger fish to fry than terrorists. To be President of the United States meant having a target on your back and on the Most Wanted list of two-dozen countries. Robert considered this threat and decided that the prestige of being Commander-in-Chief of the world's most powerful country outweighed this minor inconvenience and threat.

Soon the pictures of these men would be splashed on every television screen, in every newspaper in the country, and every transportation center would be on the lookout for the first camel jockey who stood out like a sore thumb. The president had also made a good point that Robert was very perceptive and the man for the job. If need be, these terrorists could be blamed for any number of accidents and disasters which could serve America's purposes.

Chapter Thirty-Six

With the entire academy in tow their bus was soon in Austin in preparation for the state track meet. While Kami, Oshi, and the Kings' coaches handled the track and field athletes at the U. of Texas stadium, Dr. Kerho had another dozen boys in a weekend triathlon on the outskirts of the city. The doctor had discussed with the boys whether they wanted to be entered in the age-group divisions or the open adults. The boys choose the open division.

Being in excellent condition with daily swims, running and cycling, the boys knew their strategy from the beginning: their small silicon suit encountered 32 percent less resistance and drag through the water's current than a typical bathing suit or a unitard-like swimming suit. Though it sure gained more than enough attention no one could find fault with fourteen boys who were all in the top twenty in the swim competition moving in to the racing bike section of this triathlon. The Kings' boys were instructed to swim at 90 percent of their maximum ability in preparation for the 25-mile bike ride. They drafted off competitors and succeeded far better than those who thought this was a sprint race.

Moving to the 13-mile run all the Kings' boys were packed together. Leif felt fortunate to hang with his brothers and his cycling was the only thing that moved him past so many adults. Only three adults were even close to challenging this group of boys after the first five miles.

The boys used a varying cycling intensity—5-minute intervals, alternating between 80 percent and 100 percent effort. It increased their leg power during the run by lowering their muscles' lactate levels. In the end, despite some well-trained triathletes in the Dallas area, all the boys came in bunched, with no one particular boy wanting to share the glory.

Across town the track and field competition was in its second day. The decathletes had started on Friday. Aki Musashi had done well enough in the sprints to run 10.32, but only good enough to finish third behind two fast sprinters from Houston. The 800 meters and 1500 meters proved a hi-light for the Kings' boys. In the 800, they went one-two, and the 1500 meters they swept all three top places.

With three boys who could run under 4:10 for the mile, no one in Texas could really challenge this group of talented distance kids. The 2-mile was no less victorious with the Kings taking again the top three places.

Michael Locke, the only black athlete at the academy, ran his best time of 13.54 and finished second in the high hurdles, but then turned around and won the 300 intermediate hurdles. Brett Byron came in fourth in the 400 meters with his best time of 46.88, a new national record for a fifteen-year old. Michael's progenitor was a talented Rhodes Scholar and an outstanding professor during the early Twentieth Century.

One of the big strengths for Kings Academy outside of their distance runners was in the pole vault. Their gymnastic program assisted Dane Nureyev and Rom Kerho to battle each other for how many inches they could vault over 17 feet. On their heels was Tony Simoni, a late grower, but one boy who was determined to reach the same heights as his brothers. When Tony went over seventeen feet, Dane and Rom got real serious. Dane won the event with a new state record of 18 feet one-half inch.

There were coaches who inquired about the dazzling white suits of the Kings' team. The main meet director read the rules and the suit was perfectly legal. It had to cover the anus and the genitals. This it did. What caused major fan reaction was when Wendy ran the 800 meters and the 1500.

Whistles, cheers, and a throng of males lined the track, which gave quite a demonstration of support for a girl of such beauty. The silicon suit acted both as a sports bra and a uniform, but the garment kept no secrecy of the body's contours.

One thing that Bobby Newton added to Wendy's suit was an extra coating around the pubic area. The vulva usually had been quite defined, but an extra layer still exhibited enough female sex to have males drool with lust.

An aspect Dr. Kerho knew was coming was the reaction of spectators and athletes to the boys kissing and hugging after their victories. The blatant show of approval wasn't something common in the sports world of high school athletes. Often there were a few boos, homophobic remarks, or stares that were meant to demean or intimidate. This, Dr. Kerho had prepared the boys well in his lectures.

In his pre-meet speech he prepared the boys for what was coming. "The good will always defeat the bad. Sometimes it doesn't look like you're winning by remaining silent and taking this type of abuse, but as long as you're on your feet, talking about it and finding allies in your brothers, you're beating these people who rely on intimidation, on your apathy, your silence, on you eventually giving up. This we won't do."

The boys cheered. Their results couldn't be ignored and displaying their gold medals around their necks left many a jaw locked. Coaches and athletes actually began to gather around the Kings' athletes, asking them questions on their training methods and what they ate before the meet.

These boys held no secrets and readily discussed what they did at the academy in learning their sport and preparing. Aki talked about his zigzag sprint routine on the beach by the lake, his frog squat jumps, reverse sprints, and stride frequency work using high intensity speed work.

Aki had three coaches all listening to this fifteen-year old Japanese sprinter. "See, an important tool for increasing speed is repetition, which translates into efficiency. When your fast-twitch fibers have been trained repeatedly to fire at a rapid rate, a motor pattern is established, and the muscles work in unison more efficiently. In other words, you can teach your nervous system to fire only the essential muscles and to quiet those muscles not needed, resulting in less overall fatigue."

"And your coach teaches you that?" a Dallas coach asked.

"Sure. Don't you teach your sprinters that? I can tell you something, after watching a lot of fast guys out here today, too many make the mistake of planting their feet in front of their hips. It causes a braking action that slows you down. Someone is giving them bad advice that extending their stride will make them faster. That's just not so."

The team championship had pretty much been decided by the time the mile relay was announced. Thanks to the sweeps in the distance races, Kings Academy had a lock on the team title. Never in the history of Texas has a 1A school dominated a state meet. The Kings' relay had four strong 400 meter runners, not necessarily the best or the faster boys in the state, but as a whole, they proved to be the strongest.

A Texas tradition in both high school and college teams, the mile relay always caused the most excitement. To see three white boys and a black youth beat some of the best relay teams in the country was quite a treat, but it was also an embarrassment to Dallas and Houston high schools.

All three track and field coaches were swamped after the meet by other coaches throughout the state. Even athletes from all over the State of Texas came up to the Kings' boys and congratulated their performances and sought information on how they could transfer to this school.

Dane was so happy when Niqui ran up to him with a huge smile on his face. Dane took him in his arms and kissed the boy flat on the lips. That opened quite a few eyes and a number of lips that shouted, "Fags!" Niqui no longer felt self-conscious about kissing a boy in public.

There was something overpowering in being with a winner and a teenager that eluded confidence and pride in who he was. Sure, Niqui could have told everyone he wasn't gay, but who he had sex with meant less than the joy and contentment he felt when he was in the arms of Dane or Jacob.

Jay was on overload with pride when his eyes caught Wendy's. They kissed in front of dozens of envious boys. Like his brothers, he had a sense of pride in being with the Kings' students. Jay joined with the other boys in hugging and congratulating the individual successes and the team victory.

"Shit! How'd that squirt get so lucky?" Jay had heard one boy say and decided to be the assertive one in taking Wendy's hand and escorting her back to the bus.

With the entire team back at the Holiday Inn there was a wild celebration that included several college recruiters and a dozen coaches, envious of this small academy's results. They wanted to gather as much information as they could to help their own future.

Dr. Kerho and Dr. Marion decided to allow the boys and Wendy to explore Austin's ambience, but only if they were in pairs and met at a central area every hour on the hour.

On the main boulevard of Austin City's Limits, where music thrived and college kids vented their energies, Leif and Bode Lawrence were having a time of their lives by mingling with older kids and being overwhelmed by the loudness generated by thousands of college kids and adults partying on this main thoroughfare. At their age there were few places they could just stroll into. Instead they walked the sidewalks and gawked at the strange characters with tattoos, body piercings, and sometimes risqué outfits that hinged on nudity.

A man approached the boys and asked if they knew of a great jazz place, but Bode admitted they were visitors too and weren't all that familiar with Austin. Their white LaCoste shirts and royal blue shorts gave them an appearance of neatness and cordial teens. When the man followed them a little too close, Bode and Leif speeded up.

Across the street they sprinted, but then Leif saw Sveinn standing on the next street corner waiting for them to walk that way. Leif grabbed Bode's hand and the boys dashed down an alley and looked for another exit. A car turned into this same alley and stopped as four men stepped out. The boys could have run up some iron steps to the upper floors of a building, but they didn't see this as an emergency.

All the men were dressed in coats and ties and appeared to only want to question why the boys were in this alley. Only after one of them was close did he pull out of his pocket some type of gadget, then pointed it at Leif.

Bode kicked the object from the man's hand before it fired, but by this time the other three men were on them.

"Run, Leif!" Bode yelled, but Leif was not as quick as the command to avoid being encircled by a man's arms.

Leif kicked upward and back into the man's groin. With the

release another attacker zapped Leif with a Taser gun which collapsed Leif to the pavement. Bode was in his own turmoil, dropping his first attacker with an Aikido technique, then foot sweeping another, only to have the third wrap an arm around his neck. This man was thrown with a shoulder throw. Bode was tackled from behind before he recovered from his throw; any attempt now to keep the men in focus and at a distance was hopeless. With Bode on the pavement he struggled to stand up with the weight of a 200-pounder sprawled out over a 5'8", 118-pound dynamo. Bode kicked, clawed, and screamed but he couldn't overpower the strength of three men who handcuffed his wrists behind his back. A punch to the stomach and a Taser to the neck incapacitated the boy for the night.

Practically dragged to the vehicle, the Kings' boys were thrown in the back seat with a man on both sides of them. For the spectators who had gathered on the sidewalk this appeared to be another incident of college kids getting too drunk and rowdy, if not too licentious down this darkened alleyway.

****

Toy and Kami were enjoying their own victory celebration with a few of their coaches in the hotel room. The ring of Toy's cell phone had become a nuisance with the amount of college coaches and sportswriters requesting his attention for interviews. He realized only too late that it had been a mistake to give even one person access to this number. Toy was almost delighted it was one of his security men.

"Dr. Kerho, there's that gentleman, Sveinn Eldjarn, down here and he says he desperately needs to talk to you."

Toy didn't need this type of distraction. "Tell him he's not getting Leif back and not to bother me again."

There were a few seconds of delay, then, "Sir, he says it's about the possibility that the boy was abducted tonight with another Kings' student."

Toy's alertness went sky high. "Bill, check the scanner now! If this is one of his tricks I'll fly him back to Iceland myself. I'll be right down."

Toy hustled from the room without so much as a word of panic.

Kami knew from the expression that something serious was up, excused himself from the conversation and sprang after his significant other. The men didn't bother with the busy elevator but ran down the twelve flights of stairs to the lobby. The security man's frantic display of his GPS scanner was not good news."

"I have Bode Lawrence heading north in a fast moving vehicle. I can only assume Leif is with him. We haven't put a GPS chip in the boy's heel, yet."

"Are we sure Leif was with Bode?"

"Yes, sir. The two agreed to go out together. The vehicle is moving at a high rate of speed out of town," Bill said upon examining the screen again.

Toy went straight to Sveinn. "Is this part of your antics? If it is I'll ring your neck right here."

"Believe me it's not, Dr. Kerho. I admit I was looking for Leif, but when I spotted him they ran down an alley. This black Lincoln Navigator pulled up and four men ran toward the boys. I got there too late, but I can tell you one of your boys put up quite a fight for a little guy. He almost defeated all four of those men until they ganged up and overpowered him."

"Why didn't you help them, Mr. Eldjarn?"

"I'm not much of a fighter, sir, and I think they were armed."

Toy shook his head in disbelief, turned back to his security men and blurted out immediate instructions. "Round up all the kids now, get 'em on the bus and head back to the academy. Have Oshi meet me in my room in five minutes. I want a scanner in my possession in two. Whatever you do, keep Mr. Eldjarn away from the boys."

Sveinn heard the remark. "Hey, it wasn't my responsibility to keep an eye on the boy. He's yours now."

Toy swung back to this abrasive remark. "If you had an ounce of respect and love for that boy you would have given your life to save him!" Toy let the words hang.

In minutes Toy had a swarm of his assistants and security personnel at his disposal. He was regretting certain decisions over the past few hours: one was allowing the boys to wear their shirts with the Kings' logo on the breast; yet, he knew how proud the boys were of their school; secondly, Toy figured it would have been better not to spend the extra night at the hotel.

The men watched the scanner of the two boys who were kidnapped. If there was any positive thing about this incident they knew exactly where the boys were, their condition, and, thanks to Sveinn, how many abductors. The reasons why and by whom were yet to be answered.

"Do you think this is a ransom thing?" a coach asked.

"Tough call. If I felt that Eldjarn was responsible I'd say yes. I'm not sure what this is all about. Bill, you're a helicopter pilot, right?"

"Yes sir! Flew in Vietnam, and I keep my license up."

"Head for the airport and rent us a chopper. We don't have time to return to the academy for weapons. Also, find us a martial arts store that is open. It's not that late. Buy us a few katana, throwing stars, and whatever else looks appetizing."

"Dr. Kerho!" Kelsey, the academy's chief security man, had his eyes locked on the scanner while opening a map of Texas on the bedspread. "They've pulled off the road. I've located them here at Crawford."

"Crawford?" Kami questioned. "That's where the president's ranch is. Are you telling me that our president is in on kidnapping our boys?"

"I'm not sure what to tell you, sir."

"Sorry, Kelsey. I didn't mean to imply you knew all this," Kami said.

Bill stepped forward. "Sir, I've been out to the ranch a few years ago before the man became president. It's about sixteen-hundred acres, flat terrain and well secured. I believe it is just south of Waco."

Toy thought about this. "I'd bet you the academy the president is not there. There's no way he would want to be seen or be tied to a kidnapping. If anything this abduction was okayed by him, but it's the CIA or FBI who's doing the dirty work. They declare war on me they have no idea what I have in store for 'em. Bill, while you're at the airport, secure three parachutes and buy dark clothing at the store. Make it ninja wear, if they sell them."

"You're not thinking what I think you're thinking, are you?" Kami asked.

"Can you devise a better plan? Remember when we went skydiving a few years ago? You about gave me a heart attack with this new adventure of yours, but I actually enjoyed it."

Bill added, "It might be the only way. If the president isn't there they will only have a skeleton crew surrounding the main quarters. By the looks of where they've stopped I'm guessing they're at the quarters where their agents are housed while the president is staying there during one of his retreats. The grounds have major detection devices, so there's no use in trying to drive up or walk in at ground level."

"Let's get to it, gentlemen!" Toy ordered.

It was disappointing and quite disconcerting to the students and staff to have their evening interrupted. There wasn't any question that something serious had happened and it only took an observation and face identification from the other boys and girl to realize that Leif and Bode were missing. The word spread quickly and silence became worry as their bus left Austin.

To save time Kelsey drove Toy and Kami to a martial arts store downtown. The owner was a Korean man and was waiting their arrival for the purchase of weapons he had stored in the back. This was alarming to hear of men wanting to purchase dangerous weapons. The owner was cautious in dealing with gangs and hid most of these swords in a vault.

He certainly wasn't prepared for who walked through the door, instantly recognizing Oshi as an Olympic judo champion from eight years earlier. When Kami turned around to glance at a dai-katana, the Korean also saw the tattoo on the man's neck. His bow to Oshi and Kami showed the ultimate respect, not to mention the snap of his fingers to an assistant to bring the finest of his weaponry for the men's inspection.

With the purchase of three outfits, assorted weaponry, and a new friendship from a Korean martial arts expert, Toy was on his cell phone verifying the allocation of one Kiowa/Jet Ranger helicopter and three parachutes. Toy began to settle down and tried to envision what methods or intentions the government would use with the two boys. He knew he didn't have the skills that Kami and Oshi possessed, but he had a black belt of his own and wasn't afraid to get his hands dirty.

In the rental vehicle Oshi and Kami whispered in their discussion at to what strategy to use, but also took in consideration to keep Toy out of harm's way and casualties to a minimum. Between the two men they had devised their own plan.

The transition to the helicopter went smoothly, as Bill had emphasized the emergency of the transaction; plus, money always has a way of speeding up transactions and the paperwork.

The helicopter's radar scanned the way ahead, feeding back the data that allowed Bill Spurges to follow the contours of the landscape. Low-profile flying kept the speeding craft below ground radar. His input into the chopper's development had conceived a machine built around his skills and personality, not to mention the machine's performance proved his worth as a skilled Army aviator. He had no idea what Crawford Ranch had in regards to locating and identifying incoming aircraft; though Bill was confident that these kidnappers would have no idea that they were already found out and that one of the boys had a GPS device built right into his foot.

With precise coordinates Bill gained altitude and swung to the right to avoid passing directly over the ranch. The free fall would allow the chutists to direct themselves back toward the ranch. He hovered at 7500', then shook hands with the three men before they jumped into the night sky in a most daring rescue attempt involving a government who was not used to being invaded on their own soil or being surprised with their own evil methods.

In the first few seconds the men plummeted very fast; their eyes locked on a smaller house where lights were seen through windows, compared to a larger ranch house where no vehicles were seen. Their chutes opened in silence and allowed the three to float to the surface as quiet as a feather, fifty yards to the rear of this station.

They hunched in expectation of being discovered, three men dressed from head to toe in black, with but a two-inch square opening over both eyes. Oshi and Kami had katana hanging from a sling on their back; their eyes peered through the blackness for any movement. The best thing about this outward post was that they were away from the main ranch house which was lit up like a Christmas tree.

It was understood that Toy would stay back and watch the perimeter. If need be, and without the ear pieces and mics necessary, Toy would simply call Kami on his cell phone. The men did not want this to happen at the worst of times. Toy had been gifted by the Korean store owner with usage of the man's SIG-Sauer P-226, a handgun that fitted a 20-round magazine, plus an extra magazine in case twenty bullets weren't enough. Toy didn't know whether to laugh or decide this might be out of his league all together.

With hand signals Oshi and Kami began to inch toward the rear of the house, splitting when they were within thirty-feet. As Oshi was approaching the far corner of the structure he heard footsteps before he saw the guard. The surprised agent began to turn the corner before the last few moments of his life were in wonder of how this big tree was standing in front of him. Before he raised his walkie-talkie to his mouth, two throwing stars had entered his forehead. A momentary freeze of the guard's body recognized that something was terribly wrong; he tilted forward into the dirt. Oshi also froze to wait until he was assured no one had heard the falling body.

Any previous plan of getting the boys back without casualties was now amiss. Oshi continued to the front and knew Kami was on the farthest corner, waiting for a signal. Another guard was standing by the front door, cigarette in his fingertips and not real concerned that this little kidnapping was yet known by anybody.

When Oshi saw Kami take off his hood and begin to approach the guard's location, he hesitated until he saw the guard take notice of a man approaching him, one who obviously was indistinct. The agent only had to turn his head before Oshi was at his side and taking a knife to the man's throat.

"Are you sure you wanted to do that?" Kami asked his friend with a soft voice.

"One's down in the backyard. Had to."

The two of them found the front door ajar, swung it open slowly, and then saw two men at a table staring right at them. Their hands couldn't reach for their weapons fast enough before throwing stars caught them in the throat and forehead. Their gasps and struggles were quickly ceased with gags as they were tied up. One of them was likely to survive.

With the upper rooms searched Kami began locating a door that would lead them to a basement. It was quite possible they were in the wrong house.

Off the kitchen area they saw a door that did not lead outside. Not only did the door squeak but the steps made a sound as the men began to descend. When they heard, "Max, is that you?", the two martial arts experts leaped from the stairwell, pulled their blades from their sheaths and hurdled throwing stars at the first agent who reached for his gun. The other man had a knife to Leif's penis, stretched upward as a means to make the boy confess to anything they wanted to hear.

"Drop your weapons or the boy's penis is gone," the guy threatened as his only means of retaliation.

With as quick of an underhand motion as he could deliver, Oshi put a shooting star between the man's eyes. The agent dropped like a rock.

Neither Oshi nor Kami gave the shooting stars a chance to determine the men's incapacities. Oshi sliced the first man's arm in half as the agent lifted his weapon upward. Kami then pointed his sword an inch from the agent's throat before the man made a decision that was in his best interest by raising his other arm.

Lying on a metal table with cords hanging over his bare thigh was Leif. His eyes showed both a panic and a relief when Oshi raised his hood and smiled at his admirer. He was barely conscious but gave a weak smile at this hero to his eyes.

"You okay, buddy?" Oshi asked. Leif nodded. "Where's Bode?"

"In a cell over there," Leif whispered but couldn't point because his hands were tied to the table. He was bleeding from a gash in his mouth and dazed by the torture.

"How many men do you remember?" Oshi asked and patted the boy's cheeks to keep him awake.

"Four kidnapped us, but I saw two others when we arrived."

For the time being, Oshi felt confident that they had covered their bases. He scanned the naked body of Leif and saw where these men had clamped two metal clips to the boy's testicles, with an electrical extension to a machine. Furor flooded his mind.

Looking at this wounded boy touched what Oshi had wanted to deny. He was madly in love with this boy, a young man the same age as Kami when Oshi had fallen in love with him. It was like a samurai curse to fall for a handsome male and Oshi knew he was smitten with desire.

"Did they hurt you, Leif?"

Leif's wonderment of being rescued drained from his face. Tears began to flow and a lack of words to describe what punishment the men had inflicted. As Kami finished tying the two men up, they found the key to undo a lock on a 4x6 foot room that was designed as a prison. Bode stepped out completely naked and bruised from a beating.

"I didn't tell them a thing," Bode said with pride and was wrapped up into Kami's arms. He began to cry with relief in his father's protection.

"I'm bleeding to death!" The cry for help was from the agent with his arm missing. Oshi decided not to help the guy. Instead he threw a shooting star at the agent, right into the man's upper chest, a wound that hurt like hell but wasn't fatal.

"You're lucky I'm keeping you alive, asshole! Tell your president if he wants to kidnap boys he's going to deal with people who make your Navy Seals look like Boy Scouts. Got that?"

The man nodded, then had his wrist and elbow broken. If there were more than six men guarding this site they would have heard the screams of agony.

****

Kami ran to the rear of the house and waved to Toy. Five miles away Bill did a quick liftoff and stayed no more than thirty feet above ground level to swoop in for a landing in front of the house. If the ranch had surface-to-air missiles they would soon be toast. The three men and two boys scrambled aboard the chopper as it lifted off without

399

receiving any gunfire or floodlights. A few seconds later there was a request over the pilot's radio to identify himself.

"This is Senator Franco. Have a nice day," Kami replied when he grabbed the mike from Bill. It got a hearty laugh from inside the helicopter. Oshi held Leif in his arms as if he was his own son. This Japanese psychologist sensed his love for this boy was beyond what he had realized and was prepared to quit his position as the school's therapist at this great academy.

****

Bill brought the helicopter down on the outskirts of Austin, as Oshi was giving his ninja outfit to Leif, before he changed back to his other clothing. Kami had done the same thing for Bode. They landed in a field behind a bowling alley and a plan was put in place that would be the safest for the boys and the men.

Bill didn't know what to expect when he returned the helo back to the airport. If the feds were on to this they'd likely be waiting. If not, Bill was to drive back, pick up the fellas and return to the academy.

It was precarious situation for this Vietnam pilot who hadn't had this much fun in thirty years. He was tempted to ditch this chopper and make a run for it. The leasing company had been very compliant and eager to help, so it was this that Bill made his decision to give himself up and take the consequences, if need be.

From the landing pad to the flight office Bill had his eyes and ears peeled to any sight or sound. The surprising part, everything went smoothly. He even drove the SUV around the parking lot three times to make sure no one was following him. His erratic turns and going around the block convinced him that, for whatever reason, the feds weren't about to put the president in a kidnapping plot or have his ranch disclosed as the hostage center.

****

Though Toy was reluctant the guys stopped at a fast food restaurant take-out and bought everyone a refreshing meal. Bode ate like a famished orphan through a sore mouth and a squinting eye. Never leaving Oshi's side Leif felt safe and blabbered about what they made him say, apologizing for saying it, until Toy told him a boy wouldn't be expected to defy torture just to keep a secret they'd likely to find out anyway.

Dr. Marion had examined both boys in the helicopter and determined that no immediate medical care was needed. He patched up their cuts and ceased the bleeding, then examined Leif's testicles.

"He won't miss a beat," Kami said and had Leif giggling.

As an afterthought Leif mentioned something that he hoped wouldn't come back to haunt his mother. "I had to tell them about my mother and where she lived. Do you think that will be a problem?"

Toy pulled out his cell phone and asked Leif what his mother's home number was. Their own time was now 12:47 a.m., so, in Iceland, it was 7:47 a.m.

"Yes, Mrs. Bjornsson? I don't mean to disturb you at such an early time.......This is Dr. Kerho.......Well, yes, it is an emergency. I'd like you right now to pack a few things and get out of the house. I can't tell you how important it is that you do exactly what I tell you. Do you have a place to go?...............Yes, Leif has mentioned you have a man in your life. Please go to him and wait for further instructions. For your safety and Leif's, go now."

In Reykjavik, Iceland, Mrs. Bjornson had just stepped out of her front door when two men were coming through the front gate. She was neither a naïve woman nor one who wished to put herself or her son in jeopardy.

"Mrs. Bjornson?"

"No, I'm Mrs. Bjornson's nurse and housekeeper. Just like that torrid woman to hire someone to help her get around the house, then make me do all the housework, as well!"

The agent faked a slight smile. "Well, ma'am, I need you to go back in and tell the lady there are two men who would like to talk to her."

"The hell I will! I put in my eight hours of work and look what time it is, quarter-after eight. Why that woman keeps me after my shift and now you expect me to go back in there because you're too damn lazy to ring the doorbell! You whippersnappers outta be ashamed of yourselves!"

The agent raised his hand and stopped the sermon. He motioned his peer on and let the woman proceed down the walkway. At the door they rang the bell and waited, and waited, and waited. The fellow CIA agent moved to different windows until he found one he could jimmy open, then crawled in. In moments he had opened the front door and sprinted out into the street, searching both ends. Their target had completely disappeared.

Chapter Thirty-Seven

In the corridor outside the East room of the White House the Director of the FBI was on a mission, an unexpected meeting with the president that he wouldn't take no for an answer. He was finally shown the respect he thought he was due and given a quick minute with the president.

"I lost five men!" were the first words out of the director's mouth.

The president didn't want to hear this; actually, this was Senator Franco's caper, not his, but the director didn't want to report to a senator. "How could anybody respond to the operation unless it was leaked?" The president wanted to know, as if he was now the disgruntled one.

"Whoever it was made my men look like amateurs, then left one of them alive to let you know that Navy Seals were Boy Scouts compared with the people we're dealing with. I doubt very much if this academy has its own SWAT assassins, let alone a response team that knew exactly where these boys were taken."

"Did the boys tell us anything?" the president asked.

There was an expression of disbelief from the director. It shocked him that the president was more concerned with the two boys instead of losing five of his best agents. He held his anger.

"The only boy we had time to interrogate said he was new to the academy. Don't ask, we don't know how or why a boy would be admitted late. The kid said that he came from Charles the Twelfth of Sweden, but said, if anything, he was Charles the Thirteenth, but his real name is Leif Bjornsson. Apparently one of my agents didn't like the kid's humor and hurt the boy a little. I doubt if this academy is going to tell the world where they found their students. First, they killed five men, which we can deny we knew nothing about and didn't authorize any covert mission to be at the ranch while you were absent. Secondly, why would they announce to the world now that they have these clones?

Oh yeah, the other is Bode Lawrence. We think this could be T. E. Lawrence. The boy was as stubborn as a mule, my agent said."

"Huh? Who's this character?" the president asked.

"Lawrence of Arabia, Mr. President."

The president rested back in his chair with both disbelief and a subdued grin. "I'm sorry about our losses. We underestimated this academy, apparently. I'll have to have a long talk with a senator."

"Thank you, Mr. President, I appreciate your sentiment. We did have what appeared to be a major breakthrough. One of the boys spoke of having a mother in Iceland. We put two of our agents in the vicinity after our surviving agent talked to us at the hospital. He was banged up beyond what a man should have to endure. Anyway, before we could question this woman she slipped our noose. Someone tipped her off."

"I have a feeling this is going to blow up in our faces if we don't get a grip on it. Yesterday we had a room full of reporters wanting answers, prominent members of Congress asking questions if one of our agencies has a secret operation to clone, for military uses, the great minds of history. This has become a sideshow that will destroy my administration. Do we have an understanding here?!"

The FBI Director never did like this president, didn't care for conservatives, period; he didn't like people, actually, any evangelist who thought they spoke for their type of Christians, or any other type of hypocrite.

He leaned into the president and replied, "This is not my spectacle, nor do I like losing five agents to a muffed kidnap of two boys who haven't reached puberty."

"These boys, according to Senator Franco, are supposed to be fifteen-years of age. Of course they've reached puberty!"

"My surviving agent said that neither one had a stitch of hair on them except on their heads, but they had the longest dicks he'd ever seen. Know what I think? You've got something far deeper than cloning going on here. My agent also admitted that a boy, no bigger than a rodent, had manhandled all four of them with some sort of martial arts moves—like he was a junior Bruce Lee. Now you tell me, Mr. President, who are we dealing with?!"

The president paused with a grim expression, stayed taciturn and, for the first time, realized that the Middle East crisis appeared far less a threat than an academy in Texas. If he could do it he'd send this academy to Afghanistan and North Korea to straighten out the mess over there.

"Okay, this is what we're going to do," the president began with one of his mental farts. "We put Senator Franco in the middle of this. After all, this was his little brain child that he confronted us with. If anyone sinks I'll sacrifice this man. I've heard the senator has a teenybopper on the side that he's been fucking. So much for his family values platform. I'm willing to overlook his sexual habits if he can clean up this mess. I've already given him carte blanche to investigate these people and, yes, I know I concurred with his plan, but on paper it looked fool proof. We'd get our information, terminate the boys and no one would ever report that Lawrence of Arabia and Charles the Twelfth ever existed again. Technically they don't. Let's give the senator another two weeks to clean this up, if not, maybe we can make this whole scene vanish with one explosion—accidental, of course. Shit happens all the time, and we'll make sure it doesn't look like another Waco."

"And here I thought you didn't have balls, Mr. President." The FBI Director had a momentary feeling of appreciation, but he also quickly snuffed this momentary praise out before he gave the president any inkling that he actually liked him.

＊＊＊＊＊＊＊＊＊＊＊＊＊

Senator Franco was briefed by the president's press secretary and close friend, Bryan Bell. It was further suggested for the senator to continue his investigative efforts with means that didn't require kidnapping for the time being. The president wanted to be in the shadows on this one—way in the shadows.

"I've heard this song a hundred times before, which is more than enough," Robert said after this briefing. "I become the fall guy if this thing screws up. Let me tell you something, Bryan, your FBI guys fucked up! Them and that imbecile they call their director. And, as a result, an initiative to which I've dedicated tremendous effort has been sidelined because those agents couldn't question two boys."

The press secretary raised his hand like a traffic cop. "Hold on, Senator. I don't know anything about two boys. This isn't my fight and I'm just the messenger. Children scare me. You know the thing about don't get caught in bed with a dead woman or a live boy? I try to make that my creed. Two boys sound way too desperate."

Robert didn't press the point. This was only a temporary detour. He'd come up with a more sure fire way to expose this academy without putting agents at risk. The visit of his sons was but a week away; certainly he could convince them of what was morally right and was in the best interests of their country.

The senator was more in the mood for a romp in the hay and diverted his mind to the thought of a slim brunette in his arms. With his key he let himself in and discovered that Lauri was probably out shopping again. She had more than enough shoes and handbags to compete with Amanda Marco, and Robert was paying for the condo and the utility bills. What he hadn't told her, yet, the lease was now in his name.

Lauri was excited to see the senator naked on the sofa. She dropped her latest purchase of lace panties and threw herself into the senator's arms. He disrobed her as they kissed. The first round was always heated sex with little foreplay.

It was the second round when Robert let either his frustration of his work or his demeaning of the opposite sex determine what kind of games he was in the mood for. While he had the little mistress kissing his wrinkled workhorse, until it was ready to respond again, Robert gave her the news.

"Lauri, babe, I want you to show my sons a good time when they come next week. If their pants jump you're welcome to see that they don't use their hand for a change. I'm thinking of upping your allowance to a grand a month for your efforts."

"Is that legal, Robby? I mean, helping 'em with their little problem?"

"I'll get ya diplomatic immunity. They're built like their daddy, anyway."

"Really?" Lauri asked in her best giddy naiveté. "I'll teach them to be just like their daddy."

Chapter Thirty-Eight

Three days of intense anxiety and round-the-clock surveillance was finally relaxed without one word from an investigative source or law enforcement, arrest warrant, or invasion by a swarm of governmental agents. Toy and Kami had considered every possible option in their continual review of the Austin debacle. The one thing they did agree on, they kept from their sons and daughter the magnitude of all that had happened. It wasn't necessary to inform or create unnecessary anxiety with fear of the unknown.

Toy had not even conveyed to Leif and Bode that they were hit with 50,000 volts of electricity from a Taser gun, and Leif didn't discuss with his peers that those clips on his testicles had sent a jolt of pain, so intense through his groin, that he would have welcomed to have been kicked in the balls than endure the agony those men had put him through. He was seconds from having his penis cut off when Oshi saved him.

The two directors didn't agree on how to approach Dr. Makawa. The man had gone into isolation, mediated for hours at a time and did katas to settle his mind. His only visits to the dorm were to meet with Bode and Leif. Toy thought like a psychologist and decided the man should have company, while Kami sensed the deeper emotional impact of killing men, destroying life out of pure anger with the understanding of the inhumanity that some men could treat another human being. Maybe Oshi had lost his sense of control, but the man was also a martial artist who internalized his loss of peace and destruction of men who wouldn't have hesitated a second to kill Oshi or Kami, if given the chance.

To assist this transition Brittany took the kids to grandma and grandpa's home in Dallas. She had Debbie meet her to go shopping and discuss the problems of raising husbands and children. With Oshi alone, Toy and Kami permitted Leif to visit the doctor at his home, provided that the doctor agreed.

Oshi concurred and had Leif meditate with him to regain a spiritual union, before they began a form of weapons training to clean the mind and allow energies to swirl and surface. In the past few weeks Leif had become as devoted of a martial artist as Oshi had ever taught. The boy grasped technique and the essence behind every move. He was a dream to coach.

It was slightly noted by Kami that Leif spent the night at Oshi's home, a fact that was really immaterial, and it wasn't important to question the immense love between them. Oshi knew his true passion was for the male sex, and Leif had put an arrow through his heart since day one.

There was one dilemma that was undecided; Mrs. Franco had told Brittany that the senator had requested that Niqui and Jay be flown to Washington D.C. If there was any consolation to this planned trip it was the fact that Senator Franco wouldn't kidnap his own kids, though Mrs. Franco was beginning to wonder.

In the meantime she had brought her sons and daughter to the academy with the suggestion of Dr. Kerho. He was concerned to what extent the senator might go to obtain information and compliance from the academy.

The Franco kids felt right at home in the dorm and with the people they loved the most in life. Bobby had a surprise for the Franco boys when he injected them with a genetic enhancer to increase their penis length. The boys didn't bother asking their mother for permission and were more than happy to join their peers in this department. In two days the growth began and Bobby was once again congratulated for his scientific ingenuity.

While the academy was kept under tight scrutiny without alarming the students, their schedule never varied or became so paranoid that the kids couldn't relax and enjoy the warmer temperatures as summer was upon them. Guards were posted around the fenced area off grounds to assure no boy was at risk during their morning runs.

This air of suspicion—as sensible and real as it was—gave the impression to everyone that the matter had been resolved in some way. If anyone was spying on the academy they could now witness almost three dozen naked teenagers enjoying windsurfing and other various sailing craft on the lake.

The slides gave everyone the initial queasy feeling in their stomachs; a euphoria of a roller coaster. It was interesting that, at the boys' ages, this huge slide they once were in wonder and fear of, felt more like something out of a playground.

At the Kings Academy there wasn't what was considered an end to the school year. Their usual summer break was a vacation to a country they hadn't been to before, though Japan was often the exception.

The boys and Wendy had been to Japan no less than four times. Though it was something to look forward to and plan, the kids knew these vacations were an extension to their education and social understanding of various cultures. Leif, especially, couldn't wait to travel and explore a world he had yet to witness.

****

While Travis spent several days at the academy as a guest of Alex, his brothers were driven to the airport by Mrs. Makawa and their mother. There were no instructions or warnings; Debbie tried her best to put on a good face and pretend this was all in the best interests for her sons to visit their father. Jay was reluctant to go and Niqui kept questioning why Travis wasn't invited.

Only a few nights before at their home, Travis had found what he'd been looking for. Underneath his lamp was a listening device. He quickly went to his siblings' rooms and discovered such devices in these two rooms. Jay picked his father's office door and the boys found the recorder and other paraphernalia to snoop into other people's lives. Travis confiscated all this equipment and devised his own plan.

Senator Franco met his boys right on time, huge hugs at the airport and a smile that proved he really loved his sons in front of a crowd. Though the boys expected a hotel, or wherever their father stayed while he was back here doing his job, they found themselves being introduced to his assistant campaign manager named Lauri. The condo had two bedrooms and the boys were given the one that had all new furniture and bedding. They didn't dare ask where their father stayed.

The tour began immediately after Lauri fixed the boys grilled cheese sandwiches for lunch. Both boys refused the potato chips, Oreo cookies, and the sugary juice, but they welcomed the milk and fresh fruit. Even Senator Franco raised his eyebrows to the boys' healthy eating habits.

They viewed the Supreme Court building first, then proceeded to the Library of Congress and began their way down to the Washington Monument. Their father passed the National Air and Space Museum, promising this for a later time. When they strolled inside the U.S. Capitol the senator appeared very smug in bringing his sons to his place of work.

Another senator from Georgia approached the trio, grinned a wide smile at his colleague and then eyed the two handsome boys. He put his hand out in front of Niqui.

"Mr. Franco, I presume," the senator greeted.

Niqui stretched out his hand. "No, I'm Niqui; this is my brother, Jay."

The senator was lost in confusion for a second, then began to laugh. Niqui's father was about ready to blow a casket. "Now that's funny! I think y'all will be real pleased with your visit, boys. Senator, ya got a fine lookin' pair here. Don't let the women catch an eye on 'em."

Robert thanked the senator and moved his boys to see Alexander Hamilton's statue in the Capitol Rotunda. The boys noted that he was the nation's first treasury secretary, a vocal supporter of a strong central government, and was shot by Thomas Jefferson's vice-president, Aaron Burr, in a duel in New Jersey on July 11th, 1804.

Underneath the Rotunda was an empty crypt that was set aside for the remains of George and Martha Washington, but never used. Portraits of the speakers of the House of Representatives lined the wall of the Speaker's Lobby, just outside the House chamber.

"What in the world are you two boys doing?" the senator asked his sons who were busy typing in notes on their laptops.

"Mom wanted us to keep track of our visits," Niqui answered.

Robert decided not to comment.

411

Jay brought out his small digital camera and snapped a picture of Constantino Brumidi's 185 fresco, Apotheosis, constructed in two rings—the inner representing the thirteen original states of the Union, and the outer depicted four hundred years of American history.

They saw the Senate Chamber, and again viewed the many frescos and magnificent paintings. It was quite a coincidence to meet the current Secretary of Treasury passing through with some of his relatives. The senator introduced his sons and the greeting was harmonious.

"Enjoying your stay, boys?" the secretary asked and received an affirmative. "Anything you'd like to see while in Washington?"

"Ft. Knox," Jay answered truthfully and received laughter from everyone but his father.

As soon as the cabinet member departed Robert briefed his sons on demeanor he didn't care for. "This is not a comical shop, boys. If you've learned to be clowns at school, leave it there."

"Come on, Dad, lighten up. Jay didn't know Ft. Knox wasn't around here," Niqui said and received a healthy cuff to the back of the head. It did appear that the enthusiasm for seeing the Capitol's Statuary Hall had disappeared. Niqui stayed quiet and reserved, and Jay felt just like his brother. Their visit to the U.S. Holocaust Memorial Museum was very similar to their mood—depressing.

For their first night the senator and his, supposedly, assistant campaign manager, took the boys to the colorful Adams-Morgan multicultural neighborhood. Lauri loved French food and Jay and Niqui agreed, so they decided on the La Fourchette. When polite and boring conversation had caused an eerie silence she asked a question that wouldn't have been on the senator's mind.

"How do you boys like Kings Academy? I hear it's all boys. Must be real rowdy without girls around."

"Not really," Jay spoke up. "We have a lot of violence at St. Bonaventure—that's my school. They're taught at Kings to talk out their differences and recognize that we all have similar goals and aspirations, and even similar personalities, which should be a source of unity, rather than a source of conflict. I think we all need to acknowledge the anguish we cause each other and to work in a constructive and honest way to understand the feelings of the other person."

"Wow!" Lauri said. "You talk almost like a politician. That's so smart."

The senator gave an annoying look at his girlfriend but held his tongue.

When the four of them returned to the condo their father made it a point to say he usually stayed at the Willard Hotel, but, to be with his sons, he asked Lauri to make up the sofa for him. He didn't want to give the impression he'd slept there before.

Around midnight Niqui turned out their room light and huddled next to Jay underneath the blankets. They heard soft whispers outside of their bedroom and feinted sleep.

"Fast asleep. Aren't they so cute?" they heard Lauri's whisper to their father before the door was closed softly. The boys chuckled and threw back the blankets in suspicion of their father's relationship with this girl. They tiptoed out of their room and peeked around the corner into the living area.

Lauri was dressed in a Cat Woman costume with far less fabric than the movie character. Her black boots and pointed ears almost had the boys laughing out loud. It was what she was doing that interested the boys.

Their father's briefcase was open and Lauri was checking paperwork, copying something on another piece of paper, then sliding the contents back into its sleeve. Something right out of a spy movie, Jay thought. Whatever she was doing, Jay noticed that after the girl appeared satisfied she closed the case ever so softly. A voice came from her own bedroom a second later. One thing for sure this sofa wasn't prepared for anyone to sleep in.

"You're tormenting me here, Honeybuns. When are you going to start licking me off?" The voice came from the bedroom and was certainly their father's.

Lauri dashed right by the boys who were crouched down on the carpet to her left. Once Lauri disappeared in the bedroom the boys crawled toward the half-open door and eyed their father in some type of harness, a mask around his eyes, dressed in a Batman outfit with a huge hole in the crotch. Their father had a hard-on with a bottle of chocolate syrup dangling above him and dripping onto his groin. Lauri dove right in and began slurping the man's balls and cock of this tasty goo.

This might have been hilarious if it wasn't so embarrassing for the boys. Watching this hairy groin wasn't as appealing as what they were used to. Lauri was an attractive young woman, but viewing her from the bottom didn't exactly excite the boys either.

Niqui whispered to Jay that a woman's twat resembled some farm animal's face. He had to put his hand over Jay's mouth to stop his giggling.

Jay ran back into their bedroom and came out with the two listening devices that their brother Travis had given them with specific instructions.

Niqui took one and planted it in his father's briefcase while Jay tried to find a way to attach it to his father's suit coat, hanging in the closet. When his brother told him to hurry, Jay simply dropped it in the coat pocket.

****

There was a fear their father would make them get up early the next morning. Niqui was awakened by the smell of eggs and bacon, his watch showed 9:30. Waking Jay up the boys slipped on a pair of shorts and ventured into the kitchen where Lauri was kissing their father on the lips.

"Oops!" Lauri saw them first. "I was just thanking your father for being so nice. I'm so romantic. You boys understand, don't you?"

"Our mother does the same thing," Jay said and held his smile.

"Your father was so generous by sleeping on the sofa all night," Lauri said as a means to distract the boys' curiosity.

"My father is used to sleeping everywhere and anytime," Niqui admitted. "I bet he could sleep in a bat cave and be right at home."

The senator gave his son an odd stare but then dismissed it. "I have to run. You boys behave yourself and do everything Lauri says. Got it?"

"Right, Dad," Niqui confirmed. The boys sat down and devoured the delicious breakfast while maintaining their secrets.

"Your father is such a wonderful man, boys. He's so funny, especially if he has some wine in him." Lauri tried to make conversation about this man in her life.

"Yeah, some people are like that," Niqui said. "They like themselves better when they're drunk."

Lauri agreed, though she didn't know why. "He's sometimes not very comforting, and he often thinks just of himself. I guess that's his job."

"Sounds like a lawyer," Jay said in remembering what Dr. Marion had once mentioned.

"You're funny," Lauri told Jay, which made the youngster beam, until he remembered the girl was screwing his father. She added, "What do you want to do when you grow up?"

Jay thought about the question. "Anything without the word boy in it; you know, bag boy, grocery boy, whipping boy, errand boy…"

"We got it, dude!" Niqui spoke and the three of them busted up. Being at the academy had given them a sense of humor together that had long been amiss.

Lauri cleaned up after and discovered that the two boys weren't about ready to let her do all the work. Part of their responsibilities at Kings was to clean their own messes, and the habits they were developing extended to being guests at another person's place.

When Lauri went to her room and the boys heard the shower running they searched a book where they'd seen the girl place a piece of paper after she'd rummaged through their father's briefcase.

Behind the jacket cover was the newspaper and written on it was a list of names that had been forcefully disclosed by Leif during his interrogation. In other inserted scraps of paper were other notes and an address and phone number of a magazine. Jay and Niqui grinned at each other and closed the book.

With their laptops under their arms the boys followed Lauri to a very popular mall filled with shops and restaurants. Inside and out it was a place of fun. They discovered it was an Old Post office.

Their female companion for the day bought them each a polo shirt and a pair of shorts, then insisted she would take them to lunch. It was while they were eating that she began to drop hints about the academy and the little knowledge she'd gained from talking to their father.

"I heard they have some very interesting people at that academy," she said as if talking to grade school kids. She received a shrug of shoulders and a remark that they were all boys just the same.

"Do you believe in all this cloning gossip that's going around? I bet it's really hard on a lot of those students," Lauri persisted.

"Sure is, especially for guys like Babe and Joe," Niqui spoke up and had his brother perk to attention.

"Babe and Joe? Who would those boys be?" Lauri asked.

"Well, let's see, they have…you know, I shouldn't be telling you this but it's not going to be a secret forever; anyway. They have Babe Ruth, Joe DiMaggio, Honus Wagner, Pee Wee Reese, Lou Gehrig, Dizzy Dean…"

"Wait, wasn't he an actor?' Lauri asked.

"I think you're thinking of James Dean, Lauri," Niqui offered. "Dizzy pitches for the Kings." Niqui had to kick his brother underneath the table for giggling.

"Do you think you could make a list for me later? I'd love to meet these boys sometime," Lauri admitted. "I'll have to keep the list from your father, though. He's so noisy."

"My mom once said that if we want to keep something hidden from Dad, put it on the ironing board or the washing machine. He has an aversion to certain items," Niqui revealed. Lauri laughed.

They walked up to Pershing Park which offered the serene view of the Willard Hotel. The hotel has stood at this location since 1816. Jay made mention that this must be where his father's apartment was.

"Must be," Niqui agreed, then whispered, "I bet he's never cheated on a girlfriend."

Jay had to think on that one, but grinned. "When Lincoln stayed here it cost him only four dollars a night, and the word lobbyist came from here when politicians used to influence peddlers hanging around the lobby of the hotel. Julia Ward Howe wrote The Battle Hymn of the

Republic at the Willard, also."

"You're so smart," Lauri complimented and pinched the boy's cheek. "Your father used to spend four-hundred dollars a night when he stayed here."

"Used to?" Niqui asked.

Lauri's red turned all red. "Well, what I mean is, sometimes he likes my sofa."

"Because it's nice and soft and makes his pecker stand up," Niqui said with a smile, as not to embarrass the girl too much.

"You boys are definitely way too smart," she replied.

It was near evening when the three returned after walking many miles and taking in so many sights. The boys went to their room and began listening to their father's conversation, wherever he was. The laptop acted as a receiver for the communication and the software printed any voice activation from the listening device right on the screen.

There were a multitude of voices coming from the Senate Chamber where men were arguing over a bill to increase Medicare benefits. Jay began to wrestle with his brother out of boredom, tugging on each other's clothes until they were both nude. Niqui had his favorite object of mass destruction when it came to brotherly fun; a pillow fight of hefty swings and plenty of laughter. Lauri opened the door.

"Oh, sorry, guys," she said and closed the door.

The boys paused and cracked up, then resumed their comical wrestling antics. The crackle of so many men had settled on the computer to only two voices, one being their father. They had no idea who the other man might be, but they were discussing the Kings Academy and a boy who did not have a student visa; thus, this student was illegally in the country.

Jay mentioned that they were talking about Leif. As the boys sprang to the side of the bed to listen more closely they heard their father discussing a plan to raid the academy, garnish the records and take DNA samples off of sheets, clothing and cosmetic products, to compare to the records that were being compiled from what other countries had sent to the FBI crime department.

The boys' eyes met again but there was far more seriousness behind them. The last thing of importance that was said by their father was that the man had an appointment with the president to discuss the planning of this warrant.

Both boys jumped to their feet, threw on their shorts and walked out, only to see Lauri in a see-through nightie. "Hey, guys! I see you two really like each other. This is like an apology for seeing you naked, so you deserve this. What do you think?"

Jay eyed the female, slightly better built than his sister, but not as thin or as pretty. "I like the color," Jay admitted.

"Come with me my little sweeties. I have something on the bed I want to show you," Lauri said with a finger curling to follow her. When she was in her room Lauri swept off her only garment and laid back. "I'm your present."

"You're kidding?" Niqui asked with more annoyance than being impressed. His mind raced back to a night when another girl threw herself on him.

Jay had moved toward the headboard and was examining the straps and harnesses. "What are these?"

"Oh, you're just like your daddy. I mean, like my other boyfriends. If you want to play with those we can."

Lauri sat up and unbuttoned Jay's shorts. He was naked underneath. "Oh my! How did you get a dick like that? Why, you haven't even reached puberty yet." She motioned for Niqui over, but he stepped backwards.

"No thanks," Niqui flat out told her.

Jay moved his shorts back up, then turned to all these restraints. "Can we put you in these?"

"That would really be kinky," Lauri expressed but allowed her wrists and ankles to be enclosed in the Velcro straps. Her legs were spread and her bubbly nature couldn't wait for these hot teens to jump her bones.

When a squawking sound emerged from the boys' bedroom the two of them ran to hear what was going on. In the background Lauri was saying, "But what about me?"

The static coming from one of the planted listening devices had the boys suspiciously hanging onto their hopes that nothing had gone wrong. There were several moments of silence before Niqui recommended they turn on the other device in their father's coat.   Instantly the voices started back again.

"Yes, Senator, we found a bug in your briefcase.   It wasn't professionally planted as it was simply shoved in the corner with this brown tape."

The Secret Service agent showed the senator the deactivated device, found as a routine measure when people enter the White House.

"This looks like one of mine," Robert confessed and then clarified. "I mean, I have these at home for security reasons, but my…., I think I know what has happened and I'll make sure this does not happen again."

In their bedroom Jay and Niqui gulped and knew their asses were in a sling now.   They looked at their cell phone and expected a call at any second.   Instead there was a clutter of voices.

"Senator, the president will see you in his limo.   He has another engagement that he's heading to and he wants to speak with you for a moment."

Fortunately for the senator he avoided the further clothing scan of the metal detector as he hastily departed to the driveway.   There was a spatter of white noise, a car door closing, an informal greeting, and the president's voice became clear to two Franco boys.

"Senator, sorry for the quick change of plans. I've just been told there was a bug in your briefcase.   Do we have a problem?"

"No sir.   Actually, I believe my sons are playing a practical joke that has gotten them in hot water when I return to my apartment. They're in town to see me for a few days, but this is beyond embarrassment."

Robert reached in his back pocket and pulled out his handkerchief to wipe his forehead.   To be caught with some type of spy hardware in political circles was a near death warrant.   He pushed this cloth into his coat pocket.

"Boys don't always use the best sense, Senator.   God knows I have two daughters who I have to keep an eye on. Anyway, Robert, there was

419

a meeting today at the United Nations Security Council. The Secretary of State was there and was shocked that someone had put on the agenda this investigation into these clones."

"They can't do that, Mr. President. No one even consulted our government about this," Robert protested.

"Someone took the liberty of invoking a little-used and rather esoteric regulation in the Security Council's rules of order. A 1957 addendum to Article Thirty-nine of the Security council's Affairs of Conduct allows either side of any dispute before the Security Council to provide periodic progress reports of any council-ordered investigation."

"It's so early in our investigation, Mr. President. They must give us time," Senator Franco insisted.

"Yes that all makes sense, but several countries are pointing fingers at us, accusing our numerous agencies of stalling and making excuses. We're getting our balls pinched here and I'm not sure we have any more rounds to play with."

"I know I'm onto something with the Attorney General, Mr. President. Within a few days we should have enough information to close this academy and allow the other countries to take care of their own."

"I'm not sure Russia or Sweden want to deal with kings and czars who have come back from the dead. Can you imagine the public relations nightmare, and who's to say their people might not wish to honor these heroes with the right to resume power in some way or another? I've put a lot of thought into this, Senator, and I feel it's time to count our losses and move forward. You have more important things in your future than this circus. I also think you should consider how to rid yourself of an extra-curricular sex life. If you're going to run with the ball here, Senator, you need a clean slate. Get my message?"

Robert was sweating profusely. "Yes, Mr. President. I'll get right on that."

The senator pulled the white cloth from his coat pocket and wiped his face again. What he didn't see nor feel was the tiny, circular device that spilled out and pinched itself into the leather seat.

"Good. Just in time. My press secretary is waiting on the other side of the door. Talk to your sons for me and, if they need a scare to show them the error of their ways, let my men know."

"That won't be necessary, Mr. President. I'll handle them both myself. It's been awhile since my hand has met their backside."

The boys heard the car door open, then slam again, but the two voices had changed in their appearance and intensity. Jay almost switched off the laptop but Niqui decided to listen in a few minutes more. Their cell phone rang immediately-- the boys made the best decision not to answer it.

"Bryan, I'm tired of jacking around with this place in Texas. Put the Crash Operation we talked about earlier in motion. I'd like you to also speak with Senator Franco before he leaves the grounds and mention that he should leave the girl to her own freedom tonight. Make it look like a rape of a beautiful woman who didn't know enough to lock her door at night. We'll tie up loose ends this way."

"I'll inform the Air Force General, Mr. President. I'll tell them it's a go for eight o'clock tomorrow morning. I'll also notify our plumbers' crew to dispose of this trash. I'll get the address from the senator. Do we have a story and investigation ready?"

"We'll have an Air Force Disaster Preparedness Team enter the facility right after the Fuel Air bomb is dropped. It'll look like a B-52 accidentally crashed with full tanks of fuel. Those beasts are getting old and unreliable anyway. There won't be much left of the grounds, I'm afraid."

"Enjoy your night out with the missus, Mr. President. I better run and catch the senator."

****

Jay and Niqui did more than just stare at each other; they sprinted back to Lauri's room and released her restraints. She was terribly disappointed that they wanted to stop this fun.

"Lauri, we're all in deep shit. You, too! Get your clothes on and we're out of here," Niqui explained and didn't wait for a reply.

"What do you mean, we? Who's mad at you? Come on, guys, let me in on this."

The boys scrambled to shove their clothes into their suitcases, then helped Lauri do the same, though she constantly complained and wanted to fight this decision every second. The cell phone rang ten more times before Jay grabbed it, pressed it in his shorts, and the three of them hustled out in front of the plaza to determine what their next move should be.

"Let's eat first," Lauri suggested. "I need some explanation here why I have to run from my own condo."

Niqui wasn't paying attention to this whining and flagged down a taxi. "You have cash?"

Lauri nodded and they hopped in, then decided to head for Georgetown where a restaurant became their next destination. No more were they out of the taxi when Niqui whipped out his cell phone to call Dr. Kerho. He went through Tomas to get a transfer to the doctor's office.

"Dr. Kerho? Hi, it's me, Niqui."

"Well, this is a pleasant surprise. How's Washington, young man?"

"Sir, Jay and I are in big trouble and so is the academy. We sort of bugged my dad's briefcase."

"What does sort of mean, Niqui?"

"Okay, we did. Travis found where our father had these little devices put in our room so he could listen in on our conversations and spy on us."

"Bizarre, but it does sound like your father. So what happened? Did he find these?"

"I think when he went to the White House, they found 'em, sir. At least one."

"Bad news. Why do you think the academy is in trouble? Do they think it was us who did this?"

"No, they know who did it. Jay and I overheard our father tell the president that they're going to raid the academy because Leif doesn't have a permit to go to class."

Toy couldn't help but chuckle. "That's a student visa, Niqui. We're proving to them that we have adoption papers from fifteen years ago. Where are you right now? Wait a second! Does your father know you have a cell phone?"

"He sure does and he's been trying to call us for the last hour." Niqui moved further out on the sidewalk to take a look at the restaurant's sign before giving the name to the doctor. "Can we just come home?"

"This is what I want you to do, Niqui. I'm going to see what I can do from this end to get you home, but I want you to hang up now because your father can triangulate your cell phone and find you. Call me back in about ten minutes."

"But there's more about the academy, sir," Niqui pleaded but the doctor insisted the boy hang up, so he did.

In the restaurant Lauri had ordered them three sandwiches and chips. They all felt like they were fugitives.

"I have a confession to make to you guys. I've been in communication with this tabloid, which is real interested in this academy. I gave them those names you gave me this morning and they laughed; plus, they haven't heard from those two guys who visited me. Your father talks a lot about this school, but all he wants is revenge. I saw on one of his memos that he's had the academy's phones tapped."

"He's what?!" Niqui asked with panic in his voice. He ran back outside and dialed the doctor again.

"Sir, I know I'm a little early, but I think your phones are wiretapped. My father's girlfriend says so."

"Your father's girl…? I don't want to know. Okay, Niqui, I want you to leave there immediately and go to…let me think….you know where the academy is about to vacation to? Go there and I'll call you in thirty minutes exactly. Turn your cell phone off now!"

Niqui clicked off his pone, ran back in the restaurant just as their sandwiches were brought. "Come on, guys, we're outta here!"

They barely had walked to the other side of the street when two black unmarked cars screeched up in front of this same restaurant and three men went running in. Niqui, Jay and Lauri slipped inside an ice cream shop and out the back.

****

Toy had quickly got Kami by his side as they made contingency plans for getting the boys out of Washington D.C. He used his own cell phone to call his father and have a helicopter pick him and Kami up from the academy grounds. Kings Academy belonged to a jet investment firm which made available any type of aircraft depending on the business needs. It saved any corporation from the major expense of purchase and upkeep.

Toy's father was well aware of the academy's secretive nature, and any time his son desired his or Mrs. Kerho's participation they were more than delighted. It was fortunate that Toy was able to contact his father when they went on a cruise or vacationing in Europe.

The thirty minutes were almost up between the time Toy was supposed to call Niqui when the helicopter was heard in the distance. They were picked up and headed for the airport. With a quick call to Mrs. Franco to get her son's number, Toy dialed Washington.

****

Two teenagers and their chaperon walked under the colorful, seventy-five foot wide Friendship Arch, which spanned H Street and Seventh Street, not far from the site of the MCI Center sports arena. Washington's Chinatown is small but vigorous. Niqui checked his watch every few minutes, then every thirty seconds until he was positive of the time expiration. Within a minute of him turning the phone on, it rang.

Dr. Kerho didn't waste much time. "Okay, Niqui, here's what we're going to do. I'm on my way to Washington with a Gulfstream, that's a private jet to pick you boys up. I'll be in contact with you in the air to tell you where we're going to land. Are you safe and secure?"

"Yes sir, but I gotta tell you what the president said. He talked about the Air Force making it look like an accident."

"Relax, Niqui. Think! What exactly was said?" Dr. Kerho emphasized.

Niqui's mind was confused and spinning. He wondered if the laptop had recorded it all on disc but this wasn't the time to check.

"I think I remember most of it, sir. The president was talking to some guy who wasn't my father. He said to go ahead with a plan at eight o'clock tomorrow morning and to contact the Air Force. Air something, I forget, but it would be made to look like a plane crashed and some team would go in and verify this. He said there wouldn't be much left."

"Holy……., forget that last comment, Niqui. Does the president know that you know this?"

"I'm not sure, sir. I mean, we had two devices, so I don't know how the other heard this without my father there, unless he really didn't leave, or maybe because Jay just slipped it in my father's jacket and it just fell out, or something."

"Before they can triangulate where you are, turn off that phone again and I'll call you one hour from now."

Niqui obeyed immediately and felt a sense of protection. He ran to his brother with a smile of accomplishment. Chinatown was a little scary as night settled in. The three fugitives from justice looked for any secure place that wouldn't put them on notice or in plain sight of anyone in search of them. Another restaurant fit perfectly into this plan.

****

Dr. Kerho explained to Kami what information was given to him by Niqui. Two boys eavesdropping on the President of the United States was mind-boggling by itself. The idea that the government was ready to flatten Kings Academy to eradicate all evidence of a most mysterious collection of young men was absolute genocide.

Both directors ran from the helicopter and right in to the awaiting jet. The flight plan given was a direct route to Dulles, but Toy knew the possibility of a hundred federal agents waiting for them when the jet landed.

Kami quickly contacted Oshi and directed him to go immediately to Toy's parents' house and wait for further instructions. Kami tried his best to remain calm and not give his own sense of frantic away. He told Oshi to contact him once he was outside the academy's gates. With this accomplished Kami waited the few minutes until his phone rang inside the plane.

"Oshi, we'll be leaving for China in the early morning hours. This might alone create a problem and probably confrontation. We have a serious problem."

"I'll handle it, Kami-san. I have Leif with me. Is that a problem?"

"No, you might use him to communicate with the other boys," Kami suggested.

Dr. Kerho couldn't stay seated on the plane. Even Kami was busy preparing another plane for the morning, a 747 to be readied for sixty students and staff. The huge 747-8 was available only because the Governor of Texas had recently returned from his vacation with several of his staff. Normally this 747 would carry as many as 467 passengers in a three-class arrangement.

For the Kings Academy's purposes the third tier had been arranged with all sorts of games, from a pool table to foosball. The second tier had a sleeping quarters for all the passengers, while the first tier was all first class seats. For all the traveling the academy did, this international airliner had a maximum range of 8,000 nautical miles.

Toy watched the time on his Palm Pilot and wished he'd told the boy thirty minutes instead of an hour. At exactly the one hour mark he dialed Niqui, thankful the boy answered.

"Are you okay?" were Toy's first words.

"Yes sir, we're just reading our fortune cookies and telling Lauri that the President of the United States wants her dead. Other than that, Jay and I are fine but Lauri's crying."

"Here's what I want you to do, Niqui. I have a friend I met in San Francisco when I was your age. His name is Artie Stenson and he works at the Washington Post. He writes a column for the gay community. It's important that you go there right now and bring him with you. Just mention my name and tell him it's very important I see him immediately. If you have to discuss the academy, so be it, but get him with you! Do you understand me?"

"Artie who?"

Toy repeated the instructions and set thirty-minute intervals for the next contacts. When the doctor hung up he decided the lavatory was the first place he needed to go.

Oshi called Kami after he and Leif had arrived at Dr. Kerho's home. Brittany and the children were there and very surprised to see him. Debbie was also a guest and frantic to learn what the emergency was. Travis was at the academy himself for another day until he was to fly out to Stanford for his recruiting visit.

It didn't surprise Kami that his friend was all but in a rage at this new incident and ready to destroy any threat to the academy. This was a little bigger than Oshi was prepared for, tackling the United States Air Force. Leif was given the assignment to call Bode, talk in Icelandic, and allow the Kings' boys to decipher on their computer what was said. Even then Leif talked in circles and gave clues that would eventually make sense to them—GET READY! WE'RE LEAVING FOR CHINA TOMORROW MORNING! SIX O'CLOCK P.M. CHARLES THE TWELFTH TIME!

****

Niqui, in his haste, had forgotten to shut off his cell phone; his mind had focused on memorizing facts about this Artie Stenson and the Washington Post. It did not help to have a bawling woman on their hands in disbelief that her man would allow this to happen to her. Another twenty-dollar cab fare and the three arrived at the Washington Post where they hustled to the front desk and were directed to another staff member who pointed them toward a roomful of desks.

"I'm looking for a Mr. Stenson," Niqui asked the first reporter he saw.

"Kinda late to get Artie, kid," the man said.

"It's an emergency. It has to do with the clones you've been writing about in the paper and how other countries think our government is covering it up," Niqui explained.

"Really? I suppose you're one of the clones."

"No sir, but I know of this academy and….well, I need to talk to Mr. Stenson."

"You can discuss anything with me. Artie and I are good friends. Actually, Artie writes for his own community and isn't involved in international incidents."

"I'm instructed to only talk to Mr. Stenson," Niqui insisted.

427

"Okay, kid. You got your work cut out for ya, though. Artie sometimes goes to this gay bar in Greenwich Village called the Modest Moose. It's on Eleventh Street. Let me page him, though he doesn't always answer his pager."

Niqui sat with his brother and Lauri while they waited. Two men in black suits were at the front desk, acting way too suspicious and asking questions of the receptionist. Niqui grabbed Jay's arm and the three darted off to the rear of the office.

The two federal agents saw their three targets take off to the back of the building, so they didn't wait for permission and chased right after these suspects. The elevator had a brightly lit arrow pointing upward. As one agent took the other elevator and headed up, the second agent dashed for the stairs in case it was just a trick and these kids had headed to the ground floor.

When this agent bounded for the stairs he flew by Niqui, hiding behind a wall. The boy made a dash for the door with Lauri and Jay in close pursuit. Five steps up in his pursuit the agent saw the motion of bodies behind him. Jay had already shut the stairway door, but waited and pushed it open again, just as the agent reached for it. The man stumbled backward, lost his balance and tumbled down the stairs going to the lower floors.

There was no time to waste when the three fugitives ran to the exit and out into the street. Jay stepped almost in front of a moving cab, with his arms waving, but it served its purpose when the driver slammed on his brakes. Fortunately it was empty, but for a pissed off cab driver. Before he could bitch, the three were in the back seat of the taxi.

"Sorry," Jay said and received a stare that could kill.

"You must be in one fuckin' hurry, kid! Where to?"

"Greenwich Village, sir," Niqui spoke. They drove for fifteen minutes before Niqui asked, "Do you know where a bar is that's called the Modest Mouse?"

"You mean, Modest Moose. This might be a real shocker to ya, kid, but that's a gay bar."

"That's okay. We're high school kids who are doing a report on gay living. Thought we'd check this place out."

"School's out. What kind of high school do you go to?" the driver asked.

"Summer school," Niqui quickly answered.

The cab pulled to a stop in front of a brightly lit sign that pointed to an alley. "Don't even think you'll get by the front door."

Lauri paid the driver and Niqui felt in command as they stepped out. "You and Jay go to that bookstore across the street. All three of us can't possibly get in."

"I should go with you, Niqui. You might get hurt," Jay said in respect for his older brother.

Niqui reached into Jay's shorts and held his penis. "Look, bud, I promise I'll be okay. Keep an eye on Lauri; she needs someone by her side."

"Done deal," Jay said and held onto his brother's as their bond of brotherhood. The connection caught a few curious eyes, but the boys didn't care.

"That's pretty kinky," Lauri said as Niqui breezed down the alley to the bar's entrance.

"Not as kinky as the things you do with my dad," Jay said with sternness.

Lauri wanted to argue that she should have gone with Niqui, but then wondered how successful she would be in a man's gay bar.

The first person Niqui confronted was a bald headed bouncer with leather from his neck to his ankles; a spike collar was, by itself, terrifying.

"You eighteen?" the man asked this black-headed Spaniard.

Niqui had once watched one of his mother's favorite groups on DVD-- the Village People. Though he wouldn't claim to be a fan of this old group, this guy in front of him resembled all of them. Niqui wasn't about to lie because even the metal snaps on the guy's leather jacket scared him.

"No sir."

"You eighteen?"

"No sir"

This bouncer knew an attractive teenager and a welcome addition when he saw one. He wrote the number eighteen in large letters on a piece of paper. "What's this say?"

"I'm eighteen," Niqui answered.

"Good. You said it, not me. That'll be thirty dollars."

"But I only want to find someone. Why thirty dollars?"

"Hanging charge," the bouncer said.

"Don't you mean cover charge?"

"No, tonight it's how you're hung. Five dollars for every inch comes off the charge. Show me your meat and maybe we can make a deal."

Niqui unzipped his shorts and out sprang his penis. The man just stared, then measured. "Shit, kid! You got seven inches soft. I owe you five."

Sure enough the bouncer handed Niqui five dollars and pointed to a red line ten feet away. "Ya got to walk to the line with your cock out; shows everybody why I had to pay you five."

Niqui shrugged his shoulders and started to move forward, only to have a siren and bright red lights spinning behind him. The bar was packed like a scene from Star Wars, and Niqui realized the man behind him had hit a switch to notify everyone in the bar to look at this long loop sticking out from his shorts. It sure livened the place up.

One man dashed over, his hands springing up as if he was practicing a piano that wasn't there. "Oh, my Lordy, are you hurt, precious one?"

"Why should I be hurt?" Niqui asked.

"We're not used to having angels drop from heaven," the man said and had his friends frolic in laughter.

Niqui smiled and tucked his penis back in, then asked if the man knew an Artie Stenson.

"Oh, Arlene! Arlene, are you out there somewhere?" the man gleefully requested.

A creature with a purple wig jumped high from the dance floor and waved his hand. "That's her!" the man pointed out.

"Her?" Niqui asked but moved through the couples dancing with their arms wrapped tightly around each other in this slow number from a

430

juke box. The boy felt two grabs to his butt and one pinch.

When he was near the person with the purple wig and yellow dress he wasn't sure what to do, but the reporter was telling a joke, so he listened.

"And so the cannibal had his mother over for dinner and the mother said, 'Marco, I've noticed the many boyfriends you hang around with and I don't approve of them.' So her son says, 'Then try the vegetables, Mother.'"

The group busted out in laughter and even Niqui thought it was funny. "Artie Stenson?"

Artie eyed this bombshell in front of him and raped the boy with two hands that swept down and over each curve. "Well, what do we have here? If you're eighteen I'll kiss your ass. On second thought I'd like to kiss your ass no matter how old you are. What can I do for you, sweet thing?"

"I'm one of Dr. Kerho's students. Not really, yet, but maybe soon. He told me to contact you."

"Toy Kerho?! The one and only Toy of my life?! Lordy be, It's been ages since I've seen that man, actually, held him in my arms once—held everything actually, but that's a story best left to when I can hold you."

"Toy?"

"Oops. I gave away a boyhood secret, I think.   Listen, Honey, what is my boy up to?"

"It's about that cloning thing in the paper.   Dr. Kerho wants you to come with me.   You're the only one he trusts."

Artie changed from the flighty transvestite to a serious investigative reporter in one second. "Should have figured that one out," Artie said as he escorted Niqui away from his friends and outside the bar. The ring of the cell phone surprised both of them.

"Oh shit!   I forgot to turn it off," Niqui said and answered it, but he didn't dare tell Dr. Kerho.

"Yes sir, he's right here," Niqui said and handed the phone to Artie.

The two men went through their preliminary nostalgia before Toy offered his longtime friend the story of the century if they could get out of Washington in one piece. Artie was well aware of how the feds worked, knew that this cell phone had been left on, then suggested to the doctor to call him on his own cell phone in five minutes. Artie hung up and tossed the phone inside the bar.

"Let's get out of here!" Artie insisted and the two ran across the street.

For the third time Niqui tried to spot Jay and Lauri; this time they were browsing through a clothing store. The screeching of tires caused their heads to turn and see a string of police cars and black suited men dash toward the Modest Moose. It was a raid the gay community wouldn't soon forget. Artie had a BMW parked down the street and hurried everyone to his car. They were soon driving out away from Greenwich Village and toward Virginia.

Artie picked up his own phone and began discussing strategy with Dr. Kerho, 25,000 feet in the air and approaching Virginia as they spoke. Artie knew how quickly government agents put two and two together. He also figured they had a few minutes until his car was brought up and put on a stolen car list to draw the attention of any patrol vehicle in a hundred square miles.

His cell phone would be the next thing they'd want to track. Artie used a few things only he and Toy knew about from years earlier to try to make sense of a location Artie had in mind. If they were going to escape the long reach of the federal government it had to be done quickly and evasively.

As they drove away from the main metropolis Artie tossed off his wig and put it on Jay's head, sitting next to him. The boy laughed and wasn't shocked being next to a gay guy; after all, most of his friends now were gay, including his brother. The man had a sense of humor which kept the car's occupants from worrying about this pursuit to do them harm.

It's when they headed out in the country that everyone wondered what was happening. Artie parked his vehicle in a grove of trees and had his passengers take only what they couldn't live without. They ran toward an eight-foot fence with barbwire across its top. There were a few colored lights that lit up a field, when a small plane taxied across it a quarter mile away.

To scale this fence was a major obstacle in their way, but Jay ran down a few yards and found that going under the fence would be a whole lot easier. There was enough space that, with someone lifting the fence slightly up, they could all crawl under.

Artie decided on one more call and confirmed that everything was in place. "Have the jet blink its lights on approach," Artie suggested. Ten more minutes went past before the whine of a jet was heard in the distance. Artie was the first to stand up and order everyone to run with him.

At the far end of the tarmac the jet settled down. In Niqui's opinion it didn't appear that the jet was going to stop in time without running right into a forest of trees at the end of the runway. Twenty yards from the first tree the Gulfstream settled down and began to turn. Its door swung open and steps protruded downward. Four bodies sprinted right up to the plane and hurried onboard.

Without hesitation the plane was secured and began to roll down the runway before gaining speed and lifting into the air.

"The tower is screaming at us," the pilot told Dr. Kerho from the cockpit. "I apologized for the inconvenience, but they saw our four guests get on the plane."

"So we made a mistake and landed at the wrong airport. I'm sure it happens all the time," Toy said and they all chuckled with the relief that their escape was secure. "I'm sure the FAA will figure it out eventually with the help of the White House."

Chapter Thirty-Nine

Senator Franco was keeping abreast of the evening's cloak and dagger dragnet of the city in pursuit of his two sons and the disappearance of his girlfriend.   When he'd returned to the apartment he was livid with the discovery that everyone's suitcase was gone, with no message of their whereabouts.

After notifying the agents Robert soon learned that tracking these three would be extremely arduous work.   Three different times in the last hour this trio had been located, only to have them slip through the agents' hands.   It was a SNAFU that even the CIA couldn't explain. These kids were worse than Soviet agents escaping a net of surveillance.

The senator didn't appear too concerned, except for the fact that his girlfriend wasn't there to give him pleasure for a few more minutes until her demise would less complicate his life.   He could still plan his raid on the academy, even if his sons had overheard any conversation and told the director of the impending raid.   It was the fact that his sons were getting away with this that really pissed him off, and then having a bug in his briefcase at the White House, no less, was a mark on his record that he didn't need.   The president had promised him it would be overlooked.

It was when the FAA filed a report that a Gulfstream had landed and taken off without permission at an airfield across the Potomac that confirmed the senator's worse fears.   The Kings Academy had somehow rescued his sons.   The tower had reported that four people had dashed across the tarmac and into the plane; this confused the senator on who the fourth person might have been.

His last resort for information came from the president's press secretary.   The man appeared way too calm when he mentioned that the senator's sons had requested and no doubt found a reporter from the Washington Post.   What the FBI found amusing, the reporter usually just did articles for the gay community—movie and theatre reviews and critiques. Robert was still panicky because the guy was a reporter; yet, the press secretary said things were handled and they didn't foresee a backlash. Robert breathed a sigh of relief; he knew nothing was worse than a reporter with a hot story that could bring down an administration quicker than, well, a live boy in bed.

****

By four in the morning Oshi had the academy bus readied with forty-three children and twelve staff, including his own family and Debbie Franco's. He had notified the teachers and security personnel to vacate the premises until further notice. His last responsibility was to notify the Chinese Consulate and secure permission to arrive in China a few weeks early. The Chinese seemed pleased that Kings Academy would be extending their stay in China.

When Oshi received word from Dr. Kerho they boarded the bus and left the gates of the academy grounds without the knowledge that, had they not left, their bodies would have been evaporated at eight o'clock that very morning.

Another individual had also noticed this departure. Sveinn had not given up on getting more information about the academy and unifying with Leif. Ten thousand dollars had been hardly enough incentive to give up his dreams. In fact he'd left Dallas the previous night in an attempt to sneak back on campus by whatever means and get into the dormitory at any cost. He'd managed to reach the hillside when he noticed a bus being filled with students and staff. He ran back down the hill and returned to his car in preparation to follow the bus to its destination. This would be much easier than the plan he had already put into place.

A few miles outside the academy, as the bus went zipping by, Oshi stepped out from the edge of this forest and planted a track of jagged metal, which was designed to give car tires a bad day. It was only a few seconds later that a vehicle was seen, obviously following this bus. The car's tires burst upon impact, with the tract causing the vehicle to stop fifty yards down the road.

Oshi stepped onto the road and, through the open window, knocked

435

the first agent unconscious with one single punch to the head. He reached through to the other side just as the FBI man was grabbing for his gun. Oshi dragged this agent over across his partner, through the window and out onto the pavement. A rather fast pulverizing incapacitated this driver for a period of time that the agent would be no use to the department for a long time.

A quarter-mile sprint and the driver of the bus had waited alongside the road for Oshi's return. His smile to his top security man confirmed the success.

What Oshi didn't notice was the second car without headlights that watched the federal car have its difficulties. Sveinn simply drove around the tract of metal and continued following the bus.

Oshi had to assume those agents phoned ahead that an academy bus had departed the grounds, though they could only speculate on its destination. As the bus approached the Dallas-Ft. Worth International Airport there was already a plan that had to be quick and efficient. Around a few hangers were kept dozens upon dozens of private planes. Kami was already waiting, having stripped two large gates of their locks to allow entry for the bus onto the tarmac.

Timing was critical and the 747 had clear passage to enter the runway for takeoff.

All the students, staff, and family members scrambled off the bus and quickly moved to board the aircraft. With the slight delay a private plane managed to get in front of this 747 to be the first to takeoff. So far, so good, as the Kings' plane moved behind the small Cessna and waited its turn. The smaller craft took off, only to have a car with flashing red lights on its hood drive right up in the middle of the runway.

"The feds know," Dr. Kerho said as he peered out through the cockpit window over the heads of his pilot and co-pilot. "This guy was sent to stall us until they got their backup here."

Toy told his pilot to ignore the tower's orders to exit the runway.

"I'll handle it," Oshi said and began to traipse back to where he could jump from the plane.

"Hold it, Oshi!" Kami yelled. Oshi ran back and saw that another unmarked car had pulled up alongside the police vehicle. The man who exited looked quite familiar.

Sveinn didn't know what was happening. He made his own assumptions and realized that the Kings Academy had been found out and the U.S. government had more than dishonest intentions in stopping the students' escape. When he saw that this police vehicle was blocking the plane's path he went up to the cop and ordered him to move off the runway.

"Let me see some ID," the cop said. "I have orders to stop this plane from leaving."

Sveinn saw the holstered revolver and wrestled it from the policeman. The airport security man was overweight and surprised at this aggression. His weapon was taken away with relative ease. Sveinn then ordered him out of the vehicle, jumped in and drove the car into a grassy area; whereupon, he tossed the gun as far as he could throw it.

Oshi had made the decision to open the plane's door to allow an opportunity for Mr. Eldjarn to escape to. When the emergency chute was deployed Sveinn began to run toward the plane. Unbeknownst to him the officer retrieved his weapon and yelled, "Stop!" just as Sveinn reached the chute.

Sveinn began climbing the chute until the crack of a revolver was simultaneous with the bullet he felt in his lower back. Weak and bleeding profusely Sveinn kept trying to crawl up the slippery surface to reach the hands of Oshi and Leif in wait. He glanced back and saw the officer approaching the base of the chute, then slid back down and landed right on top of the cop, tumbling both of them to the tarmac. This time an accidental shot rang out, which would be the last thing Sveinn would ever hear.

This officer had never expected to hit anyone from 50 yards away. The shot was meant more to scare than to do harm. Seeing the bloodied man on the chute was enough shock to just let the guy escape. His gun discharged into the man's chest with the collision.

In those few seconds Sveinn sprawled on the tarmac with the realization that his death was imminent and the only love of his life, a love he had disgraced with his own greed and possessiveness, was lost forever.   His last words were heard only by a thirty-year airport officer who had never hurt anyone in his entire career; "I love you, Leif."

The plane began to roll forward, the emergency chute was cut loose by Oshi and the door closed.   With the runway clear the jet soon gained speed and lifted upward into the sky, its escape helped by a man they never considered as one whose conscience would find the courage to do something so heroic with a love he needed to be expressed, even in death.

Chapter Forty

A C-130 took off from Shepard Air Force Base about the same time the Kings' 747 departed the Dallas-Ft.Worth Airport.   Its crew of two pilots, a navigator, an engineer and a load master knew of their cargo, a fuel-air bomb strapped in the cargo hold of their craft.
Their mission was to drop this bomb in a valley several miles northwest of Dallas near the Oklahoma border.

The old war plane had been around for sixty years in such duties from cargo carrier to reconnaissance.   This durable aircraft was so reliable, airlift C-130 crews found many uses since its inception that were above and beyond the call of duty of a typical tactical airlift mission.

The deployment of this bomb would be delivered in the area of container delivery known as the "TALC age."   The release would be assisted by a 15-foot extraction parachute that would cut the cotton and nylon gate that held the container in the airplane until the final moment.

When the crew approached the drop zone, Colonel Winters climbed to 500 feet, slowing to 150 mph for the drop.   Up to this point the terrain had been flat and desolate, with an occasional ranch house or arid dwelling that could survive the brisk winds across the Texas plain. Cotton fields were scattered in the fields as they approached the valley.

The aircraft's entire weapon system was rigged on a modular airdrop pallet, the bomb mounted on a wooden cradle secured to this pallet.   One minute prior to drop, the extraction parachute was released into the slipstream, where it blossomed fully open while the pallet was restrained by the right-hand locks on the 463L dual-rail cargo-handling system.

As soon as the bomb would be released of all its restraints, the weapon would fall to earth, stabilized by a small triangular parachute. A tree-penetrator fuse was designed to detonate the bomb before it burrowed into the ground. In the event that fuse failed, a delayed-action fuse activated by a timer at the rear of the bomb would hopefully detonate the weapon. The bomb was designed to flatten, if not vaporize, every structure and living creature in a half-mile square radius.

Colonel Winters had his navigator check the coordinates as he eyed the valley just ahead, a green, lush carpet with beautiful glass structures poking upward like ice crystals. A lake sparkled in the morning sun dotted with various water craft on its shore. The colonel had to blink when he saw dolphins come to the surface and jump through the air. Sweeping into this bowl-like creation was a sight few could believe when seeing it from above.

"We drop on my count," Colonel Winters announced and started down, "Five, four, three......."

Both he and his co-pilot leaned forward in their awe of the beauty below them. So amazed with this idyllic valley the colonel's voice literally froze from his count.

"I'm never seen such a place as awesome as this, Tom. This has to be a mistake," Colonel Winters announced to his co-pilot. The pilot's command to drop the bomb was temporarily suspended to the bewilderment of his crew, as the plane slightly gained altitude to re-circle this valley.

"Captain Haworth, double-check with command on those coordinates. There has to be some kind of error," the colonel ordered.

"The coordinates are correct, sir," the navigator reported.

As the second pass commenced every eye on that C-130 was glued to the serenity below them. Even if the colonel had wanted to begin his countdown, he couldn't.

"Tom, there were things that I had to do in Iraq that weighed on my conscience for years, but I excused those and forgave myself because I justified it as an act of war. This is too idiocy to fathom why anyone would want to destroy this place."

"You have your orders, Colonel. Whatever you decide, but our careers are at stake," his co-pilot responded.

On the third pass the orders to abort came through from base. Colonel Winters breathed a heavy sign of relief. "I believe this will be my last mission, Tom. Twenty-four years and my disillusionment with our government has bewildered me for the last time."

As a direct result of a late report verifying that a 747 with sixty-five people aboard from the Kings Academy had taken off and was flying due west, the president had given his decision to save the academy a few seconds of thought. He then determined that such a complex could be used for some governmental entity as a vacation resort, if not another federal prison they badly needed.

****

The 747 was barely thirty minutes from Dallas when Dr. Kerho was called to the cockpit. The doctor had been busy with Oshi, comforting an upset teenager who had just witnessed someone very special in his life gunned down in cold blood. If there was one thing that the academy had tried its best to keep to a minimum was the visual aspects of violence, whether in video games or in the entertainment of movies. To see a murder from their windows had put the passengers almost in a state of shock.

"Doctor Kerho, we have fighters on both sides of our wings," Captain Martin said as Toy entered the cabin.

"Has there been communication?" Toy asked.

"Yes sir. The pilot demands that we turn back immediately or he's been instructed to shoot us down."

Toy was too bewildered to find the words that expressed this incredulous order to shoot their plane down. The pilot turned on an open mike and allowed the doctor to hear the fighter pilot's words because there hadn't been a response from the 747.

"Four-five-zero, do we have a reply on this request?" the voice was from one of the F-22 pilots.

Toy was staring out at this menacing fighter jet, which was threatening this plane and its passengers with their very lives. The doctor relayed what to tell this pilot to give him time to think.

"Roger, this is four-five-zero. We have a flight plan to Los Angeles for refuel. We are a commercial flight from Dallas."

"Roger, four-five-zero, I show no flight plan. Please relay type of aircraft again, departure base, destination, time in flight, hours of fuel, and persons on board, please."

Toy took control of the conversation. "Whom am I talking with? This is Dr. Kerho, director of the Kings Academy."

"Doctor, this is Major Davis, United States Air Force. It is requested that your aircraft turn around immediately."

"We can't do that, Major. Who gave you the authority to harass us?"

"Sir, I don't have the authority to disclose that. If you do not turn your plane around now I have orders to down your aircraft."

"Major, I know you follow orders, but before you fire on this airplane I want you to know we have forty-five boys and girls on this plane, including Senator Franco's children and wife. I have no reason to lie to you, just look at the faces of the kids staring at you through the windows. We also have a reporter with the Washington Post, who is in direct contact with his newspaper as you are threatening us."

"I have been told, Doctor, that you have terrorists on your plane and a possible nuclear weapon that they intend to destroy downtown Los Angeles. You are possibly unable to discuss this because of your hostage status."

"Oh pleeeaassseeee!" Dr. Kerho said in disgust. "I assure you we have no terrorists on board this aircraft! The truth is, your president wants us all dead. I have thirty boys on board this aircraft who have been cloned from the world's finest leaders in science, art, composing, writing, military, and athletes from the past two thousand years. That's right, it's true what you've been reading. On this plane the boys have come from Peter the Great, Alexander the Great, Isaac Newton, Frederick the Great, Bill Tilden, Lawrence of Arabia and many others— do I need to continue? For whatever reason our government is afraid of these boys, so much so, they're asking you to commit genocide!"

There were many seconds of silence between the F-22 and the cockpit of the 747. The jet fighter appeared to gain altitude, only to come that much closer to this target and peer at the scared and curious faces of so many future leaders. To many of these youth, this was excitement, an escort that they weren't sure where to or why, but they weren't aware of their precarious position. What they didn't realize was the magnitude of the danger; their very lives depended on the common sense and diehard obedience of one Major Robert Davis, the flight commander over these two deadly aircraft.

****

SITUATION ROOM
White House

There were numerous faces in front of a NORAD screen that showed the 747 flying into New Mexico air space and two F-22s on its wings. The president was concerned on why the plane was still in the air.

"What appears to be the problem, General?"

General Shavers was General of the Air Force and was equally concerned with a message that Shepard AFB had just received from the command pilot. Senator Franco's children and wife were, apparently, aboard the plane.

"Mr. President, we have another problem. It appears, as your report says there, that the hostages include the senator's family. Are you sure of the intelligence that this plane is indeed controlled by terrorists?"

One usually didn't challenge confidential sources and the president could have flown in a rage with someone questioning his directives. "The CIA says it's highly reliable, so we have to take this one very seriously," the president imparted and liked his wording, not too disclosing to involve his decisions, but enough blame to include everyone but himself. "I believe we'll have to contact Senator Franco and make this his call."

A phone was lifted by the press secretary, who wasn't usually seen in such a secure area, but the president's confidant rarely left the president's side in the past few days. While the man had Senator Franco on the phone, he ran down the options, which were few. Robert asked to speak to the president.

"Yes, Robert, we've been told that your four children and the

443

missus are aboard the aircraft.   May I also add that your little problem is onboard, as well.   It's your call.   The way I see it, if we allow the terrorists to crash this plane into Los Angeles, your family are goners anyway.   It doesn't appear that anyone on the plane is able to solve the problem for us.   Naturally I empathize with your grief, Robert."

A few moments for the senator to ponder and respond, then the phone was handed back to the press secretary.

"Bring the plane down, General," the president ordered.

"We can't at the present time, Mr. President.   The pilot has chosen to circle Albuquerque.   He's calling our bluff that, to bring this 747 down, it will cause a lot of destruction and death in the city."

This news update enraged the president.   The whole situation would have been finished if the academy had not decided to leave on this particular morning.   There had to be a leak somewhere, but he couldn't pin the blame on anyone for the time being.   This was way too coincidental, though.

****

It was in fact Kami who suggested to the pilot to begin circling the City of Albuquerque.   Toy agreed and realized it would buy them some time to negotiate or talk some logic into two F-22 pilots who were ordered to shoot them down.

"Major Davis, this is Dr. Kerho again.   I apologize that this is difficult for both of us.   It's a mistake to think a Commander-in-Chief is any smarter than the rest of us.   If this is what you want on your conscience the rest of your life, make the call.   Your orders don't make a notion of right and wrong.   You're to believe what's good for the country is what's good for you, yet we both know that there's a God looking down at us with a notebook.   Your leaders might not believe this but every man, woman and child in this plane places their trust in God, and the children of God—which you are one."

"Doc, is your mike open to everyone in the cockpit?"   Major Davis asked.

"Yes sir, it is.   Would you like a bit of privacy?"

Toy received confirmation and made the communication just

between the pilot of the F-22 and himself.

"Doc, if you are being controlled by terrorists aboard that plane, you have five seconds to call me every name in the book and tell me to do what I have to do. If not I want five seconds of silence."

There was a longer delay as the major received not a sound for ten seconds. He took an incoming message from his base that, even with the senator's family aboard, he was ordered to attack, but not while the plane was circling the city.

As the planes continued in this tight circle with the 747, the major had to ask, "You wouldn't be the Kerho youngster back eighteen years ago that made me gay when I was in seventh grade, would you?"

Toy chuckled, which wasn't actually in accord with the mood of everyone else in that cockpit. "Yes, but I allowed the formula to be reversed, don't forget, to everyone's original orientation. Are you still holding a grudge?"

"Not really. I got to explore both worlds. Is it true you have Abraham Lincoln onboard?"

"Not the original, but a close copy. A fine volleyball and basketball player, if I don't say so myself." Toy began to wonder if this was idle chit chat while their 747 ran out of fuel in a few hours.

There was a blink of light from the F-22 cockpit, then several more flashes.

"What's the guy doing?" Kami asked.

"Morse code," their pilot answered. "It's been too many years for me. No one uses the stuff anymore."

There was a knock at the cabin door and Kami let Johnny Turing enter. "Hey, guys, I've been reading the Morse code. Do you want to know what he's telling us?"

The men listened as Johnny informed them that the pilot wished to communicate by this method to keep all communication just between the two aircraft. "He wants us to turn thirty degrees southwest."

"I'm sure he does," the co-pilot rebuked. "Enough space to shoot us down."

It wasn't what Toy wanted to hear. The doctor asked the pilot for some type of light and he quickly had a flashlight in his hand. He gave it to Johnny. "Ask that pilot if this is some kind of directive given to him by his headquarters."

Johnny did as requested and received a negative, then another message. "He wants us to immediately drop our altitude sharply until we're at three hundred feet. Expect Gatlin fire, but not at our aircraft," Johnny said.

"What do we do, gentlemen?" the pilot asked to Toy and Kami.

"It's our only option, other than land at Albuquerque, which will be the end of us by some sort of accident as we're returned to Texas," Kami gave his opinion.

"I told the pilot who I was and he's heard of my progenitor," Johnny said and added, "He thinks Alan Turing was THE hero of World War II and saved millions of lives by the man's genius with intelligence. He even knows that Alan Turing was the Bletchley Park code breaker. This guy is really cool."

"I trust this pilot. Let's begin our turn," Toy announced and hoped this decision wasn't his last.

Johnny kept up his communication. "I asked him if he was a Christian."

"What did he say?" Kami asked.

"He said he was and he'd like to meet me."

"Okay, we can arrange that," Dr. Kerho said as the 747 began losing altitude and was still descending a few miles outside of the city. It had lost ten thousand feet when Major Davis's F-22 fired a burst of canon straight ahead into nothingness. As the F-22 fighters peeled off, the 747 was at a thousand feet and entering Mexican airspace. The last flashes of light, Johnny said, were words of good luck.

Captain Martin, the pilot of the 747, was absorbing this dereliction of duty by the two pilots. He knew the officers would pay dearly for disobeying orders. He wondered who had more to worry about now that he was flying over Mexico without flight plans and no way to land. The best thing, it wasn't like Mexico even knew they had an American commercial plane over their air space; the plane was flying too low and in a matter of minutes they'd be over the Baja Peninsula and heading toward the Pacific Ocean, with fuel a potential problem.

"What exactly just happened?" Kami asked with relief.

Captain Martin had his intensity at a maximum, decreasing the plane's altitude to only 500 feet above the terrain but prepared to gain height with hills and mountains ahead of him.

"I think their conscience kicked in. Dare say they'll tell the higher echelon that we made a run for it and they didn't receive the last transmission that gave them the go-ahead. Either way their asses are in a sling."

"So where does that leave us?" Toy asked.

"See there? That's the Gulf of California and over that strip of land is the Pacific Ocean. If we land in Baja the Mexicans won't take too kindly to our explanation. Their relationship with the United States has lately been one of cooperation, so that leaves us with that bathtub ahead. It's your move, Doc."

****

General Nicholas Ramsdell, Chairman of the Joint Chiefs of Staff, stood at ramrod attention as the President of the United States came back from his restroom break. The White House Situation Room was tense, to put it mildly. The men had watched the 747 on radar deflect its course to the southwest, only to lose altitude and then vanish from the screen.

"Do we have a crash site, yet?" the president asked.

General Ramsdell spoke up before the Air Force general tried to remedy the situation with some explanation. "The command pilot said that one of the jets had engine troubles, the other had a missile jam, so he used his Gatlin to hit the aircraft. He reported that the aircraft was smoking and heading for the ground. We have planes, as we speak, searching the area."

"Very unfortunate," the president said with a most disturbed grimace on his face. "They could have avoided all this if they'd just landed. A lot of lives wasted because these terrorists think we won't shoot down our own aircraft out of respect for a few dozen lives, compared to the thousands, no, hundreds of thousands lives that would have been taken if we'd allowed that plane to continue to Los Angeles."

"I'll inform Senator Franco," the press secretary told his boss and left the Situation Room.

Before the president departed he made sure to be kept up to date when they found the wreckage. It was important that relatives were notified and proper burial arrangements were made.

Chapter Forty-One

Dr. Kerho was contemplating the possibilities that lay before him. He allowed the pilot to fly straight over the Pacific Ocean and head for Hawaii. If no other plan could be made, at least they could make Hawaii, so Toy thought. Captain Martin advised that it would be a close call, flying just above the ocean and losing far more fuel avoiding radar than flying at thirty thousand feet.

"This is a long shot but it's the only thing I can think of," Toy said and grabbed the satellite phone, checked his wallet and dialed.

"Prince William, please..........Will! This is Toy Kerho from the land across the pond..............Yes, we were preparing for your visit in a few weeks but plans have been derailed, I'm afraid. We're in a bit of a predicament, old boy."

Prince William and Dr. Kerho went back to when they were both fifteen-years of age and met totally by accident at a San Francisco Hotel, where the prince was vacationing on his travels through America. When a friendship was sparked, the prince had taken Toy and another boy on a rafting expedition with him and his entourage.

This was the time when the federal government was chasing this Texas youngster with his golden secret that kept 40 million children biologically gay. It would be the British that would eventually rescue the boys, but not without a wound to the prince that almost killed their future king but for the lifesaving heroics of Toy. Prince William never forgot and the two had visited each other several times to each other's country.

There was a laugh, not to be disrespectful, but a consensus that Toy was one to always get himself in some type of wreck.

"Toy, you have this thing for getting your government very upset with you. Are you telling me you're behind all this cloning hysteria we're reading about in the paper? I should have known it was you."

"I was going to confess this to you during your visit, but my president has pushed my confession to you now. We're skimming the ocean and fuel appears to be a problem in due time. We can make Hawaii and points beyond, just not far enough to reach China. Have any suggestions, my dear friend?"

"Toy, what if I was able to send you a tanker? England is currently engaged in naval exercises with Australia, off the coast of New Guinea. I doubt if you have refueling capabilities in the air, but can you make some type of landing strip where we can refuel you. I'll speak with the President of Australia and get a couple of jets off the carrier to escort your plane to the China coast. From there on in, you're on your own."

Dr. Kerho discussed this probability with Captain Martin. The captain was scanning a map unfolded in front of him.

"It's possible if we can make this atoll around the thirtieth parallel."

Toy eyed the map and put the phone back to his ear. "Will, how does Midway Island sound?"

****

All the passengers on this 747 hadn't relaxed since they took off, much less, told about what was transpiring with the two jets, nor why they did circles around the City of Albuquerque. It was more confusing when they flew over the desert of New Mexico, only to fly at an altitude where they would watch humans cowering and diving when this commercial jet went zooming by.

The boys surely thought that Johnny would come back and give them an update, but he was gone for almost thirty minutes before he showed up again and gave a thumbs up.

"That jet pilot over there, he wants to meet me," Johnny said with his touch of humor and the truth.

They knew something was desperate when they began to see nothing but the blue Pacific Ocean from a position where, if they could stick their hands out the window, they might have been able to feel the splash of the waves.

Dr. Kerho finally returned from the cabin and gave a short speech. "It's been quite a morning, folks."

His fatigue of being up for 30 hours straight was showing. "I'm sorry for keeping you in the dark. I saw no reason to instill fear in any of you, as long as there was hope. We can thank our Good Lord for His blessings of moral judgment from pilots who have to wonder about our leadership in this great country—great because there are men and women who will stand up to the wrongs and the injustices of our leaders.

"We're not out of the water yet, and I don't wish to make a pun. This is not the best way to fly a 747, but we're avoiding radar and our captain thinks it might be best to grab a few things you don't need; we'll dislodge a few seats and drop these items to the ocean surface. It's a temporary maneuver to make it look like we've crashed. Let's get busy."

The assignment gave the kids something to do and a part in this escape ploy from whomever desired them harm. The three stewards handled the collection of items and soon deposited the debris on the ocean's surface, just out of sight from Baja. There was no doubt that tourists and other interested spectators would report a commercial airliner flying barely above tree level and, one would think, prepared to crash.

The engines on the 747 were literally running on fumes when the aircraft settled down on the thousand acre Sand Island, the only atoll that could handle an airplane of this size. Captain Martin was most worried about making the landing strip, then when he was sure they could at least land, he became concerned about a continual problem airplanes have when landing at Midway—albatrosses. It was a miracle that one or more of these creatures didn't end up in his engines.

There was an emergency reported to this naval station, a mere outpost with only 500 people on this atoll. Within moments of the 747 landing, a British tanker landed and came up right behind the commercial airliner. Neither plane left the tarmac, despite a military jeep coming out to investigate this odd occurrence.

Captain Martin said his aircraft was a private, English airliner, carrying diplomats from Mexico to Japan, but had navigational problems and had requested a tanker as backup. For the lack of anything exciting happening on this atoll, the two military officers were happy for the company.

A half-hour later the 747 was fueled up and followed the British tanker into the air. Dr. Kerho and Dr. Marion agreed that this evasive technique was both nerve racking and too risky to continue. At 35,000 feet everyone relaxed when they saw two Australian F-15s come along side and wiggle their wings. Now all they had to worry about was if China was still willing to welcome them with open arms for being a month early. The doctors highly doubted that China would be aware that this aircraft and everyone aboard were fugitives running from the United States government.

****

"What do you mean a 747 was seen flying over Mexican territory?!" the president screamed at his press secretary.

"There are reports from several tourists in Baja that they saw this plane plunge into the ocean. We have early satellite photos of wreckage in this area, Mr. President."

"Fuck! They better show me some pictures of some bodies, or I'll have a few generals' asses by tomorrow!" The president waved off his secretary and began to worry, just a little.

****

Dr. Kerho tried his best to be upbeat talking with his sister-in-law, Brittany, and Debbie Franco. It was difficult telling a senator's wife that, in spite of communicating that she and the children were aboard, the government still supported the downing of this plane.

Debbie was near tears, but she admitted that their lives were no more important than anyone else aboard that airplane. It would have been sad that her family, because they were a senator's family, might have made them think twice about a decision to murder innocent civilians.

The doctor slumped down next to Artie; the men shook hands and then hugged each other for the years they'd been apart. The Washington Post reporter had guessed what the threatening situation involved, and he was in contact with his paper.

The Post inquiry to the White House received a confirmed report that terrorists had taken over the aircraft at Dallas-Ft. Worth

452

International Airport and tried to ransom the plane's hostages for 500 million dollars or their nuclear device onboard would devastate the City of Los Angeles.

As hard as Artie was yelling that there weren't these terrorists onboard, his paper was cautious in their own assessment of the situation. For the moment Artie only wished to gain more insight and background to his surroundings of an abundance of teenagers. He also wanted to begin his article with information on DNA and how someone clones the dead.

Toy was spent and exhausted, but he felt an explanation was owed his friend. He began with the basics of DNA.

"Artie, life is transferred from one living organism, or species, to another by DNA. If there was no DNA, reproduction would not be possible. In a eukaryote, a single or multi-celled organism containing a nucleus, the DNA is in long threads of chromosomes. The chromosomes are formed by the coded repetition of four basic units known as nucleotides. Each nucleotide molecule has three components, a base, a sugar, and a molecule of phosphoric acid. There are four bases for the nucleotides in DNA: adenine (A), cytosine (C), guanine (G), and thymine (T). The sugar in the four nucleotides is deoxyribose."

"How do you spell those?" Artie asked and Toy had to laugh he was so tired.

Toy went into a little shock culture. "Did I mention that the scientist who did all this almost obtained a piece of the Shroud?"

Artie stopped his typing on his laptop which Niqui had allowed him to use. His jaw was near his breastbone.

"See, the chromosome formula of Jesus of Nazareth, a normal baby boy, had to be 46:XY. It is true that the Shroud contained blood samples, and all three segments of the human genes tested were positives, indicating that the blood of the Man on the Shroud came from a human male. Technically, this scientist had enough material and samples to determine anything you'd like to know about the Man on the Shroud, including, yes, his sexual orientation.

"So, let's say Jesus' DNA clone is sitting with us this minute on the plane. Is He the Son of God? Is that boy there, Peter the Great? The answer is no. I believe in reincarnation, Artie. Does the Bible say anything about reincarnation, you're about to ask? It mentions people who have been reincarnated, and there are many more verses that this theory may be implicated. The truth is, I can duplicate your physical presence; in other words, I can make another Artie Stenson. Will that person be the same Artie I know? Of course not. The person will likely have your features, your creativity, your sexual orientation, your talents, but likely not the same environment, the same friends, the same influences that made Artie decide what was best for him, or what he considered right and wrong, moral and immoral.

"To assuage your mind, Jesus Christ was not cloned, though the possibility exists. That reddish mop sitting beside Johnny is Bobby Newton. Does he look like somebody who just got hit on the top of the head by an apple? Bobby invented our silicon suit the boys and Wendy wear in competition. He's a scientific wizard, which is a direct result of his progenitor's interest. He also does a mean routine in the floor exercise and not a bad second baseman. He's a lovable teenager with his own desires and needs, just like us. And yes, he loves to play bottom."

Artie laughed despite being overwhelmed. "Who's the cute brunette with the sparkle in his eyes?" Artie asked and pointed to a boy who had obviously perked his interest.

"Robin Simoni. His predecessor or forebear, if you will, was quite the artist and sculpture. I believe he did a few paintings on ceilings during his day."

"Are you trying to tell me that boy is Michelangelo?" Artie asked.

"He's a clone of Michelangelo, not the original. If you don't believe me, we'll ask him to draw a portrait of you when we get to China. I happen to know that Dr. Jolson was in search of King David's tomb, what with all the architectural findings the last few years. What an interesting DNA that would have been; I believe the doctor might have discovered that David and Jonathan were two hot lovers for each other."

Toy strayed from the conversation when his cell phone rang—it was his father. A few minutes of listening and Dr. Kerho clicked his phone shut.

"My father. He said the academy has survived. Now that's a relief. We just got all our specimens for Project Sixteen."

"You're not going to leave that comment hanging, are you?" Artie asked.

"It's a project we decided several years ago, a promise we made to the boys that they could procreate when they turned sixteen. The sperm samples were set to go; of course, asking a fifteen-year old to abstain from sex for three days was like asking them to give up Christmas."

Artie roared. "How about the blond there with the blue eyes? Do you think you could put me in the same room as that boy, Toy?" Artie turned to see Toy snoring away.

This Washington Post reporter scooted across the aisle and sat with Kami, who appeared as tired as Toy and barely had his eyes open.

"Hi, Kami. Got a question for you if you can give me a minute." Kami nodded with a smile and saw where Artie had pointed at a brunette boy across and three rows up.

"That's Randi, Artie. Good boy and too intelligent for his own good, like Bobby. Don't let the boy do any experiments on you,"

Kami advised.

"But he's so cute and darling," Artie protested.

"He and Bobby used to experiment on each other until we put a stop to it. They each increased the nerve endings on the heads of their penises, four-fold, similar to a woman's clitoris. Their orgasms had them shaking uncontrollably and they couldn't keep their hands out of their pants. It gave a new meaning to preoccupation with sex. The difficulty being cloned from absolute geniuses is that intellectuals are truly weird people. I say that with tongue in cheek because I love both of them to death. Their lives are canny compared to the originals. Leonardo was an illegitimate son in a society that afforded little opportunity to such an individual. He is a vegetarian who found detestable the idea of becoming a feeder for dead animals. Randi is also an animal rights activist, left-handed, an exceptional painter and scientist."

Artie sat with his jaw downward. "A boy his age can't invent much these days, so we don't have to worry about that."

"There's not a university in the nation who wouldn't go on their knees to acquire a dozen of our boys. Do you realize Artie what these boys have in common? They're all geniuses; their minds absorb like sponges and our teachers have to study just to keep up with their students' demands for knowledge. Bobby knew more when he was eight than his progenitor did in his entire life. I had to kick the boy out of the laboratory because of his interest in the human body and experimentation that was detrimental to the future of sexuality.

"By his own research he learned that the presence of nitric oxide inside the corpora cavernosa caused an increase in the levels of another substance called cyclic guanosine monophosphate. He made his own Viagra and was experimenting on other boys. He's a junior doctor at fifteen, and I'd trust him to do a surgery on a patient with appendicitis."

"I wonder if I could hire that boy to give me some advice," Artie suggested.

"You have no idea how an adolescent mind works, my friend. These boys aren't always angels, let me tell you. About the time that Rom was twelve years old, one of his brothers heard Rom say how he thought women were attractive. Now Rom didn't mean that he was attracted to women, only that their bodies were great looking. Leave it to a few boys to think that their brother was violating their gay space. They had Rom come into the lab, where they had him sit and gave him a drug called Uprima. As they showed their brother nude pictures of women, the boy became nauseated. Bobby believed that Rom was supposed to develop an unconscious link between naked women and nausea. For days my son became sick every time he saw a female, dressed or not. When Oshi found out he was livid. Three boys had their asses in a sling and were excluded from the lab for 90 days. Oshi had to give the boy hypnosis to stop the nausea."

"Now that's funny," Artie replied.

"Those red marks on their butts weren't funny, and Oshi had all them in extra workouts on the mat. To work out with Oshi is a torture I wouldn't impose on my worst enemy. The boys would rather be lectured and restricted from all activities, except their bed in place of putting them at the mercy of our psychologist."

"You might want to point him out so I can avoid his stare," Artie said. "These guys know a lot about sex, don't they?"

"Artie, sex is a talent, like dancing or painting. They became very interested around age ten of their sexuality and the pleasures derived from certain mutual acts. Give a genius a lab to experiment and they always think they can make something great, greater. I've restricted Bobby, Randi, Johnny, Tony, and Michael from the lab for several months at a time. It's like taking away Christmas from these boys. They've increased pleasure to where it's become dangerous, and we can't have boys masturbating in class because they can't control their arousal."

"Maybe they're just acting like boys, Kami," Artie suggested.

"Look, my dear friend, I had hard-ons in class, too, but the tip of my glans wasn't increased with the same sensory nerves as my palms or a woman's clitoris. If these boys had their way they'd be walking hard-ons all the time. They haven't quite grasped the concept that the world doesn't revolve around their penis."

"Since when?" Artie asked and laughed. "So they're all sex geniuses?"

"They're all sex masters, but don't get your hopes up. You have nothing that would impress them that they don't already have with one of their brothers. We have brilliant scientists when they're not thinking with their dicks. Randi, at age seven, mind you, knew from second grade that at noon on the longest day of the year, the day of the summer solstice, the sun would be directly overhead. He positioned a vertical post on the academy's lawn and told his teacher that what he erected would have no shadow at that moment. He then figured that a post three hundred miles north of there would have a measurable shadow. He determined the difference in the angles between the axes of the two posts to be seven degrees. These axes, he reasoned, if extrapolated downward, would meet at the center of a spherical earth.

"That seven degrees happens to be approximately one-fiftieth of a circle. Multiplying the three-hundred mile distance between the two posts by fifty, the boy obtained the circumference of the earth as 40,000 kilometers or 25,000 miles. Finally, dividing the circumference by pie, he calculated the diameter of the earth to be 12,800 kilometers. If Columbus had had Randi with him, the boy could have told him that he hadn't landed in India or Asia, but the islands of the Caribbean, and that the much more expansive Pacific Ocean lay to the west of the continent he had discovered."

"The kid's scary," Artie decided. "Do you think he'd give me the formula so I could experiment on a few of my friends?"

"I'm afraid the boys eat journalists for lunch. Don't get your hopes up."

"I'd just like to eat them before bed," Artie wished with longing. "If you don't mind me asking, who chose who came alive and who didn't?"

"You have such a way of putting it, Mr. Stenson. I can't answer that question because, whatever went through Dr. Jolson's mind, is forever secret. Maybe he thought gay men deserved a better fate. Who's to say that Matthew Shepard in Wyoming deserved a longer life and to live to love another gay person in harmony? Did he do something miraculous for humanity? Probably not. Did he leave a lasting impression on the world that bias, discrimination, hatred, and downright cruelty aren't working to improve our lives? Yes, he did. Did Peter Tchaikovsky deserve a longer life than to die at such a young age due to cholera? Dr. Jolson thought he did. Alex, the boy sleeping on Zach's shoulders, is happier than his forebear. I don't think Dr. Jolson played God, but, for reasons we'll never understand, these are the boys he chose for us to raise."

****

The president's cabinet had spent minutes debating who would walk into the Oval Office and give the president the latest reports. Though the Secretary of Defense had felt blindsided by this whole scenario it was decided it was his turn to step up to the plate. He did his own research and confirmations, then moved forward with what he was sure were the facts.

Secretary of Defense, Paul Richards, placed the paperwork on the president's desk and took a seat. "It's the Mexican authority's response to your request on that crash site," Paul mentioned and watched the president scan the report.

"Several seats, clothing, paperwork, and pushcarts. It says nothing here about any bodies. Is it possible that the plane didn't break up completely?"

"Possible, Mr. President. The seats were definitely from a 747, but, if that was the case, then there should be bodies. I have another report that may shed light on this picture. A Naval officer on Midway Island reported that two British planes landed on Sand Island. The first plane was a commercial liner, which had apparently run out of fuel. A British tanker from Melbourne came in behind and refueled the plane."

"So, what's the problem with that? We have these difficulties every month with one of our aircraft," the president noted.

"Yes sir, but the navy's radar on Hawaii reported that they only tracked one plane in the vicinity and that was the British tanker. Upon the two planes taking off, the radar picked up two planes. A 777 from Los Angeles, en route to Oahu, reported that they saw a large plane only a thousand feet above the ocean."

"Possible oversight," the president considered. "Did you check with the British Ambassador?"

"Yes, sir. He confirmed that the British had two planes in this area."

"There's your answer, Paul. Nice try."

"It doesn't stop there, Mr. President. Guam tracked the commercial aircraft when it picked up two F-15s off an Australian aircraft carrier. The two jets escorted the aircraft within twelve miles of China's coast then peeled off. Since when does a commercial airliner need escorted?"

The Head of State sat back and considered this new information. "If I remember right, didn't that Kerho kid, you remember, the one who caused all that havoc with that project...?"

"The Hyacinthus Project, Mr. President. The reason we have an ex-president in a federal prison," Paul informed.

"Yes, that's the one. The boy saved Prince William's life on that rafting trip. I bet you someone called in a favor."

The president began to chuckle at the thought. "This might be the stroke of luck we've been looking for. We wash our hands of this academy and a potential problem with international relationships. I do hope our Dr. Kerho enjoys his stay in Red China. I have a feeling it's going to be permanent."

****

Artie Stenson enjoyed moving around the cabin and talking to the various boys and girls. They weren't exactly what he was used to conversing with when he interviewed gay teenagers in and around Georgetown. He made sure he asked their names from the start, then humored himself by guessing their forebear.

Artie, for the first time in his life, acted the mature adult, if not a humbled one, when he realized he was in the presence of Alexander the Great, Lawrence of Arabia, Sir Isaac Newton, Lord Byron, Walt Whitman, Charles the Twelfth, and Richard the Lion Hearted. He couldn't take notes fast enough and was too enthused to return to his seat when the seat belt sign came on.

Before someone got yelled at for not obeying the sign, Leif had been the boy to leap up and sit next to Dr. Kerho, while Artie sat between Bobby and Randi, as the boys tried to explain their research on stem cells. Both boys had their notebooks open and were discussing their most recent experiment.

Bobby was defining what all this meant. "We're on the verge of curing paralysis, sir. See, a spinal-cord injury can cause bone fragments, disk material, or ligaments to slice into the spinal-cord tissue, severing the connections between axons and thus cutting off the brain's instructions. Randi and I have experimented on mice and cured paralysis by coaxing stem cells into repairing the decimated axons. The key was getting the stem cells to slip under the body's immune-system radar. We then extracted a ball of stem cells from an embryo, cultivated them in a Petri dish, let them multiply in the millions, and then we injected them directly into the injured spinal cord. The cells secrete hormones that stimulate the formation of new nerve cells. As communication is restored among axons, feeling returns to the body's extremities, then movement."

"Do you have any idea how many thousands of people would kiss you boys right now?" Artie asked.

"We already have enough sex at the academy. I'm just happy to help people," Randi said like a true teenager who didn't understand the magnitude of his brilliance.

"What was it like to have all those nerve endings in your penis?" Artie whispered.

The boys smiled. "Perpetual hard-ons and orgasms that make you go nearly unconscious," Bobby admitted.

"Don't suppose you have the formula with you?" Artie asked.

"Dad would cut our penises off if we did that again," Randi warned. "See Dr. Makawa over there? He doesn't understand why he wants sex every hour. We'll have to reverse the process because his wife can't keep up."

"Leif can," Bobby said with a devious grin.

"Yeah, but he's walking funny now," Randi humored back.

When the 747's wheels squealed on the runway everyone glared out the window and forgot what the subject was. Now in Beijing, China's capital city, Artie put a call in to his office in Washington. One of his cohorts answered.

"Wallace, this is Stenson. Is the chief available?"

"Stenson, the boss has been asking for you all morning. Your ass is grass if you don't get your butt in here now!"

"Tell that homophobic wimp to kiss my ass! Wait! I'll do that myself. Put him on or I'll resign over the phone and sell this story to the highest bidder, and I think you know what I'm talking about."

"That's not how it works around here, you know that Stenson. Redford isn't going to speak to one of his peons," Wallace tried to convey.

"You have sixty seconds, Wallace, or you'll be looking for a new job yourself when they find out you didn't put me through."

There was silence, then a wait while voices could be heard in the background.   The chief editor came on the line and was less than hospitable.

"Stenson, where in hell are you?!"   Redford asked.

"Sir, we've just arrived in the capital of China, I believe.   Did the Post report how the United States Air Force almost shot down a civilian aircraft?"

"Your ass is in hot water, Stenson!   Our rapport with the White House is a perpetual teeter-totter as it is without you running around with terrorists and juvenile delinquents.   You're a reporter for the gay community, nothing else!   No one gave you permission for your cross country tour or to go snooping around in business you have no right sticking your nose into.   I have a mind to fire your ass over the phone."

"Mr. Redford, if I may.   If after what I tell you, you wish to fire me, be my guest.   I'm tired of working for bigoted people who don't appreciate a class act when they see one, and you're not getting this information for my measly forty-five thousand a year.   I can't even afford my uplifting bras and gowns my boyfriends thrive on.   Anyway, sweetheart, I'm sitting here with Abraham Lincoln, Walt Whitman, Herman Melville, Cole Porter, Peter the Great—just to name a few."

"You're stone drunk, Stenson.   You're also wasting my time."

"Hey, bozo, you're the one with a bottle of Jack Daniels in your drawer.   Just ask Senator Franco if England doesn't currently have a DNA sample of James Barrie, and if the United States just sent them another sample to match.   It came from a boy, let me see, yes, he's three rows back and he's necking with another boy.   I should be watching them instead of talking to you.   Where was I?   I want…Wow!   You have no idea how good it is to say that.   I have a story here that will make me the queen of divas, and you can get your usual pat on the ass, though you deserve a good fuck."

"Stenson!   I don't have to take this abuse!"

"Actually you do.   You know I'm onto something and whatever the White House told the press about terrorists, it was a lie.   I'm looking at Mrs. Franco as I'm talking.   She doesn't look exactly terrorized to me, yet our president wanted to do her and her family in.   Sorry about that Bobby and Randi, forget what I said.   Oh, Redford, those words

462

were meant for my new friends; you might know them as Newton and Da Vinci. Did you know as teens they fought over the same boyfriends? Or was that with Michelangelo? Oh wait, you wouldn't, that was five hundred years ago."

"Stop it, Stenson. I know you're having fun with this. I need proof, hard evidence and photographs that what you're telling me is factual. We can't exactly compare pictures of these teenagers to their, whatever you call them."

"Try their original forebear. I'm going to talk with Dr. Kerho and see if I can have DNA samples taken from each boy to compare them with the originals. I think you'll find that several countries have initiated their own investigation and have taken samples from the tombs they suspect were broken into."

"What's it going to cost me, Stenson? I want you to know I didn't blackmail the Post when I was on that Watergate fiasco."

"Yes you did. Your bonuses and book sales made you and your sidekick a mint, and you're still milking it for every dollar you can get. I want two million in an overseas account, and I'll arrange my own book deal."

"You're a crook, Stenson. Why don't I just purchase you your own gay bar? I'll have Barbara Streisand cut the ribbon on your grand opening."

"It's a deal. I'll take two million and the bar. I'm bigger than Streisand, so screw her."

"I see the samples first, then we have a deal."

"What if I bring down the president in the process?"

Redford laughed. "If you can bring down the Teflon Texan, I'll kiss your ass, Stenson."

"I'm polishing the spot as we speak. I prefer you wear peach lipstick so my boyfriends won't get jealous."

Artie hung up and received two wide grins from both sides of him. Now all he had to do was get a few strands of hair from every boy; of course, he was thinking pubic hair.

Chapter Forty-Two

The landing at Beijing International Airport was smooth and a relief to how this adventure began.   There was difficulty right off the bat with Artie and Lauri not having their passports.   When the Chinese considered putting Lauri on the next plane back to the states, she filed for political asylum.

Taken to an interrogation room by herself, and in the midst of Chinese interrogators, who looked at her like she was a spy, Lauri panicked and admitted she had slept with Senator Franco, who was now running for the next President of the United States. Only one Chinese man smiled, the one who began calling someone far more important than himself.

Artie admitted he was a journalist and his passport was coming by express within the next few hours. The rest of the Kings Academy sailed through customs and caught a bus to the China World Trade Center Hotel, approximately 20 miles away from the airport. Off the 2nd Ring Road the mammoth trade center stood as an example for the new prosperity and vision of this world power.

Dr. Kerho was prepared for surprises and found his first frustration when the group was transferred to the Shangri-La Hotel, for the convenience of all—whatever that meant.

As they drove to the hotel, Toy saw quite a change in China's Imperial City. It was always a city in a hurry, but automobiles had largely replaced the swarm of bicycles on the major boulevards and the new freeway flyovers were car-clotted. Only through the last few years did the city restrict 1.3 million cars a day by requiring the drivers to take busses. This was done to reduce the smog problem; yet a dark haze hung like a storm cloud over Beijing. High rises towered over historic courtyard homes and athletic venues could be seen in the distance, a tribute to this metropolis of 13 million people as host of the summer Olympic Games years earlier.

The climate was an oven as the boys and girls scurried into the air-conditioned comfort of the hotel and were soon assigned their rooms. The students had changed on the plane in their stunning blazers with a gold crown on their vest. Dr. Kerho had even given Artie one of these Kings' crowns as a gift from the boys for his bravery and survival.

Artie had admired this beautiful 3 by 3 inch object and asked how much it was worth.

"To our academy, invaluable," Toy said. "The boys designed it in metallurgy class. It's made from one of the world's hardest metals and is gold-plated. Bobby says it is bullet proof, if you want to take the word of a fifteen-year old." Toy chuckled.

"With Rev. Falwell dead, and Buchanan, who can't shoot straight, that leaves only Anita Bryant to worry about. I think I'm safe because she's too old to fire her gun anymore. Is this going to cost me my pension to buy this thing?"

"Only if you touch one of my boys," Toy asserted.

The time change and stress had the whole group clamor for rest and sleep. It was the next day until Toy and Kami could organize any type of trip and organization. They hadn't seen nor heard from Lauri, but Toy didn't dare call the American Embassy for answers.

A representative with the Chinese Sports Committee had met with Kami and welcomed the academy, then began to arrange dates and times for competition with this surprise early arrival. It was decided to give a few days of rest for the students to visit the sights and practice in an assigned complex that the athletes could use in the upcoming two weeks.

With spiffy blue tennis shirts and shorts they toured the Forbidden City and Tiananmen Square, over to the White Pagoda Temple and to the Peking Zoo. There was no question that the Forbidden City was their favorite, so far. Built in 1406, its 250 acres were laid out under geomantic laws, aligned on a north-south axis. Colors, names, and shapes all had meaning. Symmetry and magnificence were Ming Emperor Yong Le's goals when he began this construction.

The city on a whole lies low. Two-story houses were long prohibited, to keep ordinary Chinese from the sacrilege of looking down on the emperor if his sedan chair should pass in the street. The architects who laid out the city put the altars of the sun, moon, earth, and heaven at the cardinal points of the compass, and the Forbidden City at the center of the dial. It was all a nearly perfect checkerboard.

In the Tiananmen, the gate is the nations' symbol, and Chinese children were singing "I Love Peking's Tiananmen," just as American children might sing "My Country 'Tis of Thee." Tiananmen is the place where 24 Chinese emperors handed down their edicts. When the People's Republic of China was born, it was here that Mao Zedong hoisted the Red flag.

Chapter Forty-Three

Within days the Republic of China had made known that they wished Senator Franco of Texas to assist in the negotiations between Taiwan and the mainland. The points of contention had been a precarious situation for many years, with China determined to eventually regain control of what they considered a domicile of their own.

The president had listened to the Chinese Ambassador and was delighted that the Chinese were so impressed with this Texas senator that they would extend an invitation to this political figure. It was almost too perfect of planning that the senator would gain such a rapport with a foreign power when the Republican Party was within days of endorsing Senator Franco as the front runner in this heated contest between four other candidates. Three other men had stepped down in their vision to run for the nation's highest office, what with lack of support and dwindling financial resources.

Overall the American people did not trust China, any more than they trusted the Russians during the Cold War. For an American senator to barter a tenuous political issue, such as Taiwan, was a carrot in this ambitious leader's hat. This came as more of a surprise to Robert than it did the president, though the senator tried his best to play it off as a result of his tough, but diplomatic stance when he was head of the Foreign Relations Committee. When he met with the Chinese Ambassador that day, he was greeted with extreme warmth and discovered that the Chinese were ready to negotiate a more complacent interaction with the Taiwanese government, given the senator's concerns and suggestions. An off-the-record comment surprised the senator that much more.

The Chinese admitted that they'd heard through the political circles of gossip that the senator was close to being nominated by the Republican Party to run as the next President of the United States. A large financial contribution was forthcoming to assist in the senator's campaign.   It was only a matter of how this money would be distributed and hid from the eyes of both the Democrats and any other overseers that scrutinized the smallest of donations.   Private donations had a way of disappearing into the wood work, and cash expenses were far more easily hid and spent than corporate checks.

To an exuberant and aspiring senator, the handshake with the Chinese was gladly accepted, though the senator was adamant that no favoritism or promises were part of this financial assistance.   The Chinese felt almost insulted that the senator would think such a thing. They just wanted to make sure that the right man, one of intelligence and honor, became the next president.

****

While the political arena became convoluted the Kings' athletes were a bundle of energy waiting to explode when they reached the Beijing Shichahia Sports School.   This massive sports center was the key to the last several decades of rapid growth and world supremacy in several sports.   There were over 20,000 Chinese children and adults in their elite program, resembling a great deal what the East Germans had put together in the 1960s and 70s.

When the East German program was discovered to be nothing more than a steroid factory, its demise had been forthcoming, only to have the Chinese recruit the fallen coaches and system that developed this small nation into a world power in sport.

This academy from the United States was recognized by the Chinese because of word-of-mouth, not because of international competition in worldwide circles.   Two visits by the academy had raised eyebrows: the first was when the boys had visited Germany at ten years of age and competed against German youth of the same age in a variety of sports.

For children to have such technical proficiency, especially boys, was a remarkable feat.   To those coaches, keeping an eye on Olympic and world class talent, the rumors and acclaim quickly spread to such countries as Russia and China, who still relished their own status based on results in international events.   The second trip was when the boys were thirteen and returned to Japan for the fourth time.   The Kodokan welcomed their native hero in Dr. Makawa, and considered Dr. Marion as an adopted brother.   Their Japanese judoka were trained since the age of three, but the Kings' boys gave them stiff competition and mutual respect.   In the sports of gymnastics and swimming and diving, the academy excelled even beyond Toy's and Kami's expectations.   Their basketball and volleyball teams had little trouble with the Japanese junior teams.

During the first few hours the Kings' athletes watched the various levels of athletes train, the young children were constantly tested; their outfits were no more than youth underwear.   There was nothing overly impressive with the arena, the same mats and safety harnesses and abundance of foam rubber to try different stunts without injury.   The seriousness of the training wasn't as relaxed as at the academy because of the nationalistic expectations of pride that were expected as part of the Chinese system.

Toy was well aware of the many bonuses given by the Chinese for athletic accomplishments.   A family's honor depended a great deal on their child's success.   This program was so regimented it felt like a military institute with its bearing and discipline.

When the Kings' athletes began their own time in between the Chinese training classes, there were watchful eyes that would see a similar training pattern, but a different approach and style.   The academy's coaches were hired and trained in a system that was unique and futuristic in its approach.

Rather than work an athlete in a sport with an encompassing viewpoint of their technique, the event was split into parts, technical nuances.   Each minute detail had to be understood and incorporated into the child's mind and body before it was all implemented as a whole. The harnesses that the Chinese used were not much different than at the academy, just that they were redone so that the gymnast or diver could experience twists and turns far in excess to what they would experience in competition.   Much of this was psychological to prepare the mind for what the body would not comprehend or find within its realm to be practical.

The academy's coaches didn't mind sharing their advance

approach, though they kept the psychological side a close knit secret. Coaches, especially with world class athletes, were known to be selfish. There was little effort in international circles to share breakthroughs or suggestions that might benefit other athletes. It was truly a game of one-upmanship.

It was the second day when the Chinese allowed the American boys to workout with their youth. Artie received more scrutiny than the coaches and athletes, but he kept assuring his hosts that he was there writing about the academy's team and not spying on the Chinese, though that was not the words he used.

Several of the Chinese athletes spoke many words of English and they found delight when they discovered that a few of the Kings' boys spoke their language. Kids are kids, their smiles developed into friendships, in spite of what the coaches might have wanted in China.

The Kings worked out in their body stockings, a full length garment, so light, it felt like terrycloth but was designed to keep the body's temperature conducive to competitive readiness. To the boys it felt like they were naked and had full mobility. Skin tight it also had a built-in pouch to keep the genitals from swinging. The Chinese athletes wondered why they had to wear uncomfortable jocks and small, white shorts that were often binding, if not restrictive to their maneuvers.

Toy and Kami met with the coaches of their team at a restaurant that was only chosen after driving through the city for a while. They choose the Palace Hotel and the topic of conversation was not so much on their own athletes as it was about the tight coiled butts and calves of the Chinese youth.

"They're gene doping," Coach Larsen expressed. "They've gotten away from using roids and human growth hormone. This is the future in creating the gifted athlete."

"I noticed they weren't even tired after two hours. Our kids were ready for the showers," Reece explained to his brother.

Reece had also brought his oldest son, Reece Jr., nine years of age and a future gymnastics star. Out of Reece's four children, his first was the only one to have the gay gene. Though his son didn't stay in the dormitory the boy was well versed with what this meant and was madly in love with Oshi's son, Mikki, who was ten and gay.

Oshi's children had all come with him and their mother to Beijing, with Mikki being the third youngest and his only gay child. For Mikki, gymnastics was something he did when he wasn't doing judo and other martial arts.

Artie had tagged along with the directors and the coaches, his role was not to intervene but to listen and learn. Athletics had never excited him all that much, unless it was his fascination with Greg Louganis, a former American diver and a gay icon for the community. Anyone in Speedos was okay with Artie. He asked what this was all about.

The group's molecular geneticist gave him a quick review. "Gene doping is different from other performance-enhancing techniques. Human growth hormone, for example, occurs naturally in the body and will accelerate cell division in many types of tissue. Taken in high doses, it can provide a head-to-toe muscle boost and can even add a few extra inches of height. Anabolic steroids are used heavily in our professional sports: baseball, track and field, football, basketball, and hockey. Big muscle growth in the upper body. Synthetic erythropoietin, or EPS, a chemical naturally produced by the kidneys, is a favorite of cyclists, triathletes, marathon runners, and people who engage in long periods of aerobic activity. EPO flushes fatigued muscles with oxygen to stave off exhaustion."

Artie nodded. "So what you're saying, these kids are taking a quick fix to Olympic gold."

"It's to the point where doctors are designing the perfect athlete, and gene doping will give that extra tenth to a sprinter; an inch for a high jumper; and a foot to a long jumper who only needs that as a difference for a gold medal between him or her and the next athlete," Toy conveyed.

"Any downside?" Artie asked.

"IGF-I, that's the vector they would be using, could make precancerous cells grow faster and stronger. We've done our own tests and the vector virus can run amok. Bobby and Randi were testing this on mice when they were twelve. It's risky, at best."

"Your boys were experimenting with this at twelve years of age?"

Artie had wide eyes.

"Not on themselves. The technology is too easy. It's just graduate student science, and we're talking about boys with IQs that make for a good bowling score. Bobby even discovered a noninvasive imaging device akin to an X-ray that detects bits and pieces of leftover viruses used to introduce performance-enhancing genes. I know of no other device in the world," Kami said.

"Is there any way of getting the Chinese athletes to cooperate?" Artie asked.

Coach Larsen chuckled. "It might take a reporter like you, Artie, to ask the Chinese if they'd consider us testing their athletes. You do have your visa ready, right?"

The men laughed and Coach Randal, the gymnastics coach, changed the subject. "The combination of athletes sure had the two teams wanting to show off for each other today."

"I thought it was because Wendy and Marie were spotters for our boys. Their leotards distracted more than a few eyes," Coach Larsen admitted.

"I don't think their coaches appreciate our girls in there," Toy remembered. "Knowing the way the Chinese think, they will accuse us of using the girls to disrupt their practice. Our boys stayed motivated, which brings me to Rom and Tim. Those two made awfully quick friends with two Chinese boys."

Coach Randal knew exactly where the doctor was coming from. "Once those kids locked eyes you could smell the testosterone escalate. I gave them permission to tour the city with their hosts, though I dare say they'll tour each other given the chance."

****

Another strenuous workout followed for the Kings Academy in the morning and the team was off on planned tours to see four-thousand years of history, art, and architecture. Rom and Tim had discussed their excursion through the city with two Chinese boys, Xi and Chen. As teenagers do, the Chinese boys admitted getting shots in the rear end, but they were told these were vitamins and not to question the intent. To boys whose future depended on acceptance from a society that judged their worth by what they did for their country, they weren't about to question their coaches' training methods.

Kings Academy group flew into Shanghai for a night on Shanghai's Wall Street. They visited the Shanghai Museum where the world's greatest collection of ancient Chinese art is kept—440 bronze vessels, 517 pieces of ceramic, 417 pieces of jade and an abundance of gold. A night's rest and the students went to the Municipal children's Palace, housed in an old mansion. There they met some of China's young athletic, academic, and musical prodigies prepping for tomorrow's performance.

It was a day of working out, learning from each other and sharing knowledge and ideas. Dr. Kerho invited this unique academy like his own to Texas, if and when they were able to return.

The leaders of this prodigious school weren't political puppets and they relished being with the American children. The evening became a show and tell, of sorts, as each group or person presented a show. Members of the Shanghai Acrobatic Troupe exhibited just what the human body is capable of.

Flexibility, agility, and balance didn't begin to describe what the Kings' students saw. Being that the show was private, revealing costumes increased the fun as the Kings' boys and girls enjoyed the show in their fundoshi, a Japanese version of a loin cloth. The props used by the Chinese children were simple—hula hoops, knives, umbrellas, though the usage of these props had never seen such talent.

Oshi's ten-year old son, Mikki, and eleven year old daughter, Mishi, wearing just a belt of colorful ribbons, put on a martial arts show with knives, sun spears, and assorted blades. Their expertise and knowledge had been learned with their years in Japan and the intermittent time they spent in conditioning at the Kings Academy.

To them it was equal love between the attention of their grandfather in Tokyo and their many friends at the academy. Following these two siblings, Hikoyi and Oshiwa, twelve and eight, respectfully, teamed up and did a kata demonstration for the audience. Oshi beamed with his oldest and youngest son in syncopation.

Dr. Marion's boys were almost a year older than the other boys on the academy. Mikki and Rikki were superb tennis players and baseball players, but loved the ballet like Lance, Trent, and Trevor. They had all trained at the Dallas Ballet Company until the academy had hired their own teacher.

Toy had also been a ballet dancer in his youth, but was more delighted to have Lance take over in his father's footsteps. Toy's sister, Brandy, dedicated her time to teach dance at the academy and was very delighted when Leif came.

Leif offered humor and a new personality that invigorated this group of boys who loved to display their bodies in all forms of dance. To get Leif away from Oshi's arts and loves became a tug-of-war.

Five of this group performed a small segment of the ballet, Nijinsky's Clown of God. Mikki had been especially adamant in examining the floor of this small auditorium, but found the unfinished pine, smoothed over and perfect to perform on.

Too many school stage floors had highly polished hardwood or glossy linoleum, or wood laid over cement or linoleum. He knew dancers cannot commit themselves wholly to a dance on the wrong type of floor; this would risk injury and a career very quickly.

The mass of Chinese youngsters were enthused and in anticipation of seeing a ballet performed by this American academy. They knew ballet is a visual art, like painting or architecture. But unlike other arts, ballet is a visual spectacle that moves.

The setting of the stage was dominated by a large cross, paper, in this case, because the academy couldn't travel with a wooden one. For the first few minutes Oshi and Leif had a candle light service for the death of Sveinn Eldjarn. The boys gathered around Leif and grieved with the boy because they knew how much this teacher had meant to him. The tribute was heartbreaking, but it was also a type of closure Leif needed.

With the ceremony finished the candles were blown out. Everything was dark, then a voice announced: "The world was made by God, Man was made by God. It is impossible for man to understand God, but God understands God."

A blue light lit up the middle of a circle, creating an image that this was the beginning of the world. Inert, naked, and faceless bodies covered the stage. Under God's eyes they became animated, organized, breathing. The circle was formed. Life!

A being detached himself from the anonymous mass and placed himself in the center of the circles. God gives him a face. It is man: The Clown of God. Slowly the Creator formed him, teaches him to live, to walk, to jump, to dance.

Sergei found himself dancing this role, a role that never in his right mind he would have thought of playing in China when he started on his visit a week before. The boys of the academy adopted him as one of their own, even if he was in love with Trevor. Being center stage was his finest moment and he had even shaved down to blend in with his peers.

The Chinese kids were spellbound, riveted to the male nudity on stage in all its beauty and fascination. There was no laughter as what might be expected from American children to cover their own embarrassment, only the absorption of watching boys take pride and enjoyment in human form. Nudity is often used in ballet as a stimulating but serious ingredient which completely justifies itself artistically.

As the dancing became more progressive in God's life pronounced, familiar strains of Tchaikovsky's Pathetique Symphony came from a speaker near the stage, a substitute for a full blown orchestra, but no one really cared.

The stage grew with earthly paradise with revelations of color, sound, and light. Four other Kings' boys sprang to dance, each representing other creatures: a Rose, a Faun, Petrouchka, and a Golden Slave, symbols of the four elements: Air, Earth, Water, and Fire. Lance's craving for a great and constant love, and his tenderness forced him to create a new image of love.

A woman, played by Wendy, unreal at first, grows in his soul. Little by little she takes a tangible shape and he seeks in her the support of a creature flesh and blood. The face of his dreams suddenly becomes the face of one woman. The nude female danced to the center of this circle and the two of them swept into each other's arms.

476

"That's so beautiful," Artie whispered to Toy. "They must be in love; they melted into each other's arms so easily."

"They're brother and sister, Artie, and they fight all the time," Toy replied with a subdued smile that they were his children.

All this dance and joy led to the further advancement of all life. The man's companion was by his side, but she could not follow him in his search for the Divine, in his love of humanity. Instead Lance began to dance with his five brothers, as an unspoken way that this was his destiny, to love and know this unique love, angelic beings, in whom physical grace reflects spiritual grace, and whose technique and genius were immense because they serve God and brotherly love.

It was interesting whether the Chinese children caught this message, but when the lights came back on they stood and applauded loudly. For anyone watching the children's faces, the expression of nudity was an expression of freedom and love. This, any child understood.

Toy wasn't sure how all the nudity would be accepted, but the smiles said it all. The boys had performed this number in New Guinea, as well, and it was received by licentious hard-ons there. He told Artie this story and cracked him up.

The Chinese performers were discussing with each other the beauty of the performance and, no doubt, the male physiques that gave new meaning to anatomy.

****

No time for leisure days the group was off to Xian; 13 centuries ago it was the largest city in the world. Here they saw the mammoth underground army, built at the command of Emperor Qin Shihuang, first Son of Heaven and father of the Great Wall. Peasants digging a well in 1974 made the discovery. The terracotta soldiers, archers, and horsemen, each with a face different from the next, defy age. Some still have their brilliantly colored glaze intact. The boys were given special privilege in seeing Qin Shihuang's men face to face.

Possibly their tour guide had heard of the performance of the previous evening, as the guide took the team to a lesser known area. The location was no less surprising with the display of the Han Dynasty naked warriors.

Emperor Jingdi died in 141 B.C., but he left behind vividly molded earthenware figures originally dressed in designer silk. The clothing did not survive the journey through time, but the sculptures are exquisite au natural.

Tad, who wrote the Kings' monthly magazine for the academy, was now envisioning doing the same thing for the academy, creating mini-statues of each boy in their athletic performance to compliment the other statues around the track. Dr. Kerho said he'd have to consider this suggestion.

On their way to the Great Wall, Artie was truly glad he'd come, if not hijacked, now that he thought about it. His mind sensed he wouldn't be returning to the beat of writing and being a critic for the Georgetown gay community. The possibilities were endless, but seeing and being around these boys and girls from Kings had truly given him a new perspective of life, if not the value of truly being human and alive with his sexuality. He had his favorites, which Artie readily admitted to Kami and Toy.

Alan, one of the dancers from the previous evening, had sent an arrow through his heart. The boy was not as muscular as his peers, a backstroker that didn't require a great deal of body mass. Alan was not effeminate, only lithe with beauty, limbs that matched the length of his penis. Artie had done seven interviews in his spare time, but he knew who his next conversation was going to be with.

A standard visit to the Great Wall of China often took a week or longer. The academy didn't, nor could they afford to have the athletes take this much time off. The Wall, or Chang Cheny, as it's called in China, was quite an eye-opener. The Chinese thought of it as "the long city."

As it wiggled off through the mountains into the horizon, one can actually be on the wall and not know it; so much of it has been ruined or taken apart by peasants to build houses or other dwellings. For the boys it was a day of hiking; for Artie, he had his interview and complete attention from Alan.

Artie found Alan Marshall with three other Kings' boys at Mu Tien

Yu—the best view on the Great Wall. He discovered Alan to be soft spoken, bright and cooperative. Artie started by complimenting the boy on his dancing and how gorgeous he looked in the all-together.

Alan thanked the writer and returned the praise by admitting that he'd heard from Niqui that the reporter was glamorous in a dress. Artie gleamed with his reputation intact.

"Tell me, Alan, I'm just an ignorant Washington Post reporter; who was Alan Marshal?"

"My progenitor was Will Marshal, sir. Born in 1147, his father was John Marshal, who was a leader of Queen Matilda's forces. John was in conflict with King Stephen, when he sent out a message calling for a one-day truce. The agreement was, if the king would grant the respite, Marshal promised he would attempt to persuade Queen Matilda to surrender the castle at Newbury, just west of London. The king agreed only if John would offer a guarantee in the form of a hostage—his youngest son, William. The five-year old was duly sent into the enemy camp as security against his father's pledge.

"Sadly, John Marshal had no intention of keeping his word. All the man did was bring in fresh troops and more provisions. The king threatened to slaughter the boy at the foot of the gates, if John Marshal didn't budge. A large-siege catapult rolled up to the battle lines; the power of this beam could hurl a large boulder some two-hundred yards. The king's men had placed young William into the catapult's load basket, and the story goes that young William thought this was most amusing. He'd cheerfully called out over his shoulder, wanting to know what sort of game the men were playing.

"In the end, Stephen had no stomach for the murder of an innocent boy, but the king kept the boy for four more years before he was returned. William went on to become a famous warrior who commanded vast resources of wealth and soldiers. King Henry took quite a liking to him and knighted the man. As a young knight he became enamored with the younger Henry III. William would go on to befriend King Richard and King Philip of France, who considered William the most loyal man and true as they'd ever met. It's fair to say the man served for many kings who desired my ancestor's companionship."

"Fascinating," Artie said and photographed this stunning male against the landscape of the Great Wall. He wrote the boy's talents and goals for his future, then went in search of Peter Tchaikovsky.

Alexie Tchaikovsky was not one of the fine athletes of the academy; his pursuits were music and poetry from a child; yet, he enjoyed running and ballet, though he desired to choreograph rather than dance. His martial arts ability was impressive with the years of study, and he could swing a mean bat and play a fine third base, but the arts were truly his love.

Artie asked the boy if there was something special about his ancestor that stuck with the boy.

"Peter's life was tragic but brilliant," Alexie said. "To be free and able to love without fear of persecution is something I relish in his name. I use his music a great deal, though I'm also an admirer of jazz and the blues. I can admire Peter's efforts to keep his sexuality hidden, if not try to convince himself that he could live a life as a heterosexual. He discovered this was all but an impossible endeavor. I'm only happy to extend this man's vision and live a future of being happy with who I am."

"So what's in Alexie's future and, if I may ask, who's your boyfriend?"

Alexie flashed a bright smile. "I love Marc Alger, maybe it's because we share so much of the same interests and our ancestors both tried their best to cover their secret. I'm using my electronic keyboard to do a certain amount of improvisation. The dual musical/dance numbers of Leif and Mikki make for a future of contrast. I want to score and make a movie, which my brothers will act in. Maybe when we return to Texas."

With each individual Artie met and talked with, he saw a boy who shared his own vision; they were no more their forebear than Artie was his father. Artie figured that what distorted mind-set he knew he had when the truth was revealed must be what others would think if they heard that a clone was made from someone else. Outside of some physical resemblance, these boys were America's future and not the recreations of past heroes as others would view them.

Artie remembered why he desired to be a reporter, an instinct that gave him an opportunity to transcend some of his personal failings. He might not have probed the troubling aspects of his personality in the same depth as a novelist does when depicting the flaws and strengths of a fictional character, but even his most mundane, day-to-day existence as a gay man gave him the chance to present himself as a better person than his father had so labeled him. These boys brought out the best in Artie, and he was giving second thoughts to a lifestyle that was sometimes demeaning by being a prissy for others to use.

****

While touring the Great Wall, Toy had received a most interesting call from a Major Robert Davis. The Air Force pilot had managed to obtain the number of the senior doctor living in Dallas to get his son's number. The pilot and graduate of the Air Force Academy wasn't despondent over being grounded while an investigation ensued, only that he felt cleansed and worthy because he avoided the cold-blooded killing of so many innocent people.

There was no doubt to the major that terrorists were never involved in any plot to hijack an airplane. Major Davis admitted he'd like to meet Toy, if and when the academy returned to Texas. He also wanted Mrs. Franco to understand that the pilot did relay to the highest echelon the fact of the senator's family being aboard the flight. Even then the decision to down the aircraft was announced from Washington and the Pentagon. If the senator himself hadn't given the okay, it was promoted by the president himself.

Toy was elated with the call and promised the man that he would be vindicated of any dereliction of duty. It might take the press to do this, but this conspiracy was going to be revealed one way or another.

****

Upon the team's return to Beijing, Mrs. Franco had a conversation with Dr. Kerho. She called a lawyer in San Antonio and proceeded to file divorce papers against Senator Robert Franco. The truth sent a cold chill down her back. Her one saving grace was the company of her four healthy and intelligent children.

****

Senator Franco's announcement as a candidate for the Presidency of the United States was skyrocketed by a massive newspaper barrage, financed by more sources than the American public realized. His voting history had few errors for his competitors to ridicule, and his campaign headquarters covered the absence of his wife and children by stating they were vacationing in China as part of this country's new open-door policy that welcomed the American people to visit this modernized country. The way it was presented made it appear that Mrs. Franco was already an ambassador of friendship.

Robert had wanted to move in on the Kings Academy as abandoned property, or at least have all the time they needed to rifle through the academy's paperwork and computers, not to mention the dormitories and athletic facilities. He was seriously thinking of purchasing the grounds for another St. Bonaventure extension. It was too good to pass up.

For right now, the senator was busy formulating his views as a viable candidate. He laid out sharply contrasting opinions on Middle East policy, terrorism, and health care, taxes, abortion, and stem cell research. Wherever Robert saw the swing of voters, he adjusted his viewpoint. The man was certainly "likable" enough. The public was not familiar with the senator's family, so Robert didn't have to support equal rights, even for his own gay son.

By the third day of his announcement to run, Robert had presented a string of negative ads, personal attacks and blanket denunciations against his opponents—the American way of fighting for elective offices. What's particularly depressing is that this primitive campaign style was based on simple reality: "It works." American politics is all about image and style points.

Candidates can't reach voters personally by stumping or even through national newspapers; the country is too massive and too fragmented. Television is the way to reach Americans. Entertain them with hard-hitting tactics that make it look like political victory has the same strategy as the Super Bowl.

The candidates are fully aware that they are really not leaders or legislators, but performers, actors with a legal education that, if they fail, they can always go back to the courtroom or the senate chamber. In this election, the candidates were no different than B-list actors in an audition for the lead in a bad sitcom denoting Washington politics.

Robert had amassed several major kudos with his recent ties with the Chinese. One of these was in the settlement of an issue with Taiwan and its representation in future Olympics. Though the incident was over a simple use of a flag, the senator's suggestion of give and take had been accepted by both governments so that the athletes could participate.

The amount of money the Chinese were investing in his campaign was of some concern, but possibly the Chinese government saw leadership potential in him and a man they could work with to eliminate this 19-year ban of weapons by the European community. Anyway, the American people needed China as a wedge against North Korea and other evil despots of humanity. Just because China was building two new nuclear submarines and increasing their military weaponry didn't make them a new power to begin another cold war.

There was some disturbing news when the family attorney notified the senator of Mrs. Franco's plan in divorcing him. One thing about having a family attorney for the last fifteen years, the man was loyal. Robert had the attorney stall the paperwork for a few weeks. With all the new Chinese contacts that Robert had made over the past few days, he knew just the man who could help him with his problem.

Chapter Forty-Four

With a week to prepare advertising, ticket sales were rampant in preparation for the Chinese competition against the Kings Academy. The Chinese promoted it as America's best junior team against China's. This was supposed to be low-key, a form of friendship event which would create harmony and mutual respect between the two countries. At the new gymnastics center, created especially for the Olympic Games of 2008, there was nothing low key about the attendance that filled the arena of twelve-thousand seats.

Girls often received the most attention of the two teams because of their world championships. The Chinese boys had been highly rated in the last Olympics, but fell short of their expectations, despite winning the World Championships the year before. This team looked very familiar to their Olympic team of the future.

Though the Kings' boys were finished with their own gymnastics season, they had kept in training for this with more advanced skills. The one aspect they did represent extremely well was the psychological aspects of competing; the most important element between being world caliber and choking when push came to shove.

As the Kings' boys did their routines with less flare and degree of difficulty, they also did these routines with less errors and total precision. This alone put the Chinese team tense in front of a packed arena, expecting their own gymnasts to shine above a mere academy of unknowns.

There is one aspect of gymnastics and diving that not even a world's champion can eliminate, and this is the aspect of judging. Given this was an all-Chinese judging team, it was a foregone conclusion that even if the Chinese had flopped on the mat the scoring would still favor their own.

Bobby Newton had shined on his floor exercise and pommel horse with his new exercise routine. For a boy not accustomed to international competition he showed no nervousness or reluctance to go balls-out. Their hypnosis definitely contributed to this state of mind, a perfectly legal way for athletes to block all outside influences and totally zone in on their event.

By the end of the evening the Chinese had taken the competition by a minuscule five-tenths of a point. The crowd and the Chinese team were just glad this wasn't the American team that would be competing in the coming Olympic Games.

Only a few blocks away, the boys' swimming and diving teams were competing against the best of this particular sports school. If China had a weakness outside of their baseball team it was in swimming. Between Alex and Luke Eakins they split first and second in three different races, both bettering their times in every event they swam. For a high school team they dominated the Chinese juniors in every race except for diving.

Trent Nijinsky and Spirit Owens matched their Chinese welterweights with near splash less dives; yet, the American boys got their 9s and 9.5, while the Chinese received 9.5s to 10s. The Kings' coach could only shrug his shoulders at the scoring and compliment his divers with their perfection.

The boys' swimsuits were closely scrutinized by the judges to violations, just like the gymnasts were. Dr. Kerho reminded a Chinese coach that only a few years before it was the Chinese girls who were told to wear something more modest than the see-through leotards to impress the judges. There was nothing illegal in the Kings' suits and their coaches were in agreement that the Chinese would be wearing these soon.

****

Debbie Franco received a pleasant surprise when she returned to her room and found that an invitation had been slid underneath her door. Though it was a formal letterhead, the signature didn't reveal the organization or party that invited Debbie and her children to Houhai Lanke for lunch and tea, just outside the city limits.

Brittany was consequently invited by Debbie, but Oshi wasn't going to allow the two females and their children to go unescorted. He rented a minivan to haul the nine children—Reece's nine-year old was invited by Mikki—and the three adults. A Chinese driver was mandatory as their chauffeur.

They were barely outside of Beijing when the driver stopped for gas, then said he was called in for some emergency and it was okay for Oshi to resume the driving. The driver mumbled the directions which weren't all that difficult to follow.

Houhai Lake is a popular hangout for lovers and those wishing to escape the hub of activity within the city. Their destination was north of the lake as the elevation climbed to steep ridges overlooking the serene beauty of the Lake District.

There were steep drop-offs with but a flimsy guardrail that acted as no more than a warning that anything beyond the side of the road would be flying into space.

The kids in back were examining a pair of chopsticks, learning to balance one chopstick along the base of their thumb and the other stick on their forefinger, then picking up a piece of line by moving only the top chopstick with the forefinger. It was a lesson in dexterity.

In the meantime the adults were busy in idle chitchat and Oshi was keeping his eye on his rearview mirror when he noticed a military jeep coming up quickly with the intent to pass. The two military men in the front seat acted stiff and focused. Oshi eyed them suspiciously when the jeep came along side.

It was the men's eyes and demeanor that gave Oshi a wrong feeling about all this. When the jeep's steering wheel was yanked to the right, Oshi jammed on the brakes of the minivan to avoid having the jeep literally ram them into the guardrail on this narrow two-lane road. Even then the jeep caught part of the minivan's front bumper and ripped it from its holdings. The jeep slowed, stopped and began to back up in an attempt to finish this van and run it off the road to a 500' drop to the lake below.

There was great commotion within the van but Oshi turned the vehicle to his left and began backing up, but away from the rail. When the jeep skidded off the front bumper again, Oshi braked hard and leaped from the front seat. The driver of the jeep tried his best to hurry shifting the jeep into forward, only to be yanked backwards by this Japanese man and have his neck quickly wrapped up in the man's arms. The passenger of this jeep reached for his revolver and began to point it at this adversary choking the driver.

No one heard Oshi's twelve-year old son, but this military man took a flying side kick in the back of the head that sent him sprawling onto the jeep floor. Mikki was at his brother's rear and yanked the gun from the man's hand before tossing it over the cliff. With both boys diving on this stunned Chinese sergeant, the man's arm was snapped at the elbow.

The sound of a pop wasn't a gun but the driver's neck as Oshi finished what he started. He tapped his twelve-year old to release the choke hold on this passenger, then a quick punch to the back of the neck left this man unconscious.

"Good work, sons! Now get back into the van!" Oshi directed.

He saw a pair of handcuffs in the jeep and put both men's hands as part of the steering wheel. Directing the jeep toward the guardrail, he put the jeep in neutral and rolled the vehicle through the wooded rail until it disappeared off the cliff below.

"Maybe they didn't think I gave them enough room to pass," Oshi said as he re-entered the van.

It did get a giggle from the scared teens in the back. "Personally I don't think they're waiting for us at this restaurant."

"I'm inclined to agree," Debbie said. "I'm so shaken I think its best we turn around."

Brittany had no disagreement with this idea as the pensive group stayed silent all the way back to Beijing. Rather than return to the hotel, Oshi drove everyone to the afternoon baseball game that was planned between the two junior teams—junior for the Chinese meant 19 and under. The kids were more than delighted to be at this game than a luncheon and tea party at some old, dumb lake.

＊＊＊＊＊＊＊＊＊＊＊＊＊＊

The incident was the talk around the stands at the ball game. Oshi

487

spent several minutes in discussion with Dr. Kerho and Dr. Marion on what this all meant.

Hikoyi and Mikki received all the praise they could handle from the other boys when they heard what heroics the boys had done. To these two martial arts experts they had inherited their father's courage and total lack of fear. Their successes were equaled by Alex's pitching. He threw a nifty three-hitter against the Chinese team as the American boys won seven to nothing.

There was a cautious exuberance at the hotel when the team returned and prepared for an evening of volleyball.

Toy and Kami weren't enjoying themselves as much as their students. The adults met in the restaurant to discuss this latest incident and its ramifications.

"Who'd you call to start the divorce?" Kami asked Debbie. When he was told the family attorney, he obviously asked how close the guy was to her husband.

"They've known each other for twenty years," Debbie replied and had more than enough reason to suspect she'd been undercut there.

"I'm not going to make assumptions," Toy remarked. "We've all read the English version of their paper here and know that Robert is odd-man out with the other candidates having full support of their families. I find it difficult to believe this was an idea to remove this barrier. I'm wary of the senator's association with the Chinese on this recent Taiwan complication."

Dr. Kerho decided not to elaborate.

"From now on we all stick together," Kami decided. "We play basketball tomorrow night and leave for Japan on Sunday. We don't have to worry about this happening in Japan, believe me."

****

Five floors up the boys and girls were in celebration of all that had happened throughout the day. The two youngster boys had bought squirt guns and ran in one of the boys' rooms to nail Leif in the shower with blue water. He chased them out into the hall and into Alex's and

Travis's room while they were lounging on the bed. There was soon a mild melee of nude wrestling and swinging lightweight youngsters from bed to bed.

One thing the Franco children had learned through the past weeks was that, at the academy, monogamous relationships didn't really exist. There weren't the petty jealousies and possessive mentality like at their high school. Boys loved each other both in pairings and in all forms of activity, including sexual. No one got upset because their best friend had sex with another brother, or as a group.

This was difficult for Jay and Niqui to decipher. Jay had already walked in and found the love of his life having sex with his sister without the boys there. It just seemed that there was some kind of violation to their own love in seeing two girls making out, but his sister had told him weeks before that the epitome of sex was two girls, not a girl and a guy. Jay now knew what this was all about. He began to realize that he didn't own Wendy, any more than she owned him.

Jay began to contemplate his own sexual opinions. He'd had sex with Lance, the time the four of them were in bed together. His mind had been so much on making out with another boy that he hadn't realized how enjoyable the union was.

Lance was truly an expert and had Jay squirming in bed with the greatest of pleasures. Come to think of it, Jay thought, Lance was far more seductive in bed than Wendy was. Maybe it took a boy to really love another boy to its fullest.

The night before his own brother Niqui had come to ask Jay if it would be okay to have sex with Wendy, only because she suggested it. "Only if I can have Dane," Jay replied and knew he was asking a lot from Niqui with this request.

"Of course! Dane thinks you're gorgeous."

Niqui finally had his dream come true. Wendy was as sweet and gentle with her foreplay as she was in heated sex. As much experience as Niqui had gained through his friendship with Dane he still only lasted one minute in his first intercourse.

By the night's end he was rocking with splendor for a whole twelve minutes. He desperately wanted to tell Wendy he was in love with her;

yet, he sensed everyone there was in love with each other.

Jay found more than Dane in his pursuit to cover his mood. Both double beds were slid together and five boys were shooting the bull when Jay moseyed in with a droopy face.

"Get your butt in here!" Dane shouted when Jay showed resistance. "We were just discussing who the best athletes in the world are."

Jacob scooted over and gave room for Jay to sit. All of them were wearing their fundoshi, while focusing on this intense debate.

"Tennis players are the greatest athletes because they possess high levels of coordination, balance, finesse and instinct. They're mentally tough and have excellent physical conditioning," Jacob surmised and got punched by Aki, who was the fastest runner at Kings.

"Play with your fuzzy balls, dude! Basketball players have it all: speed and strength, eye-hand coordination and stop-and-start quickness, mental toughness and physical stamina."

"Cyclists have all that!" Leif spoke up. "We'd leave your sprint-and-stop athletes in the dirt. Try going hours non-stop. What do you think, Jay?"

"My sport requires total fitness, balance, speed, courage, and tremendous flexibility. The mental aspect is no doubt the hardest part of the competition. If you're not mentally tough, forget it! Plus, we have cute bodies."

"I agree with the cute part," Trent jumped in this debate. "Ballet is like sex. You have to be versatile, flexible, and have stamina and endurance. I've been a top and bottom, had one in my mouth and two in my hands, all at the same time. Now that's multi-tasking. No other sport as that much focus and concentration. I've been able to bring everyone off at the same time. Can you imagine the timing that takes?"

"That sounds like fun," Jay added to this picture.

"Let's see!" Dane shouted and they all mobbed the youngest of the six boys there.

The light was soon turned off, if not knocked off by a flying leg. Jay had never had so many bodies and body parts in and around him. If he wasn't laughing he was kissing someone and something—it didn't make much of a difference in the end. To hell with girls was Jay's final

490

thought before his second orgasm in less than twenty minutes.

The adult colloquy in the restaurant downstairs was only dissolved with the report that two young boys were seen sprinting naked through the lobby and into an elevator. Brittany and Reece had a good idea who these two boys might be and dismissed themselves to find the culprits. This hotel would never be quite the same with this American group as their guests.

Chapter Forty-Five

The basketball game between the Kings Academy and the Chinese junior team was a mismatch as far as height was concerned. The Chinese had a seven-foot center and two forwards at six-eight. The tallest boy on the Kings' squad was Jon at six-six, then Trevor who had grown two inches over the last month to six-five.

The same arena used in the Olympic Games was packed for this demonstrational event between two junior teams. What the Chinese witnessed was not the version of American basketball they'd seen in international and Olympic competition, let alone the NBA, which was often televised.

The Kings' coach developed from the very beginning a team concept of passing and something new to basketball—pure shooting. Basketball had evolved into a fast break, inside the paint, lay-up type of game, composed of people who could jump and run.

Volleyball had occupied the previous evening and the Kings' squad was also seen as the underdog. Their unique game of two outside hitters and two back court hitters in the center completely had the Chinese teens rattled. In the end the academy won 30-25, 30-28, and 30-22. The boys liked to think of it as a get-back for the Chinese scoring fiasco in gymnastics and diving.

With basketball there was much the same frustration for the Chinese. They were used to the American game of run and gun, but these American boys weaved around the perimeter until someone got open for the 3-pointer. Kings forced the Chinese to come outside and away from the key.

Jon, Trevor and Michael Locke were hitting the bottom of the net with 3-point shots to match the easier buckets the Chinese were getting from feeding their big men. There wasn't much the Kings could do but double team the tall men and hope the jump shots weren't all successful.

No matter how many easy lay-ups were made, or fifteen-foot jump shots, their three-point baskets raised the score up faster than the Chinese could find a remedy.

Frustrated with this American way of playing, the Chinese went to a wide zone, but then the Kings began passing inside for easy lay-ups after spreading the defense, occasionally moving their fastest boys down court for quick breakaways after a defensive rebound. This was not a style of basketball that any team, including their Olympic men was used to playing. Balls-out basketball was tiring and the Chinese were not in the condition that the American boys were.

There would be no letting up as the Kings found their range from the onset and never faltered. Their victory of 86 to 65 had the Chinese shaking their heads as they exited the Olympic arena.

With a quick shower and change of clothes the academy moved across town to an auditorium where the Beijing Opera was usually held, in addition to other different styles of dance and theatre. From what was supposed to be another recital for Chinese junior aspirants in the art of ballet, the Kings Academy found a full house to witness this taste of ballet.

On this night there were politicians and authority figures that were there to scrutinize this strange academy with so many talented boys. They were also ready to censor any homoerotic style of dance, which was rumored after the academy performed in Shanghai.

Dr. Kerho had a suspicion of what the Chinese might do if they found the boys in violation of China's homophobic philosophy. On this evening the kids would keep their clothes on.

The dance that Trent and Dane choreographed for the Beijing performance was called Force of Rhythm. It was decided upon because of the combination of styles of classical ballet, modern and ethnic dance. The idea was to capture the essence of ballet and ethnic dancing styles in such a way as to display the rapport, beauty, and relationship between the two.

When the curtain rose the twelve dancers heard a voice above the music: "We're in a difficult situation. We're going to have to live together."

Wendy wore a white cotton dress, an attire reminiscent of the old South. The boys wore tights, their physiques magnificently defined without the usual ballet jock.

Once again it was Tchaikovsky symphonies to folk, blues, rock, and jazz; the dances revealed variations on these musical themes in dance expression. Tchaikovsky's music: "Distress," for four boys; "Shout," a solo for one boy, Trevor, and demanding rhythm came from "Forces of Rhythm," a dance for three. There were recollections of minstrel days, and the Marion brothers did a "Roots and Rhythm," in red loincloths, as a display of native African dance.

It was an excellent performance, yet cautious, because the dancers had an unusual floor to dance on and it lacked any type of pitch, which most dance floors slant from the rear to the front. Nonetheless, the dance made for closure to their Chinese vacation—so they thought.

****

When the academy arrived back at their hotel, everyone's cell phone had been taken and their rooms had been rummaged through. Toy was fit to be tied and discovered that the desk clerks, who had been so hospitable before, didn't quite understand English as well as they did prior, nor comprehend what the complaint was about.

To heighten the tension two Chinese policemen came to Artie's door and said he had to leave the country. He was being accused of spying or some type of illegal reporting. The departure was so hurried Toy had barely enough time to slip a piece of paper in Artie's pants as he hugged the man. He knew Artie would be flown to Tokyo first, then to San Francisco.

By morning the two directors were informed that their aircraft would not be leaving Beijing that day. In actuality, the Chinese government was requesting that the academy play more basketball games in demonstration of good will. Of course they promised that the team would be allowed to leave after these exhibitions.

Most Americans would have called, if not run, to the American Embassy. This was not an option for Dr. Kerho. Though it appeared they were under house arrest, Kami was the first to explore the idea of wondering outside the hotel unmolested. The ploy was to verify he was being followed and his reasoning and assumptions were valid.

When the Chinese Sports Commission official arrived to discuss these proposed basketball games, that were never scheduled to begin with, Toy listened to the proposition. This was more than basketball games, Toy and Kami recognized. How long they would be stalled in China was what they needed to know. They did come to one conclusion, they couldn't keep 40-plus boys and girls confined to their rooms for too long.

In the afternoon there came another surprise. A Chinese officer informed the two directors that Oshi's and Reece's families, all the coaches and staff were to depart China immediately. Though shocking in its appearance, Oshi and Kami decided it might serve their best interests for the time being. If Artie couldn't accomplish what was given him, Oshi knew exactly what to do and who to see.

Nine children and ten adults left that evening for Tokyo. All Oshi could do was wink and relay to Toy and Kami that he wouldn't let them down. Toy had gone back to the hotel and begged the Chinese for the use of the pool to give the kids something to do besides sit in their rooms and read Manga comics. He was told at midnight that a plane would be leaving for Hong Kong the next morning, and the Kings would be playing another national team from South China.

Toy had only one request, as he knew to protest would only cause more sanctions or harassment; he insisted that two Chinese Sports Commission representatives travel with his team. For some reason Toy thought a plane crash would be lessened with the addition of important Chinese onboard.

The mood was sour and angry at breakfast. Toy, Kami and Debbie ate at the same table and discussed this new China reverting back to its old ways of simply denying human rights to foreigners. The men had sent Sergei to the Russia Embassy that morning with the hope they would hear the boy out and allow him immunity. Sergei received far more than access to the embassy, the Kremlin was now listening very closely to the fact that there were three of their most famous men of all time being held hostage in China.

A waiter came up to their breakfast table, accidentally dropped a napkin on Kami's lap, and then replaced it with a piece of paper shoved slightly between Kami's legs. Kami pretended this was a minor inconvenience and a mistake, as the waiter apologized for his clumsiness.

It wasn't until Kami went to his room that he read the note. His eyes widened when he realized this was probably the work of Mr. Kamito, by way of the Japanese Embassy in Beijing. The instructions were direct and to the letter. Kami shot a glimpse at his watch and realized he had six minutes to find the Starbucks down the street. He sprinted down the hall to pull Johnny Turing from Trent's arms in bed. Still sporting an erection from a morning woody, Dr. Marion tossed the boy a pair of shorts.

"Sorry for the interruption. I need you right now."

With Toy's penlight and a cosmetic mirror from Debbie's handbag the two meandered down the street to this coffee house.

****

Artie Stenson had landed at Nikona International Airport in Tokyo and immediately called a phone number Kami had given him. It was definitely an old man who answered the phone, but he was very interested in what Artie had to say. This senior citizen said he'd take it from there, so Artie decided not to hurry his departure for the states and stick around to see what transpired.

Now that he had volumes of information and pictures from the Beijing trip Artie began to organize and put into effect his goal of letting the world know of this great creation of human beings. On a laptop that Jay Franco had loaned him, his pictures of these Kings' boys also included more nudity and sex play than legally acceptable.

In his visits to rooms he'd walked in on orgies and erotic love scenes that were indescribable to his friends in Washington. Even then, they'd likely never believe him. He had honored his work to Toy by keeping his hands off these beauties. For the time being the pictures were enough to keep him in fantasies for years.

Artie also had his own perspective of the attack on Mrs. Franco and her family, as well as on Oshi's. There were political repercussions that were a puzzle for the time being, but Artie was determined to sort through these to find the truth. What he had was enough to bring this to a world stage that might just protect the people still held in Beijing. He decided to call the Washington Post with what he had.

"Mr. Bernstein is not going to give you two million, Stenson," the chief editor informed Artie as a renege on their previous deal.

"No problem, I've turned independent. I'm running with the story today; if you want what I have it will cost you one million and you won't have sole copy or first refusal."

"I highly doubt if another paper is going to give you a second of their time unless you have pictures and valid proof," Redford replied.

"That I have. You are wasting my time, Redford. You have five seconds."

The editor took two. "Okay, I'll see your million. Send what you have."

With verification that the money had been wired to his account, Artie sent copies of his pictures, interviews, DNA samples and reports of the China trip. He also gave details of everything he had suspected had happened between the U.S. government and the Kings Academy. With seven straight calls he had managed to consummate business arrangements with the Los Angeles Times and all the major cities from San Francisco to New York and Miami.

When Artie counted his transactions he was now a wealthy business man with eight million in the bank. Artie patted the gold crown on his vest pocket, a crown he'd worn all over China with pride. It was quite a surprise when two Japanese men smiled with their acknowledgment of this gold emblem. Toy hadn't mentioned that the academy was that well known in the Japanese culture.

****

The Chinese government wasn't aware of the magnitude of letting a reporter go who was bent on bringing awareness to the world of China's egregious behavior. What the Chinese were more concerned about was keeping their guests for an extended period of time. The information given to them by Senator Franco was more than shocking, but a leverage they were looking for in dealing with other countries like Russia and various European nations who could sell to China the latest in military wares. Whose citizens were these boys when the United States wouldn't even protect them?

While they began negotiations with several countries to return human property to the rightful owners, it would serve a purpose to keep the Kings Academy confused by making them a traveling circus for the entertainment of the Chinese people. Sending this team to Hong Kong was a way the men's team could dominate an American team at their own sport. It wasn't important to advertise that these boys were only fifteen.

What the Chinese government didn't expect was the rapid backlash from countries like Russia and England for even thinking that negotiations were possible. Holding human beings for ransom, even cloned ones, didn't set well with current government leaders.

Russia was to the point of demanding the return of Peter the Great, Peter Tchaikovsky and Vaslav Nijinsky. It became a pissing contest that threatened world peace with the rhetoric that commenced a day after Sergei was flown to Moscow.

****

By the time the American boys were in the air flying to Hong Kong, the New York Times had on their front page:

'THE GREATS ARE CLONED!'

The headlines alone woke up a nation that, indeed, an American academy in Texas had taken responsibility for an awry experiment in genetics a little over sixteen years previous and had hidden this secret from general knowledge and the world at large. It was a four-page spread with pictures and bios on a number of boys whom Artie had time to interview.

The New York Times beat everyone to this announcement, though the LA Times was but three hours behind. They hadn't taken the extra day like the Washington Post to interview key figures in government or get a denial from the White House that they had no idea the existence of these clones right underneath the nation's nose.

The Chicago Tribune stuck their neck out ever so slightly and mentioned the possible kidnapping of two of the academy's boys, who were rescued from Crawford Ranch.   The article implied that several federal agents had been killed in the rescue of these boys in recent weeks, but there had been no confirmation or denial that federal agents were involved in this fiasco.

The White House fumed and responded to this outrageous accusation, calling it irresponsible journalism.

The Tribune and other major city newspapers had attempted contact with the Kings Academy in China, only to have their Consulate General of the People's Republic of China announce that the members of the academy and their families had sought political asylum in their country. To verify this news was also refused, since the students were currently on a country wide sports tour.

If there was anything enlightening for the Chinese by reading the New York Times article, they now had an abundance of names to which they could begin blackmailing more countries.   For a meager sum of one billion per clone and a promise to discard the current 19-year moratorium on selling weapons technology to the Republic of China the clones would be returned unharmed.

The Kings Academy, instead of being a focus for extermination, had now become a major asset to China's future financial wealth and political advantage.

Another major tabloid in Florida felt like they had had their legs cut off when the major newspapers of the country had somehow discovered this 21st Century miracle.   They had just explored the death of the one man they thought would be their source of all this information.   Their investigation had taken them to Dallas, Texas, where the magazine interviewed numerous bystanders and personnel in the tower who had witnessed the death of this reporter.

There was an immense apology from a now retired airport security man who admitted in tears that he had no intention to hurt anyone.   He was currently under psychological care for depression.

Based on supposition and speculation, the tabloid used the information about Crawford Ranch with their own findings to speculate the interference of the United States Intelligence Services with the disappearance and the asylum of thirty clones from world figures.

Senator Franco played the indifferent candidate; one who was shocked to hear of this news, yet intrigued about this experiment that would cause international chaos.   He firmly denied any involvement in any cover-up or knowledge of government interference.   He did admit knowing of this academy and his pursuit on legal matters concerning moral turpitude.   Robert didn't happen to mention that this legal maneuver had long been decided in the Kings' favor.

Campaigning through the Midwest, he hadn't had a chance to hear the final results of a plan to extinguish a potential problem overseas. This would have to wait until he returned to Washington, but, in this case, no news was good news.   The sympathy that would be gained by such an unfortunate accident would likely increase his votes with women who could empathize with his grief and loss.   Men would see his bachelorhood as favorable, since families and First Ladies were often a nuisance.

****

The Washington Post had headlines much different than their competition, as it were. *'CLONES KEPT HOSTAGE?'*

They quoted their correspondence with a reporter in Japan, who was recently asked to leave China, immediately.   Of course the Chinese Embassy in Washington denied that the man had been under surveillance as a CIA operative and had escaped the country before being arrested.

The Post quoted the reporter as saying that the Kings Academy visas and passports were confiscated hours before their departure to Japan.   Their rooms had also been ransacked with medicines and cell phones confiscated.

This crisis put the American people's eyes on China, a nation with a poor reputation in human rights and a track record of deceit and dishonesty.   The news article also included the recent arrival of children and adults who weren't directly cloned, but related somehow to the academy as being forced to leave China, while forty-five staff and students were required to stay.

The two-page coverage raised many questions for the United States' government in addressing this issue of who the boys really belonged to and if they actually represented these icons in history.

There had been one E-mail from a spouse of a federal agent who was in the vicinity of Crawford Ranch, as so reported by the Chicago Tribune.   The wife of this agent had been told her husband died in preventing a bank robbery.   When the Post went to further interview this woman she had retracted her interest as confusion and a misunderstanding.   What the Post didn't know, but suspected, the woman had been paid a healthy, seven-figure settlement to forget the particulars of her husband's demise.

****

The president wasn't extremely pleased with the headlines that shocked the nation.   He played the surprised role, one of empathy and ownership.   This wasn't so much a case of anyone stealing another country's treasures, any more than an immigrant would damage their homeland.   He certainly didn't wish to rile the Chinese; yet, to look soft to the American people wasn't a wise approach either.

Without accusing, the president mentioned that he hoped China wasn't exploiting this opportunity for their own gains.   The boys were still human beings, American citizens who were under the protection of the United States and thus deserved respect and human dignity like any other tourist.

It was an interesting press conference to cover his ass, though asserting that these boys were American citizens was a mistake he told his press secretary.   His speech did draw a line in the sand with Russia, England, Italy, Germany and several other countries that these clones were now U.S. property.   The United Nations was just about to become a very ugly scene.

There was some mention of getting Senator Franco to discuss this problem with the Chinese Ambassador, except for the recent activity by the Chinese in showing a unique interest in the senator's campaign for president.   It was the amount of money that had raised the president's eyebrows, including the abundance of foreign corporations giving substantial financial contributions to the senator's campaign.

The obvious viewpoints taken by the president and his cabinet were kept under wraps. It was most pleasant that the Chinese were the ones to have to deal with a group of clones representing the world powers and the legal quagmire of what nation they really belonged to. The president was betting the Chinese just might have the same solution he had come up with—an unfortunate plane crash. He was quite willing to force China's hand, knowing full well they wouldn't back down. In China, world leaders and journalists would never get an answer or an internal investigation.

****

At the Tokyo Narita Airport the departure of 45 happy, go-lucky Japanese wasn't an unusual scene. Their clothing and friendships only fit in to their planned five-day visit to Hong Kong. Cameras hung from every neck and the talk centered on gambling to bars and women. A logo print of a rock band adorned one of the male's sweatshirts; his raucous behavior had a few drinks behind his exuberance for this excursion.

"Omoiyare!" one Japanese man yelled to a woman who was waving fanatically to this departing friend.

It was fair to say that this departure was so routine no one really gave a rat's ass whether the men had a good time or not. If someone had examined the men's necks they might have seen something that stood out. The coincidence of all these men wearing the same tattoo would have made a person question what kind of organization this was. There again, the answer was merited in the tattoo—Seijitsu. (Loyalty)

Chapter Forty-Six

The Chinese Air Bus was equivalent to an American A-321 and had two low-level diplomats flying with the academy as it headed for Hong Kong. Dr. Kerho felt relaxed with the smiling Chinese executives, who considered this a great honor to escort this American group of boys to this sports event.

As this flight was approaching Hong Kong the vast city sprawled out below the many eyes peeking out through the plane's windows. It was when the plane flew out over the sea without the slightest hint of circling back when the doctors looked at each other with curious expressions.

Toy and Kami stood up at the same time and looked for a stewardess. In broken English, they assured the doctors that the pilots knew what they were doing. Five minutes went by when nothing but water was below the plane. This time the doctors went directly to the cockpit and knocked on the door. Nothing.

Dr. Marion wasted no time, took a step back and kicked in the locked door. In both the pilot's and co-pilot's seat were two men completely unconscious. Toy checked their pulse and found them both deceased.

"Probably poisoned," Kami admitted. He stuck his head around the opening, over two startled women and two very concerned Chinese diplomats.

"Will! Alexie! Get your butts up here!" In seconds the two boys ran between the adults to their father's wish.

Toy selected them because they were beyond doubt the two finest pilots in the academy and had won the most prizes in the boys' contests of dog fighting. Will was even called the Red Baron by his peers.

Both boys took the place of the dead pilots, once their fathers had removed their bodies from the seats.

"This plane is on autopilot, Dad. We'll have it headed in the right direction in no time," Alexie stated.

"How are we on fuel?" Toy asked.

"It'll be tight, but I might be able to fly this bird in on glider mode," Will said to the very concerned looks on all the adults' faces.

"It appears we've been through this before," Kami stated with frustration. "If we had the fuel I'd have this plane flown to Taiwan. Notice we're not exactly surrounded by Chinese fighters for not landing in Hong Kong. How coincidental."

There wasn't panic, only a whole lot of boys fastening their seatbelts and listening to the flight attendants explain what to do if they had to ditch in the sea.

Gradually the A321 turned and headed back for Hong Kong. Within minutes Will had the plane lined up at the International Airport after he had made contact and was put on a direct course for an emergency landing. Through the last mile of flight the plane literally dropped like a wounded bird with power. Though the landing was bumpy, Will brought the big airliner to a dead stop to the applause of his peers and several adults.

Dr. Kerho didn't expect that his entourage would have a welcoming committee. Not surprisingly it took another hour of waiting in the terminal before a bus showed up with very apologetic officials unaware of this scheduled arrival. If Toy could have sworn in Mandarin, he would have. One thing for sure, there would be no more flights or transportation anywhere outside of this city.

Wendy snuggled up to her father on the bus, oblivious of all that had transpired. She asked, "Daddy, do you think we can purchase any birth control pills in this city?"

"What happened to the supply you had?" Dr. Kerho asked, almost delighted that a different subject helped take his mind off the Chinese.

"The Chinese took them. They confiscated everybody's herbs and pills, like they were looking for steroids," Wendy assumed.

"I'll see what I can do," the doctor promised. "Have you considered keeping your legs together for a few days?" It was at least said with a smile.

"Oh, Daddy," Wendy whined. "You know I'm in love."

Wendy smiled and kissed her father on the cheek, but she was only slightly worried that she and Maria had missed the last two days. She didn't want to admit that her father was right. There was no sense in risking pregnancy for the next few days until she and Maria had their birth control pills again. Lance had admitted he no longer took his male birth control pills and Jay had come to the conclusion that hanging around a girl was about as boring as watching gold fish in a bowl.

Sitting next to his sons on the plane Kami wasn't sure if he'd done everything that the message he'd received from the waiter had asked him to do. He and Johnny had gone to the Starbucks, sat outside at one of the tables, while this teenage genius held a mirror between his legs and flashed Morse code words to reflect upward. The whole scene was comical, if not bizarre. If anyone was watching, it looked like a boy in play while the two waited for their croissants and coffee.

Kami understood space technology to a point; to think that there was a Japanese satellite, 80 miles in space, which could look down on a boy's crotch was scary if not out of some Star Wars movie. If anyone could do it, the Japanese could.

It was disturbing that they all had to pack their belongings and equipment for this jaunt to Hong Kong. Toy made sure all the kids wore their white cowboy hats with the Kings' emblem stenciled in front. They all dressed in sharp outfits, which made the students appear more like representative of the United States at the Olympic Games than teenagers from a private institution in Texas.

Whatever was happening in Japan Toy knew better than to question the directions. If the kids were to wear hats, they'd wear hats 24 hours a day. Any attempt at rescue was now imperative with their lives at risk every hour.

The academy had almost not boarded the plane to Hong Kong, a determination that Toy had now wished he had been adamant against. The Chinese had implied giving the Franco family dignitary status and insisted that they be escorted down on a private jet. Not only did Debbie vehemently oppose this idea but Toy and Kami halted the boarding of their own students. The Chinese finally relented and allowed everyone to board together. Toy didn't want to think that the Chinese had decided that everyone was expendable, and not just the Franco family.

Debbie was both serious and attempting to lighten the ambience when she commented, "I hope this doesn't mean we all die together because I refused to be a dignitary."

"With politics you get strange bedfellows and company you could do without," Kami replied overhearing the conversation. "Your divorce has now exempted you from political assassination."

Debbie had hoped Kami was right, but little did any of them know.

The academy's staff and students were taken to the Intercontinental Hong Kong Hotel, a nicely furnished hotel and above average in luxury. There were no phones in the room and Toy was instructed that there would be no outside calls allowed. They certainly weren't expected.

As a group they ventured through the city's Central district to the commercial heart of Hong Kong and boarded a cable car to the top of Victoria Peak, the quintessential colonial millionaire hill station, where a panoramic view of Hong Kong could be seen.

Every time Toy or Kami attempted communication with a tourist or an American, a Chinese guide intervened. Dr. Kerho wasn't sure he knew who to call anyway, nor did he feel it necessary, but it was fun teasing their so-called hosts.

The group ate at Yan Toh Heen, an exceptional Cantonese restaurant. To finish this tour, they went on a sampan ride through the floating village of Aberdeen, thick with junks and sampans, houseboats for Hong Kong's fishing families. Kami couldn't help but comment that such housing projects probably didn't require a city permit. His guide smiled.

****

The Hong Kong Sports Arena rivaled the Staple's Center in Los Angeles. It was truly massive and a modernized structure, though the credit should go to the British more than the Chinese. The game between the Kings Academy and the Chinese All-Stars had been quickly assembled and advertised in this seaport in southern China by TV and radio. The academy would not see a dime of the ticket sales, and all

18,000 tickets were sold in six hours.   No one mentioned to these interested Chinese men, women and children, who relished any sporting event put on by their government that the Kings' athletes were but high school students.   The promotion certainly included that the Chinese National Junior team had taken a close loss to this team because of injuries.

When the Kings' students were shown to their seats they could've used binoculars with this nosebleed section given to them to watch their own team.   Kami sat on one side of his significant other, while Wendy and Jay, then Lance and Maria finished the row.   Toy leaned over to whisper to his daughter.

"The pharmacy wouldn't give me birth control pills without a prescription from a Chinese doctor, but I bought a package of Morning After pills.   Did you use one?   I had them delivered to your room."

"Not yet, Daddy," Wendy whispered back.   "It's only evening and that's why they are called Morning After for a reason.   Are you sure you went to college, Dad?"

Toy smiled and felt put in place by his daughter.   "Excuse me, my precocious one. You should remember, when you were six-years old you loved to sit on my lap and pretend you were helping me with my homework."

"Pretending?   Pleeeaaassssseeee!   A six-year old can certainly understand molecular genetics.   It just wasn't as much fun as crayons and colored blocks.   Anyway, Kami's homework was easier.   He fed me to keep quiet."

"I heard that," Kami commented from the other side of Toy.

Father and daughter chuckled.   Wendy gave her father a hug and thanked him for his thoughtfulness on her needs.   Earlier, to celebrate the pills arrival, Jay and her had passionate sex in the shower, then left the tub for Maria and Lance to take a bath with very romantic sitting positions in the bathtub.   She and Jay had sat on the stool and admired the gymnastics.   For Jay it felt good to be accepted again.

The Kings' basketball team resembled the towel boys next to this Chinese men's group, a quickly assembled bunch of wannabe pros from assorted teams.   Each of the starting Chinese was over six-five.

Dr. Marion watched the boys warm up, then proceeded to move down to the bench and take his place to substitute for their basketball coach who was sent to Japan—forced, that is.   To Kami it was ridiculous to allow a team of fifteen-year olds to play grown men with international experience and referees who looked the other way as muscular men checked boys with 28 and 30 inch waists.

Three minutes into the game the Kings had these Chinese men eyeing each other.   The boys had made four-three-pointers in a row to lead 12 to 10; the score was but a temporary moment of hope.   All the Chinese did was bring out their three six-eight men to guard the perimeter and put a stop to this long-range bombing.

The Chinese were too quick, too big and too experienced, using the post and paint to shoot over the smaller Kings' inside men.   The seven-three center for the Chinese used his ass like Shaq O'Neal to send the boy guarding him stumbling backward a few steps.   A twist to the inside and the man had an easy lay-up or dunk over young Lincoln.

There was one positive aspect to several of the Kings' boys who spoke Cantonese; they were able to trash talk with the Chinese.

The Kings had their moments of ball control and waited for the open man to avoid the taller Chinese.   The score wasn't as lopsided as it could have been: 65 to 52; a respectable showing for a group of teens against a semi-pro like team in the states.

As the two teams lined up to shake hands a beverage vendor came up to Dr. Marion's row with five drinks.   Kami began to protest that he didn't order this until the man said, "Compliments of Mr. Kamito." Kami was handed an empty cup and at the bottom was a message. (Everyone to Locker Room—be ready to move—spit in cup)   To Wendy's peculiar stare Kami sent a glob of spit into the cup which made the words dissolve.

"We're all going to the locker room," Dr. Marion said. "Pass it on."

This large group of American kids paraded through the exiting spectators who weren't sure what they saw that evening. A team of five boys, clean shaven and too young to drive, yet they all shot from a distance that had the fans holding their breaths with each shot.   If the shots weren't blocked by their Chinese counterparts with arms like tree limbs, the ball usually found net. In the final half the majority of spectators were rooting for the youngsters.

Kami and his followers found that the team had beaten them to the locker room. The nine basketball players were all showering and surprised to see their classmates. No more was the last student inside when a man sprang out from behind an equipment door and braced his back against the door.

"Reece?" Kami asked in disbelief. The two brothers gripped each other with a mighty hug as dozens of Japanese men swarmed into the area.

Reece quickly explained the proposed plan in detail as the Japanese began to change clothes with all the students. By the time everyone was dressed and the cowboy hats were pulled down over their faces, both doctors stood back and gave it a fifty-fifty chance of working. There were positives: the bus was directly parked by the backdoor; it was dark; and the Japanese were close to the size of the athletes, except for the tallest of boys. If there was a problem the Chinese would need guns—lots of gun—if they were still alive to obtain them.

The two brothers also switched clothing, then hugged again with a promise to see each other soon. A losing team is never happy in their post-game mood.

The Japanese played the role with perfection: quiet, heads hung low, they boarded the bus past two Chinese agents who, in some ways, had to feel sorry for this losing group against their giants. One thing for sure, they knew how to dress sharp, these boys from America with their cowboy hats on, the agents agreed with each other.

With the bus headed back to the hotel, Toy and Kami waited patiently inside the locker room with forty-plus bodies holding their breath. A custodian surprised the group when he arrived to do his job. Kami didn't wish to hurt the man, so he tied him up and promised he'd be freed in the morning. He also left him more money in his pocket than the janitor would have made in six months of work.

The casually dressed men, boys, and three women, all wearing Mao caps with red stars on the crown, nonchalantly traipsed out the back door and blended in with the remaining crowd leaving the arena.

Two blocks away Toy saw the multitude of bicycles that had been previously arranged as part of their getaway. Each person was given precise instructions, leaving 30-seconds apart and in pairs, except for the three females who were each accompanied by a male. Slow and relaxed the riders were guided by a distant tower with an air traffic light on its tip. The reference would take them in the direction of their destination: the East Lamma Channel.

Only a mile away a Chinese man was dressed up as a street vendor, selling meat on a grill by a street corner. When each cyclist approached, the man offered a piece of meat, then a nod toward a path that swept down between groves of trees. The bikes were quickly discarded one by one, the rider stripped naked and shown one of hundreds of dark shadows out on the harbor. The only thing different about this one junk, it had a blue light on its deck. There were so many junks out in the channel it was like a Christmas parade of boats.

Toy had trusted Kami to go first, to lead this string of escapees to the proper location of extraction. It wasn't easy for Kami to be the first to arrive at the water. He tried to persuade his brother that he should stay and assist, but Reece pointed to another shadow by a tree.

The Kings' director took a closer look but there was no mistaking a man Kami trusted above all—Oshi. Kami slid in the harbor waters and began swimming toward the distant junk. In pitch darkness he heard the next swimmer splash in behind him as they churned through water that smelled like diesel fuel, with occasional substances and artifacts in the water that were best left for the darkness to keep as its secrets.

Within a half-hour there was a line of swimmers making their way through this pool of slime and harbor wastes. Freedom had its price so no one was complaining with this means to an end. Sometime between the 20th and 21st rider, a strolling policeman noticed a pattern of bicyclists exiting off the main boulevard and getting fed when no money appeared to change hands. The Chinese man's curiosity had him investigating this odd occurrence, but it would be a mistake that this man wouldn't live to regret.

Kami was the first to approach the junk with the blue bulb on its stern. He was mentally preparing himself to have to climb up a rope ladder, an effort his tired arms didn't look forward to. He found himself literally swooped out of the water by a net; a net with a black coating inside that, if anyone was looking even a short distance away, they wouldn't have noticed anything rising from the water. Over the next hour every swimmer found themselves hauled up and put into a cargo area, where they were showered and wrapped in a blanket, then sat and ordered to wait for everyone to gather.

Thanks to Alex and Travis, Mrs. Franco made it in spite of the tiring swim. Debbie knew her heavy breasts would come in handy sooner or later. She just never envisioned herself skinny dipping in a Chinese harbor to save her children's lives and her own.

The junk soon began to sail northwest to what was known as Deep Water Bay. The group of blanketed and happy survivors were directed to move below deck to another cargo hold, where they each stepped down six metal steps and taken down a narrow aisle to bunks lined up on the side. No one had the slightest idea that they'd just boarded a Japanese sub, which had its own unique docking station within the bowels of this fishing ship. The sub wasn't designed for an attack sub but one for rescue and reconnaissance.

Toy and Kami were directed to the con tower where a man with his back turned toward them greeted, "Dr. Marion and Dr. Kerho, I presume?"

"Watakushi wa nihongo ga wakarimasu," Kami replied by saying he understood Japanese, if the man wished to speak it.

Reece turned around and pointed to where Oshi was standing by the periscope. "I'm sure you do, brother. I always hoped I could save your asses someday because you've sure been good to Oshi and me."

****

The submarine docked a day and a-half later in a clandestine docking platform south of Tokyo. Everyone boarded a bus and was driven to the impeccable Park Hyatt Tokyo Hotel in the tiny West Shinjuku district. The staff greeted the group dressed in gray sweats with bows and smiles. Brittany was there to give Debbie a hug and kiss and welcome her brother and all the Kings' kids. Debbie knew the bucket of water dropped over her body on the fishing trawler didn't quite prepare her for such a greeting. She ran straight to the showers when she stepped foot in her room. At $400 dollars a night she was going to find comfort and a good night's sleep.

Brittany had done prior shopping to give the teens a clean set of clothes before she had a chance to take them out for shopping and a good meal. Most of the kids flopped down on their beds after a hot shower and slept until morning. No rest for the weary, Oshi had them up early. They shopped for running shoes and gear at the Art Sports store in Shibuya, then had his pledges jog four miles up Roppongi-dori to where they could circle around the Imperial Palace and Kitanomaru Park. There was no grumbling and they all knew they could use the exercise after being coupled up for two days.

Toy and Kami had their own run, then settled down in the comfort of their room to feel their safety in what the two men considered like their second home. Toy called Artie's cell phone and found that his favorite reporter was at the Capsule Hotel Fontaine Akasaka in the western part of Tokyo. The 75 square foot capsule was only $42 a night, had TV (Japanese only) and radio but no fan. Artie was ready to kiss Toy's feet when the doctor invited him to spend the next few nights at the Hyatt--on the academy, of course.

A conference room was reserved for dinner during the evening hours. There were chairs and place settings for over one hundred, as their rescuers had arrived back from Hong Kong and there was a joyous reception with the mission's success. Laptops and other paraphernalia were politely returned to their respectful owners as these Japanese had used the students' room for a few hours until they returned to their own during the night and made their exit back to the Hong Kong International Airport.

The Franco kids were briefed by several Kings' boys on what to expect in Japanese custom: When meeting for the first time, make eye contact, then drop your eyes to the ground, bow slightly, and extend your hand in a soft handshake; mirror the actions of the Japanese—when they bow, you bow; when they lift their glasses, lift yours (they all laughed because nobody had glasses on)—"I meant drinking glasses,"

Alex replied and had to laugh at his friends' humor.

He continued to tell them that when passing a glass, cup, or a gift, always give and receive with two hands; never rest your chopsticks on separate sides of the plate, but do put them together and place them sideways on the plate; never point, spear your food, or squash items with a chopstick, and never stand chopsticks upright in your rice bowl.

"Do we ever get to eat?" Jay had to ask. He got pinched on the butt by Wendy. Jay was thinking that girls acted too much like wives, after you get to know them.

Brittany and Debbie had an afternoon to prepare with all their elegant purchases. Satisfied with their shopping obsession they returned to the hotel for the anti-aging facial and jet-lag renewal treatment. At 59,000 yen, or $545, they knew they deserved the very best after what men had put them through.

After the main course of lamb, they had chocolate mousse for dessert. The rescuers had also removed the gold crowns from the blazers on the beds of the Kings' students. These were passed back to each student and staff as a gesture of success.

Wendy and Marie realized that they were well behind their safety measures. What they would discover a few weeks later, the mistake was costly. They were both already pregnant. It was only a question of who were the fathers.

The evening would not have been complete without the appearance of a 77-year old gentleman. Mr. Kamito received a standing ovation and deserving kisses from his two favorite boys, Kami and Reece. Miyato was by his side, his niece and forever guardian. Miyato, eighteen years after meeting her half-brothers, was still glamorous and fit at 39. She had mothered Kami's two boys and was soon to be a grandmother when the boys returned to Texas.

Chapter Forty-Seven

Leave it to the British to give this Chinese blackmail to the press and let the English people have their say. The British had more to lose than any country in regards to the number of clones derived from deceased British citizens—eight, to be exact.

There was a mixed reaction to the initial London news that the clones did exist and had been growing up in the United States. There was the typical outrage of grave robbing and stealing the very significance of English history.

Other British were more logical, realizing that these boys were created by unorthodox methods, but were not these original kings, poets, writers, or honored citizens.

The greatest debate appeared to challenge this notion of eugenics; the systematic control of trying to develop a superior race of humans by genetic engineering. With the names of most of the thirty boys revealed, it no longer took a brain surgeon to figure out that these were gay individuals who were now progenitors. The question still remained which would obviously prove the genetic bearing of gay orientation— were these thirty boys themselves gay? The choice-versus-genetics debate had been mostly settled over the past decades in favor of genetics. There was no doubt that men and women still had difficulty with gender identity—how people see themselves—but sexual orientation was truly about attraction and that's something you eventually can't avoid.

The word clone was no longer a popular term, partly because of the misunderstanding of the configuration. Some people wanted to believe that Richard Kerho was still Richard the Lion Hearted and desired to be King of England. Or that Oscar Wilde would return to England to become a famous playwright and be the flaming homosexual who hustled good looking young men.

In successive articles following the worldwide news of the discovery of these thirty boys, molecular biologists put the one issue at rest; chances were that these boys were homosexual because of the direct cloning process. If a female had had the opportunity in the development of the embryo, the only hope that a child would become a homosexual was in a female region called Xq28 on the X chromosome, inherited from the mother that is statistically correlated to homosexuality. Since these boys directly inherited all the DNA of their forebear, this biological gamble was already decided.

The British had set up negotiations with the Chinese to bargain for the release of all the boys. China had begun to realize the political fallout and international pressure to release their hostages. Their solution was a plane crash.

It was a call from Dr. Kerho to Prince William that announced the escape of all the hostages which allowed the British Ambassador to make the Chinese Ambassador wait two hours for a meeting that was never going to take place.

The Chinese government became very stoic in this embarrassing news that the Kings Academy had never sought asylum, but were more paraded around China as a sideshow for someone's financial interest. They quickly paraded two female flight attendants and two diplomats to admit an attempted takeover by these Americans of a Chinese aircraft in an attempt to make their host look like kidnappers. China said they only had the best intentions for the academy.

The prince and Toy had discussed at length the foreseeable problems with how the general public and governments viewed these thirty boys. The first thing Toy did was come forward with the original brainchild, Dr. Jacob Jolson, an Israel scientist, developer and archeologist, who schemed this plot from the very beginning. Fortunately for all, the doctor had since succumbed and left a project of newborns as someone else's problem.

With the help of Prince William the London press gave a perspective that these teenagers were no different than if they were children born of these famous men, whether they were from the past or the present was little difference. The English had many famous citizens living in America, Paul McCartney, as a primary example. It didn't mean that Great Britain was ready to start a war because they wanted Paul's children in England. The boys were no more and no less than descendants of famous to some, infamous to others, but all historical men of remembrance.

****

While the Washington Post kept their headlines: *CLONES HELD HOSTAGE*; *CLONES ESCAPE CHINA*; and *CLONES SEEK A HOME,* the Japan Times viewed these boys as heroes and the world's future. All over Japan the teens were given recognition, treated like they were the Emperor's sons. One in particular, Aki, was nearly worshipped as a returning samurai from four centuries back. It was a whole other education to convince many Japanese that this boy wasn't a reincarnation.

Toy and Kami immediately put the students through a rigorous schedule of staying physically active in conjunction with their many invitations all over Japan. They attended a samurai competition with over 600 mounted warriors competing in traditional samurai skills. Many festivals that were held throughout the year at various times were demonstrated just for the Kings Academy in honor of their visit to Japan. The Naked Festival around the Konomiya Shrine in Inazawa was held as such a celebration.

The students were invited to observe the creation of handmade paper, and then stayed at a Buddhist temple in Tokushima, Shikoku. The teens entered their own kite at the Hamamatsu Kite Festival, where kites don't just fly—they fight. The teams try to cut their opponents' strings with their own to bring down the other's kite.

There was something for everyone, even the musicians at the academy took part in a performance of taiko drumming. It was a special tradition of Japanese music that required one to extend their arms smartly, spin on cue, and flip the drumsticks in the air—all the while keeping perfect time. The humor and exercise had the kids in stitches.

In the meantime Artie had lost all control of his network of news reporting. He found humor in his immediate wealth and was just happy to be part of this world breaking story. What he desired most was for the truth to come out about how his government almost destroyed so many valuable and interesting lives. Toy and he set off through Tokyo to discuss plans to return the kids to the states and be treated with respect instead of pariahs. Kami was their guide through a completely different Tokyo than Artie had explored in his own previous visit. At the end of the Arakawa tramline, there was a warren of narrow streets and neighborhood shops. They had a lunch of mitarashi-dango and an-dango, rice and bean-paste concoctions.

The men walked further into the neighborhood, past cramped, cluttered alleys, wild gardens, and wooden homes. There was no neon, no crowds, and no signs except for those that graced simple storefronts. Shopkeepers still calculated by abacus or in their heads. More bicycles than cars.

Kami led them to a bathhouse where they had to duck to enter a low-ceilinged room with cold and hot plunge pools and twin steam rooms. Naked men were in abundance and greeted Kami with the ultimate of respect. As confident as Artie felt in a gay bar, he knew he'd found heaven with all these naked Japanese bowing to him, and he in return bowed to them. Their massages were rough but delightfully refreshing. Rooms with men and boys, or any other combination, were wide open for those who wish to voyeur.

"Well, Artie, where are we?" Toy asked and he didn't mean in a bathhouse.

"I put a stop to the Washington Post using clones in their articles. I said one more time and I'd cut them off. They're starting to call the boys the thirty wonders of the world," Artie said.

"What can we expect as a reception if we return to America?" Kami asked.

"It's a toss-up," Artie replied. "Our government won't be playing their games, though that doesn't mean that something mysterious won't happen. All the boys' accusations are just that, hearsay, without any substantial evidence. Naturally our government has denied any knowledge of your existence until I broke the story. Apparently this Lauri Fisher applied for asylum in China. I believe this to be an actual request. My sources at the Post believe that the Chinese have gotten in bed with the senator, but I'd guess he doesn't know why."

"I see where you're coming from, Artie," Toy spoke up, his body melting in the hot vat. "Let's assume Lauri told the Chinese everything, including their relationship. Robert Franco becomes president and China blackmails him with the truth of the adultery. He either kow tows or the Chinese screw up his presidency. It sounds like Lauri Fisher was being treated a whole lot better than we were."

Artie moved back and sat on the edge of the tub. "I'm going to fly back to Washington tomorrow and see what I can stir up. If I shake a few trees I could get a few panicked individuals," he suggested.

"Don't get yourself in a noose, Artie," Kami warned. "Our current president and Senator Franco have too much to lose if they feel threatened. Would you like to see a Kabuki tonight? You might not want to leave after you see your competition in Japan."

Artie laughed. "I've bought the most perfect evening gown for the occasion. Did I tell you that Bobby promised me a penis like theirs when I get back?"

****

A week after missing their periods Wendy and Marie were discussing the outside possibilities that they were pregnant. They each gave the other the opinion that they were just late with their periods, though that thought didn't stop them from dashing to the nearest pharmacy to buy a pregnancy kit. The positives made them gulp and Maria only hoped that Lance had been luckier than her brother or Hikoyi. For all the assurances Maria had given her mother, to have Jay's baby would be terribly embarrassing.

As gossip went, the news spread quickly. Lance was both excited and felt empathy for Maria, if that was the case. He'd also had sex with Wendy during their orgy in Beijing, but he'd done it more with Maria. It was all too confusing to know if he was the father of two babies, or maybe Jay, or was it one each? Then there was Hikoyi who was just glad to have reached puberty and have a girl like him enough for his heroics with his father. He let it be known he wanted the baby to be his.

Though Travis didn't know the circumstances, he just haphazardly heard the news from Alex. Travis hinted this to Niqui, who told Jay, who didn't smile and hid his tears from his brother by leaving the room.

****

Dr. Kerho and Dr. Marion visited each room every night before they turned in. Their interruption of sexual unions had its humorous moments. It was also a time to listen to problems or observe the boys' growing up dilemmas with their peers. In this case there was one boy they couldn't find—Jay. When they looked for the youngster in Wendy and Maria's room it wasn't the missing teenager that was discussed but the speech Wendy had prepared for her father.

"Daddy, I'm pregnant. I have to ask if a twelve-year old has the ability to impregnate, if it's his first time?" she admitted and waited for the reaction.

"It doesn't come as a surprise that this happened. In answer to your question, just because a boy reaches puberty, his sperm is usually too immature to penetrate an egg. There are exceptions, but rare." Toy replied without emotion. "I can't blame you for being irresponsible with what happened in China. Do you think Jay has panicked?"

Maria decided not to keep any secrets, as well. "Dr. Kerho, I'm pregnant, too, and it might be Jay's, we're not sure. I think he might have panicked if he's not in his room."

Dr. Kerho considered this, knowing Jay as a fragile boy who often just went along to keep pace with two bigger brothers. He was about as vulnerable of a child as they had in the group. Trouble was, the boy didn't have a GPS chip in his foot. Toy and Kami immediately contacted Oshi, Debbie, and Brittany; all of them hit the Tokyo streets at 12:30 in the middle of the night. A city of 13 million and they were looking for a forlorn kid in a blue T-shirt and shorts. How difficult could that be?

The adults spread out in different directions with Kami and Oshi thinking the obvious places where a teenager might feel comfortable. While Kami walked the crowded Shinjuku District, where streets blared with the sound of J-pop and the clatter of pachinko parlors, Oshi traveled near the Harajuku station, which was Tokyo's teenage gathering spot for preening, strutting, or in comparing their latest collective fashion statement.

519

There were kids in foot-high platform boots staggering about in Saran Wrap tight miniskirts; Maria Antoinette wigs; Goth dress; harum-scarum, multicolored hair; tattoos and skin piercings; and makeup that rendered the boys and girls with the subtle Kabuki look.   Girls bantered animatedly, feathered boas swishing and makeup trunks open.   Boys dressed in their favorite Manga comic book characters, their boyishness being both feminine and the boy-for-sale look.

Oshi didn't expect to see Jay mixed in with any one of these groups, so he stood off to the side, not wanting to necessarily blend in, but not desiring to offend either.   Girls and boys offered their services to this handsome Japanese man who, if they only knew, was a Japanese sports hero.   Too busy to be annoyed, two boys who wanted this man as a daddy figure for a night had their ears twisted to remove their bodies from Oshi's way.

Kami scanned his area of neon scrum of love hotels and bars— Footnik, Atomic Heart Mother, Hungry Humphrey, Bar Bee.   His search took him by numerous alleys and off-streets, and when he spotted a group of teens in a half-circle staring down on something they were finding pleasure in, Kami stepped into this alley and approached.   In between one of the teen's legs Kami spotted someone sitting back against the wall.   He put in a call to Oshi, just in case.

"Jay?" Kami called out and saw a head pop up from the sitting position.   Whatever abuse or discussion he was involved in had come to a halt.

Jay started to stand up until one of these youth pushed the boy back down.   "Fuck off!" Kami heard from one of these youth.   Dr. Marion wasn't sure of the circumstances if these were friends Jay had found, or if he was being tormented in some way.

"I've come for the boy," Dr. Marion said and started forward.   He found himself quickly surrounded, but he wasn't going to start any aggressiveness as he reached to help Jay up.   One of these teens laid a hand on Kami's shoulder and that's all it took for this teen to find himself part of the wall.   The other odd-looking characters stepped back and decided to let this American have his way.   It was when they were walking out of this alley when someone kicked Jay in the seat of his pants.

Dr. Marion had the boy stand where he was, then went back and did a few slaps, two throws and a couple of roundhouse kicks that had five Japanese youth reclined on the street. For some reason no one wanted to get back up. When he turned back for Jay, the boy had disappeared.

Out on the main street Kami saw an arm waving with a captured boy in the other hand. Oshi had a habit of boys running into his arms. Dr. Marion was a trifle heated when he arrived next to Oshi.

"Thanks for waiting," Dr. Marion said harshly. Jay was too busy crying to reply.

The three males waited for Toy to arrive. He had called the women and thought it best for them to return to the hotel. There was no use embarrassing this youngster with the stares of women in all their sympathy or caring.

"Ginza ni ikimasho!" Toy expressed to his buds when he approached.

Without conversation the four males went to a bathhouse with an adjoining restaurant. With all the walking, Toy, Oshi and Kami ate while Jay stayed stubborn. They acted like this was a night out for the men and Jay did as everyone else, not sure what these three adults had in store for him. They undressed, hit the stream room, soaked, scrubbed, and then had their massages. All the time Jay did exactly as his mentors. Finally, Kami and Oshi went to a separate tub, while Toy made sure he and Jay had their own.

"Ready to talk, youngster?"

Jay cried.

"Okay, you're upset because Wendy is pregnant. Do you think I'm all upset with this? Do you think I'm upset with you?"

"I didn't mean to do that," Jay squeaked out.

"Let's understand something, Jay. We promote you kids to share, to experience, to love one another. Wendy loves you a great deal and, if there's blame here, I'm more upset with the Chinese for disrupting the measures that were in place to avoid this for the time being. Guess what? Mistakes happen, things don't always go as planned, and life isn't always fair. You running away causes me more concern than my

521

daughter's pregnancy."

"I can't take care of a baby. I don't even know how," Jay expressed with pleading eyes of being completely overwhelmed.

"Jay, I had Lance and Wendy when I was fifteen. I had no idea what to do or how to act with two babies. The truth is most humans are not equipped to raise children, even in their thirties and forties, and shouldn't be expected to be. They really haven't lived enough as adults to pass deep wisdom to their children. My parents helped raise my children and twenty-nine other boys. Your younger years are not meant for truth-teaching, but for truth-gathering. How can you teach children a truth you haven't yet gathered? You end up telling them the only truth you know, the truth of others: Your father's, your mother's, your culture's, your religion's. When Dr. Marion and I put together the Kings Academy we wanted what was best for all the boys—not to bring all our negative baggage of prior sexual embarrassment, repression, and shame—which would lead to sexual inhibition, dysfunction, and violence. We created a different, humane world. You're part of this now.

"Wendy knows and accepts the fact that her elders will raise this child, which does not exclude your love or hers. You won't be solely responsible for their care or upbringing."

"Will I have to marry Wendy?" Jay asked.

"Have to?" Dr. Kerho wanted to laugh but he resisted the urge. "Of course not. As much as I would love to have you as my son-in-law I don't have a shotgun tonight, as if this was a backwoods family—you get my girl pregnant, you're going to marry the gal. We've evolved beyond such histrionics. Maybe if you are in love with my daughter, when you both reach your twenties and have finished your education, then you two can talk about it. Until then, I want you to grow, to learn, to love and play as a teenager with all the responsibilities of children, no job or stress. Got it?"

"You're not mad at me?"

"I'm only mad at any of my boys when they don't face their issues straight up. You've made me a grandfather, so how can I be mad at you?"

"Won't the other boys be upset?"

"There's not a boy here that's perfect, Jay. In a month or so we'll be starting on another dorm for fifty or more boys, many of them twins or triplets. This is what all my sons have wanted, to begin a new generation of students. Much like they got started in the world, you'll be the big brothers who will set an example for your children. You just jumped the gun on everybody."

Jay smiled for the first time that night. "I'm done with girls, Dr. Kerho. They're too stressful and domineering. Boys are far more fun and sexy."

"Wise decision, young man, though your support for Wendy through all this would be appreciated. Give yourself time to sort out your sexual needs. Girls aren't all that bad."

Jay nodded.

"I've been meaning to ask you, Jay. If you and Niqui had anything else we might use to support the government's conspiracy against you boys or the academy?"

"No sir, just what we listened to with our security system," Jay answered.

****

Seven o'clock in the morning Toy sat straight up and nudged Jay who was sleeping in the same bed. Kami and Mikki were in other bed, as that was where Kami found the boy when they returned. As was often the case, different boys loved to sleep with an adult through the night. Mikki had got a tongue lashing from his mother for streaking through the lobby, naked. He sought comfort with someone who would have compassion for such antics.

"Jay, sorry for waking you, but I just had a thought from last night. Did you say you guys used a security disk? Like a listening and recording device?"

Jay rubbed his eyes and realized where he had spent the night. He nodded and whispered. "It was part of our laptop, but we never listened to it that way. I gave my laptop to Artie to use."

Toy swept the boy's hair and put him back to sleep. He got

523

dressed and slipped out into the lobby, where he dialed the Washington Post. It was four in the afternoon Washington time.

"Wallace here," the voice answered.

"This is Dr. Kerho, in Tokyo."

"Oh, yes, Dr. Kerho. We've written a great deal about you lately. Is there a problem I can handle? The editor is out of his office for the time being."

"Artie Stenson is flying back to Washington as we speak. If you wouldn't mind relaying a message for me I'd really appreciate it."

"No problem, Doctor. What is it you want me to tell him?"

"Tell him that behind a Velcro cover in the top of the laptop he's carrying is a security disk. Tell him it's important."

"I understand, Doctor. I'll make sure he gets the message," Wallace replied.

Toy decided to spend the morning on the streets of Tokyo, while his academy slept in through the early hours. He shopped on Omote-sando, a broad tree-lined avenue with high-end shops—Chanel, Dior—and upscale cafes, where he grabbed a breakfast for sixty-five dollars and barely felt full. He noticed there were more jeans and cloyingly cute Hello Kitty gear on Tokyo's streets than kimonos, although there were always a few of the elderly women who still wore the custom dress.

He purchased a couple dozen Manga comics at Akihabara district for his kids, and enjoyed the contrast of teens in neon-colored hair and punk and biker gear. His joy centered in what that security disk was going to have recorded from what the boys' heard. If everything was on that disk, they had a president and a senator nailed to the cross. As a celebration for his quick thinking—only six hours later did it click in—he bought several pieces of Japanese furniture for his and Kami's office and a few Japanese prints that would please the love of his life. Toy even decided to take the two new fathers and others out for a shopping spree later that day.

****

A fourteen-hour flight from Tokyo to Dulles Airport in Washington had left Artie happy to step off and smell the Capital once again after almost two months of vacation that had changed his life forever. He had a new sense of power and control; a sense of pride that he could walk into the offices of the Washington Post and not grovel for attention or play a second rate journalist because he did the gay beat.

The wait for his luggage was long and patient, his lone suitcase that he'd purchased in Tokyo was full of new clothes and artifacts obtained in Beijing and in Japan. Jay's laptop was cradled under his left arm; weeks of biographies and notes and a great deal else he hadn't yet shared with the press. He had promised Jay a newer and upgraded model for being kind enough to allow him to use the boy's computer.

As soon as he stepped out of the terminal the sun made for near blindness as Artie held his arm up in pursuit of a taxi. The impact struck his chest point blank like a sledge hammer, spinning and dropping Artie like a heavyweight punch that he didn't see coming. It was the hit on the pavement by the back of his head that added to this instantaneous shock. His eyes glazed over and the last thing he remembered was some woman asking him if he was all right.

✴✴✴✴✴✴✴✴✴✴✴✴✴✴✴

The yellow taxi had parked in a most likely spot to receive passengers; it was the driver who refused any request, though his taxi sign definitely read Vacant on the roof. No one noticed the small hole in the trunk that served as a clandestine exit for a silencer screwed into a 9mm Berretta.

When the agent spotted the target exiting the terminal his intention was a head shot until the man put his arm up to attract a taxi. With the crowd of incoming and outgoing passengers, the agent took his best shot with the right angle—a direct shot to the heart. As the target spun and dropped, another tourist quickly came to the man's aide, only to slip the laptop under his own arm while pointing to another passerby, his partner, to get help. Of course in his departure to find a policeman he carted away Artie's suitcase. Within thirty seconds the taxi finally drove off with the two gentlemen inside.

✴✴✴✴✴✴✴✴✴✴✴✴✴✴✴

It took twelve minutes for an ambulance to arrive at the spot where Artie lay. If there was any consolation, a doctor, already late for a

525

flight to Atlanta, played more than his Good Samaritan role and stopped the bleeding from a cracked skull. There wasn't much else he could do with the man unconscious and the sirens of the emergency vehicle heard in the distance.

In the Emergency Room Artie Stenson was examined thoroughly, where it was discovered that there was far more to this than a man who had accidentally slipped on the sidewalk. A 3 by 3-inch gold crown had been dented severely by a projectile that was quickly determined to be a bullet. The police were called that second.

Woozy and disoriented, Artie had regained his consciousness as he arrived at the hospital. With the stitches being finished, Artie couldn't recall what he could have possibly slipped on. This treating doctor leaned over and replied, "Mr. Stenson, you were shot in the chest by a nine-millimeter bullet built to fragment once it enters your body. Do you have any enemies?"

Artie sat stunned, a bandage around his head and he was given the gold crown that he cherished so much as evidence that he'd been very lucky. "Maybe someone didn't like my last review," Artie said but he wasn't smiling. He grabbed for his coat pocket and pulled out his cell phone to dial Tokyo.

"What do you mean you've been shot?!" Toy frantically asked.

"I'm sitting here with stitches in my head because some potshot tried to kill me. Sure I left a few boyfriends sulking, but none of them are that violent."

"Look, Artie. I called the Post and talked to a guy named Wallace. Did he give you the message? I thought you had a security disk with the voices of the president and Senator Franco."

"Wallace? The man's a weasel, Toy. Probably went right to the feds with that. The Post has its own deep-throats who will hang a guy for a dime. So you think…wait, I don't have my laptop or my luggage. Maybe they're in the waiting room or still at the airport."

"Artie, you can best bet they're long gone. We have a problem and I mean this more for your safety now. Only after I talked with you did I find out that it was Niqui's laptop that the boys used and it has the thumb drive with the recordings on it, which I have in front of me.

When whoever tried to bump you off finds an empty disk, they're coming back to check your clothes—no doubt in the morgue. You probably only have minutes to get out of there."

"I'm outta here, Toy! I'll call you back in thirty minutes."

"Don't forget to turn off your cell phone!" Toy tried to say but he wasn't sure if Artie had disconnected before he had heard the warning.

Chapter Forty-Eight

The two agents with a counterintelligence organization, which is so secret it answers only to a retired CIA agent and the president, passed through the hospital's emergency room doors and didn't bother to give a man in a white smock with a surgical mask on a second look. They held out their fake FBI badges and wanted directions to the hospital morgue.

Artie figured the two blue-suited gentlemen weren't there to give blood, so he wheeled out an empty gurney to an EMT vehicle and began sliding the cart in the back when an Emergency Med Tech asked him what he thought he was doing. Artie flicked out his wallet and offered the man and his partner five-hundred dollars in cash. The two young EMTs, operating from a private business, glanced at the five bills and saw two days of work for a taxi ride. They agreed.

A choice had to be made before the emergency vehicle left the parking lot. Artie considered the Japanese Embassy, but decided instead on the British, only because they spoke almost the same language. His arrival at the embassy caused a great deal of confusion until Artie managed to get Toy on the phone. Toy contacted Prince William and within minutes the British Embassy welcomed this wanted American citizen and journalist with ultimate protection.

The British Embassy was now privy to the information and evidence that could destroy the current administration and wreak havoc on American politics. They were quite willing to assist in whatever function was requested. The tape was played over the phone as one smiling ambassador listened and knew the transmission would be on its way to British Intelligence within seconds.

Dr. Kerho had made forty copies of the tape, giving one to each student to keep, sent one to Prince William and one to his father. It wasn't his intention to put these people at risk, only to saturate the tape

528

to so many people there would be no attempt to gather the evidence to suppress its contents. Questions of how to contact became the paramount issue when Artie considered a senator from Massachusetts, a man of his own sexual orientation, though they barely knew each other.

"Have you ever met the man, Artie?" Toy asked.

"I saw the man in a gay bar over in Georgetown. He might recognize my name if he's ever read my articles in the Post, but at the time I was dressed as Mae West, so I don't know if he was impressed. I was very good that night, though."

"I'm sure you were," Toy complimented.

Artie played the important journalist from the Post and finally did get through to the senator, who actually knew the man's name. "I find your critiques most accurate," the senator said and began talking about the last film Artie had reviewed.

"Sir, not to disrupt your praise and good tastes, but I need for you to come to the British Embassy immediately."

"If there's some kind of problem I can get someone from my office to assist or a representative from our embassy, Mr. Stenson."

"No sir. This involves something so secret that the president tried to kill me this morning—at least his people did. I'm probably the most wanted man in Washington D.C. at the moment."

"This sounds interesting," the Democratic senator said. "I'll be right there."

When the senator heard the tape he was dumbstruck. It wasn't so much that the Republicans were in deep water as much as the nefarious actions of the nation's leaders would put Washington on its head if this was indeed fact. He knew just having this information would put his own life in hot water, as well.

The Massachusetts' senator took account of where he was with all this, the electronic eavesdropping without court authorization against those suspected of domestic "subversive" activity, and the downfall if such information was leaked. He needed direction himself since this was way out of his league. With one call the senator had the Director of Intelligence Services, now heading for the British Embassy where this forum of Americans were going to decide the fate of American politics in another country's house.

Mr. Joyce had only been recently appointed in this new position to overview eleven intelligence agencies within the government; a task that was catastrophic to begin with. "We have an immediate constitutional problem that we can't indict an incumbent president. If he can't be indicted, he can't be called before a grand jury."

"What if a Federal Grand Jury begins their own investigation?" the senator suggested.

"We could disclose the tape but then the president could serve up doctored or manufactured tapes to exculpate himself and others," Joyce considered in a mumble, as if he was giving himself ideas. "I need to bring in a few of my directors. Don't panic, these are men I trust. I'll make sure we cover this leak at the Post, as well. Where is Senator Franco as we talk?"

Artie saw his future and his safety tossed between these men's hands. He wondered if these professionals could even bring the heads of government to justice. Before Joyce left the embassy he had his own copy of the tape as well as the senator. For the time being Artie was a welcomed guest to the British Embassy.

****

The Post's Editor in Chief was called upon within the hour. He was found on the Post's roof garden outside the owner's office. After a few questions and information of the morning attempted murder of one of his journalists, Mr. Redford went in search of Mr. John Wallace. The man was enjoying a coke at the water cooler with a few colleagues.

Mr. Redford pulled Wallace to the side. "I need for you to meet Stenson at this address. He says he has something important to deliver and will only trust someone from our office."

"Are you sure it was Stenson?" Wallace asked.

"Positive. He was in a panic and thinks people are after him."

Wallace lit up, took the address and gathered his car keys. The phone call from Wallace to the White House was picked up by a court-ordered wiretap. At a coffee shop only a few blocks away from the Washington Post Office Mr. Wallace was arrested a half-hour later for conspiracy to commit murder.

At another residence, that had been given Mr. Wallace, three men were arrested for attempted break-in, concealed weapons, and resisting arrest. In a laptop computer confiscated in one of their cars, they also found an abundance of child pornography of children and teenagers in sexual positions. An additional felony charge was forthcoming with this find.

Later on, Artie would admit that he had no idea how this was put on his computer. He was certain one of these people out to get him was a pervert. Anyway, Artie did request a copy of the disc for evidence as part of his investigation for the Washington Post. The Attorney General refused this, but decided to add the erotic video to his own collection at home. In addition to the porn being discovered, one of the guns on a man arrested was discovered to be the make and model of the gun in an attempted murder at Dulles International that morning.

****

Congressman Joe Hart was running a close second to Senator Franco in the last stretch toward the Republican Convention a few months away in Philadelphia, Pennsylvania. Joe was one of the die-hard campaigners who enthusiastically walked neighborhoods in every contested state, trying to win votes one person at a time.

Hart's biggest gripe about Senator Franco was the fact the man must be breaking every campaign-reform law in acquiring soft-money slush funds. There was a powerful loophole in the tax law where someone could raise unlimited money and use it for almost any political purpose, as long as he didn't explicitly endorse any candidate. So far the ads that were being funded weren't so much in favor of Franco as bashing every other candidate.

Being that Wisconsin was an important swing state for both candidates, Hart and Franco had a heated debate on issues from health care to Middle East policy. It was after the debate when the two men met back stage, shook hands again and apologized for any off-handed remarks that might dismiss either one of them in consideration for vice-president. There were challenges to one's character and then there were character assassinations. To preserve the integrity of the party the line was closely drawn.

With Joe was a Chinese man, a reporter for the San Francisco Chronicle and a traveling journalist who covered the primaries of various states for assorted Chinese speaking newspapers across the country and for the Republic of China.

"Robert, I want you to meet Xiong Wu. For a reason I can't put my finger on, Xiong gives you better ink than me," Joe said in his introduction.

The two men shook hands and Wu asked the senator for a few minutes of his time to discuss the senator's goals, trips and agenda in the upcoming weeks. Robert waved off his campaign manager and secretary, then asked Wu to join him in his dressing room. The appearance of a Chinese reporter didn't necessarily surprise Robert, but the privacy of the meeting was unusual and Robert hoped that the man had news of another sort. It was a worthwhile suspicion.

"I've been asked to inform you of the death of your wife and two of your sons," Wu blurted out before the conversation centered on politics.

Robert was honestly stunned, though he had had a great deal of introspection over the past several weeks when the news would come. His silence and grief were sincere, but he also realized this had been his decision. When the senator didn't ask how or when, Wu interjected his own questions.

"Would you like the bodies returned to American?"

"Yes, that would be best," Robert replied. "I can have a proper funeral and burial. Which son survived?"

"I believe your oldest, Senator. He is in Japan, but we have agents everywhere who may assist your wishes."

Robert had read where the Kings Academy had fled China and escaped to Japan. He just hadn't heard who all escaped or anything about his family until now. At least the divorce wouldn't be necessary or ever revealed. The way this journalist spoke Robert sensed the man was far more than a reporter. It was as if he was asking if they should also eliminate Travis. Having a vindictive gay son was not in his best interest, but having the boy murdered would certainly raise suspicion, especially amongst the Kings' people when and if they returned.

"I appreciate the news, Mr. Wu. I'll have to give this considerable thought. If there's nothing else I will need a few hours by myself."

"There is something else, Senator Franco. The Military Intelligence Department of the People's Liberation Army General Staff would like some assurance from you, if you are elected, that the sale of U.S. military technology to Taiwan will be minimized."

"Yes, I can understand their concern. Obviously I will uphold the Taiwan Security Enhancement Act. I can assure the Chinese that we will retarget strategic nuclear weapons that were aimed on China."

"That is not enough, Senator. The Chinese people do not wish harm by use of military force against Taiwan, but we do need information to our specifications that Taiwan does not possess in the future plan on having nuclear weapons. Beijing is emphasizing the mountain site at Tien Mou and what the U.S. government refers to as Taipei Air Station. Do these installations currently have a nuclear arsenal?"

Though the senator was privy to several committees in the Senate, this exact reference to nuclear weapons was not available to him at that time.

"I need to do a bit of research before I can answer that question, sir. Can we meet tomorrow? I'll be in Kenosha, just down the road."

Since the senator had been directly involved in recent negotiations between China and Taiwan he didn't receive the suspicion by the National Security Agency when he casually sought information to assist his diplomatic viewpoints and negotiating capabilities. The work at NSA is wryly said to mean "No Such Agency," the work is so secret. They were well aware that the People's Republic of China (PRC) had launched a major program to influence American politics, including Congress and the White House.

To the NSA the Chinese loved to quantify things, and their espionage and influence campaign was part of a strategic plan to revive China through the "four modernizations," which were directed at industry, agriculture, science and technology, and defense. Their biggest goal was to build the People's Liberation Army into a world-class military force with the most modern strategic nuclear missiles and a navy capable of projecting power far from Chinese shores. To quote Mao, "The mind of the enemy and the will of his leaders is a target of far more importance than the bodies of his troops."

The NSA gave a great deal of Top Secret data to the senator, though the information was fictional and intended to mislead the Chinese. Within 24 hours, Mr. Xiong Wu had the material in his possession and immediately handed it back to his superiors in the Central Intelligence Agency.

****

Major Robert Davis was enjoying his down time from flying. He was still given base privileges while this endless investigation continued into his failure to follow through with orders since his aircraft was determined to be in top rate condition. Having six children from 3 to 11, they were even more appreciative of having their father home during the weekends.

The two agents who flashed badges from the CIA wasn't a good sign of progress for the major waiting for his verdict, which he hoped would be quick. Their questions centered on what his orders were and the information the major relayed back to command about who was onboard. He was perfectly honest with the two men why he refused to blow-up a 747, which showed no sign of being controlled by terrorists.

The men of the C-130 were also interviewed to their orders and, consequently, its cancellation at the last second. Though the pilot had no reservations about carrying out the precise orders given, as he put it, he did mention that he did extra passes to assure confirmation.

****

The autopsies of five FBI agents were done more secretive, with the burial plots that had to be excavated and exhumed. It was done so clandestine the families were never notified, nor would they ever be.

****

Great Britain decided to become proactive in their pursuit to consummate the world's reaction to this discovery of 30 clones of famous figures in history. Prince William became the lead spokesperson for his country in denouncing the ridicule and harsh words from various countries whom had had their tombs tampered with. To assure a speedy assembly to address these issues England called on a Special Session in the Security Council of the United Nations. They easily received the 9 member votes for such a session to commence.

On the East River in New York City, one-hundred and fifty-six members out of the 191 members gathered in the General Assembly Building. Unusual but not unprecedented, Prince William addressed this congregation of world leaders and designees as representatives of the countries around the globe. He discussed cloning as an advancement in science and technology and how it had forced people to reconsider long-held notions of parenthood, childhood, and the meaning of life.

There were truths in his presentation how, not long ago, transplanting hearts was unnatural and against God's wishes and divine plan—much like in vitro fertilization—often called test tube babies. These were now considered a normal part of medicine and family planning.

Prince William elaborated on how the deceased John Paul II had condemned the very idea of human cloning, calling it a tragic attempt by humans to imitate God's unique and special life—giving powers. This religious objection paralleled the similar reactions that were once raised against autopsies, anesthesia, artificial insemination, and the entire genetic revolution of our day—yet enormous benefits have accrued from each of these developments.

Previously, after hours of conversing with his dear friend, Dr. Kerho, Prince William had certainly come away with a better understanding of science and cloning, which was represented in his speech. He talked about the moral Rubicon....if this was a moral catastrophe for humans.

"Is it the mark of our ultimate hubris, our need to "play God"? Yes, ladies and gentlemen, it is. Playing God is what we do for a living. We've been doing it for centuries. We arrange the natural landscape through plant and animal breeding. We have discovered vaccinations and antibiotics to defeat plagues that once decimated populations. We exterminated species because they were in our way. We created reproductive technologies to aid people who would otherwise not be able to have children. We have advanced to where we can clone people, all part of a technological imperative that is as unstoppable as the passage of time.

"I assure you that cloning a person will duplicate only his physical appearance and other genetic elements, not the person's personality and memories. Thus, the cloned person will grow up with its own personality, experiences, likes and dislikes, and personal desires and goals.

"This great organization of the United Nations has already preset protection of every citizen in the world. If such a reminder is necessary, is it not written that we reaffirm faith in fundamental human rights, in the dignity and worth of the human person, in the equal rights of men and women and of nations large and small? We practice tolerance and live together in peace and security, and we are instructed as a great body to promote social progress and better standards of life in larger freedom.

"What right do we as nations to demand the return of people we once ostracized and abused? England has been honored with eight selections who were cloned. Do we not have the right to argue that most for these boys should be returned to their rightful country? Hardly. Oscar Wilde was put on trial and condemned for being homosexual. He would die a shortened life because of this abuse. Alan Mathison Turing saved millions of lives through breaking the German code during World War Two, and then we imprisoned him in 1954 for having a homosexual affair. He died a year later in the prime of his life due to depression and the failure of his country to recognize him as a national hero.

"Do you really think we have a right to demand his child to be returned? We sent Wilfred Owens to war, from which he wouldn't return. His poetic license destroyed, but only after he told the world that war is the sickest of diseases. Lord Byron was all but chased from

England for his orientation and possible capital punishment. How many others have faced the shame of discovery? James Barrie, Richard the Second, T. E. Lawrence and Isaac Newton. What lives would they truly have lived with a past condemnation of a sexuality we failed to give validity to?

"Italy evoked harsh penalties on their homosexuals and had men like Michelangelo and Leonardo de Vinci hide their secrets with guilt and shame. America is no different in their responsibility of persecution and prejudice. The Thomas Eakins of the world could never live a free life with laws and mentality that meant hatred and torment to those proclaiming to love another male. We can honor Walt Whitman, Cole Porter, Herman Melville, Nathaniel Hawthorne, Abraham Lincoln, Alain Locke, Bill Tilden, and Horatio Alger with their great deeds and accomplishments in the past, but can we truly love their sexuality?

"Does the Soviet Union have the right to declare ownership of the children of Peter Tchaikovsky, a man who never felt comfortable with his sexuality due to the shame his countrymen forced him to endure? Rudolph Nureyev ran away from a country who wouldn't accept his style of freedom and love. What right does a country have in wanting his child back, when the child wishes freedom for his love of another? The greatest ballet dancer of all time, Nijinsky, was another man who escaped the oppression of a country that wouldn't allow him to express his feelings for another male. Do you really think even his son would want to return to Russia?"

Prince William paused and allowed a film that Dr. Kerho in Dallas had hurriedly sent the prince. The assembly viewed the most recent competition between St. Bonaventure and the Kings Academy. They were pleased to be a part of this monumental aspect of these famous boys demonstrating their variety of talents.

A taste of humor swept the crowd when Wendy took off her wrestling helmet after she defeated Jay. The short video showed this massive campus, then gave a view of talented and happy boys, which the people of the world had never witnessed before; plus, the interior included many of the scientific experiments and displays. A gasp of wonder came from the mouths of so many dignitaries when the lake glistened with naked teenagers and a dolphin had leaped out of the water to appease this cameraman.   A quick view of the inside of the dormitory's arcade caused laughter and uneasiness among some members of the United Nations.

The nudity wasn't hidden from the reality and philosophy of the academy.   Smiling faces and waves from several of the boys showed a happiness and contentment seldom witnessed in sports schools and any academic setting. When the video was finished there was a round of applause and a new appreciation for these teenagers they had yet to meet.

Prince William stepped back up to the mike.   "And these boys, ladies and gentlemen, are your future leaders that represent all of us without regard to origin, culture, or a particular religious dogma.   In conclusion, I believe in order to justify our human fears about these youth, we imagine a God who acts just like us.   Therefore, we speak of God's "promise" to His "Chosen people," and of covenants between God and those God loves, in a special way.

"We cannot stand the thought of a God who loves no one in a way which is more special than any other, so you create fictions about a God who only loves certain people for certain reasons.   And we call these fictions religions.   I call them blasphemies.   For any thought that God loves one more than another is false--any ritual which asks you to make the same statement is not a sacrament, but a sacrilege.   How can we as human beings make judgment on these boys and expect God not to judge us?

"We make God say whatever we need God to say in order to continue limiting each other, hurting each other, and killing each other in His name.   Yes, many have invoked God's name in this arena during their discussion of thirty boys, and waved flags of possession, and carried crosses in retaliation to those who raised these boys, all as proof that God loves one people more than another and would ask those to kill to prove it.   Many have created a culture based on exclusions: gays, blacks, handicap, disadvantaged, mentally ill, and supported it with a cultural myth of a savior who excludes.   I tell you today that the culture of God is based on inclusion. In God's love everyone is included. Into God's Kingdom everyone is invited. All human conventions and all human constructions are faulty to the degree that they are not unlimited, eternal, and free.

"I ask of you today to include these boys, young men of an alternative lifestyle and sexuality, as part of the world of acceptance and love. We must teach what it means to be aware of other people's feelings and respect of others' paths. Thank you."

By an overwhelming majority the United Nations voted to make these boys citizens of the world and the right to their own decisions of where they wish to live and in any manner. Prince William's speech was presented on the front page of the New York Times and was soon translated around the globe.

Chapter Forty-Nine

The Kings' boys spent the morning at the Kodokan, Japan's martial arts center for aspiring judokas. To have Oshi as their guest the Japanese were honored with his presence in working out with many of their future stars.

Jay, Niqui, and Travis were but new comers to this world of marital arts and were learning their falls in preparation to learn judo, besides Aikido and other styles of Bushido. The trip to China and Japan was an eye-opener for all three of these Franco teens. Only the day before they were taken to a local sumo talent "stable" where naked young men, save a brown loincloth, worked out under the baleful eye of an immense former champion. The future sumo had bodies rippling sheets of pink Jell-O. If their aspirations are successful, these youngsters could make millions in sponsorship and prize money.

The boys were coming to the realization that their daily lives were a combination of school, play, and sometimes work. There weren't set dates when school was expected to start or end; every day there was something to learn and appreciate.

Travis, for one, had been anticipating the start of college, but his role with Kings Academy was far more enlightening than being accepted to a university or answering the many scholarship offers he'd received after his swimming performances. Dr. Kerho was already offering the boy to continue on at Kings in his education and being able to compete at a national level in swimming.

Jay and Niqui were thrown off-balance by a call from the Director of the Intelligence Services. They had no idea the importance of such a call, nor the status of a man that coordinated the offices of so many important men and women. They were simply asked how the electronic listening device got in the president's limousine.

"I think it was by accident," Niqui honestly admitted. "My brother had slipped the device in my father's coat pocket, so, somehow, it must have gotten booted out."

The director had to chuckle over this bizarre occurrence that had undermined this Head-of-State. He thanked the boys for their cooperation and promised them that they weren't in any kind of trouble, but they might have to testify at a hearing in the future.

Though Jay was spending most all of his time with Dane and other boys, he still made it a point to talk to Wendy and Maria. There were no resentments or guilt and the feelings of being scared or apprehensive were discussed the first night into the wee hours. In their final summation they were all looking forward to being a part of raising two new children to the academy. Dr. Kerho and Dr. Marion were changing their own time tables to begin another generation of Kings' children.

The teens also took a trip to Mr. Kamito's resort, west of Kyoto. It was a relief from the big city of Tokyo, and they roamed around the resort in nothing but fundoshi, as the boys who had spent a few summers at the resort in previous years showed the Franco children this escape from the rigors of the academy. The abundance of warrior skills of kendo (sword fighting), archer (kyudo), and hand-to-hand, unarmed combat (jujitsu) were all major parts of the curriculum of the Japanese children there at the resort. The kids referred to it as the sister school of their Texas academy.

All three of the Franco boys took an interest in dance. Learning from Trevor, Dane and Lance was entertaining and funny. There was yoga, taught by Miyato—tantric, of course, and calligraphy taught by Mr. Kamito, himself, and all the beach sports to add even more variety.

Toy and Kami patiently waited word from Artie on the proceedings in Washington and the United Nations. The most recent articles in the nation's newspapers came from information Artie had given them on the sporting events. To read the results would give the impression that the Kings beat the Chinese junior teams only the day before.

The best news was the coordination between the doctors and Toy's father with Prince William to present an image of the academy to the world. Nothing was hidden or forbidden to see, and they all knew it would raise eyebrows. Within hours after the United Nations' Special assembly the paparazzi flooded into Japan in search of the "thirty clones." It didn't take long for them to locate the whereabouts of this resort and flock to its entrance. At first they were refused admission, but the long telephoto lenses began capturing these youth in their morning run without clothes on and soon there was no privacy to be had at the resort.

As a last ditch effort to compromise, Dr. Kerho allowed the press on the resort grounds to photograph the boys in play and leisure with their fundoshi on. A rare group photograph was allowed that would make worldwide coverage with this one picture. Dressed in yukata the boys smiled as one delighted team, brothers who were now under the world's scrutiny as to their normalcy.

Christmas was celebrated with fewer journalists, but no less coverage as every movement, trip, and companionship had a camera somewhere aimed at these youth. Celebrity status was now something these boys would have to tolerate the rest of their lives. Their annual Christmas concert with the choir went on as scheduled, only with the change of location being Tokyo instead of Dallas.

****

Senator Franco was in good spirits. His opponent, Joe Hart, had cancelled out of the debate the day before for personal reasons and this gave Robert that much more ammunition to demonstrate his ability to work through stress and the demands of campaigning. He was going to campaign southward until he arrived in Texas just before the convention. It was in Memphis that he received a phone call that would ruin his day, if not his life.

Robert was not concerned that the president sent a private plane to fly him back, or that he was to leave his staff in Memphis. It was especially encouraging that he was to fly to Maine, where the president was on one of his many retreats away from the Capital. The senator once complained because the president spent more than 50% of his term either at Crawford Ranch or at Walker's Point in Kennebunkport, Maine. There might be time to get in a few rounds of golf or lobster fishing.

This was a good opportunity for a break from the rigors of

542

campaigning, the senator thought. He couldn't get the death of his wife and two sons out of his mind, so he needed this breath of fresh air and the ocean's breezes. His enthusiasm was forced during his rallies as he stared out at the voters and wondered what they'd think if they knew he had all but murdered his own family. Politics was a strange business but Robert knew he had the stomach to succeed.

Inside the Lear jet was a copy of the latest Washington Post with a cover photo of thirty boys vacationing in Japan. Their smiles were appealing and contagious in the realization that they were a happy lot, considering everything they had had to endure. The article accompanying this photo was both full of moral indignation and deliberate provocation in tone. Robert was not amused with this editor who announced: "I reject this extraordinary, un-humane bias whereby the blood of a Kings' child is nothing but a clone with no country, no identity but one we remember in a history book.

" And to think there are those out there who think that the blood of you and me from a lineage of a generation is somehow holier and respected. The Christian Right calls themselves children of God; yet, these Kings have been labeled immoral, test-tube babies who need reparation therapy for their sexual orientation. I would dare say that these youth are happier and more settled in life than these hypocrites of humanity and religious dogma."

In some ways Robert empathized with these teenagers at Kings Academy. He knew a few of them and, if he hadn't known any different, they were the same likeable kids as his own children. This thought of his boys—two dead by his own orders—brought emotions of grief and guilt. The thoughts were becoming too consistent, too depressive and clouding his focus. Tears flowed from his eyes and the senator had to hide his despair from others on the plane. He knew he'd have to ask his doctor for medication or hope his memory would somehow drive this from his conscience.

The senator's arrival at Walker's Point was greeted by a solemn president, once again dressed in what the press called his "Crawford casual" ensemble—jeans with a big belt buckle, a short-sleeved button-down shirt, and cowboy boots. The man was Texas even in Maine.

Without so much as a prelude to why the president requested his appearance, the two men adjourned to a dinner of steak and lobster. A '74 Cabinet of Mondavi vintage made the steak that much more tasteful. There was no sign of the First Lady or other invitees; just the president and senator discussing the Washington baseball team, like their own forebear, the Senators--last place.

Near nightfall the president suggested to Robert that they should take a drive down to the beach and enjoy the sunset. Robert agreed and was glad this wasn't an invite to Crawford, where they would have had burgers in one-hundred degree temperatures during the day and eighty-degrees at night. It was truly the armpit of America, as far as Robert was concerned.

Senator Franco felt uncomfortable with just him and the president staring out the limo windows at the gorgeous sunset reflecting off the Atlantic Ocean. The driver of this limo had walked away while smoking a cigarette. Silence had prevailed for much of the last fifteen minutes, but Robert wasn't going to bring an issue up until he knew what the president's agenda was.

"We've really fucked-up this time, Robert," the president spoke ever so softly and never took his eyes off the ocean.

"Excuse me, Mr. President? I'm not sure I understand," Robert replied, his interest in nature was now forgotten.

"I'm afraid your teenage sons' espionage ploy was more serious than we thought. There are tapes, apparently, of discussions that put us in a conspiracy crossfire. Our political careers are over, I'm afraid."

Robert tried to remember everything that was said during that day they had found the eavesdropping device. There was nothing that could pin him to a crime. "I'm not part of any conspiracy, Mr. President. I'm as clean as a polished diamond. Possibly this is an attempt by one of my Democratic challengers to frame you and me."

The president sighed. "Robert, did you recently meet with a Chinese reporter and give him Top Secret information on Taiwan?"

The senator felt bile rise up from his stomach; the eyes of the president bore into him like a diamond drill knowing there was a tarnished spot somewhere. This question was worse than being hit by a sledge hammer and far more penetrating. He stammered for an answer, wanting to deny this outright, but someone had evidence to the contrary. "The man discussed my views on Taiwan and I gave him info that any citizen has access to."

"Nuclear capabilities of the Taiwanese are not something that is public knowledge, nor does one call the NSA to request such sensitive material, then relay this to a foreign government; except, in this case, you handed it to a CIA agent."

"Shit!" The senator felt his past, present and future drain from his life—there would be no returning to Memphis to run on the Republican ticket for the Presidency of the United States.

His mind went numb and dizzy, but it still surfaced the murders of his wife and sons. He quickly threw his head out of the window and threw up on the grass. Recapturing his composure Robert had to think that his involvement in his family's murder was just a guess by this CIA agent, or did he really know? If there was any consolation, at least he didn't tell the guy to go ahead and kill his eldest son. His only scramble had to be to save his own skin, to somehow avoid prison and complete humiliation. Words of explanation couldn't form, nor did he wish to appear as a traitor to his country.

Speechless faces made the rear seat of this limo a place of horror to both men. Robert was close to tears, not wanting to meet eye to eye with the president. When he did glance back the president had a wooden box on his lap, opening the lid to reveal the gun set Robert had given the president as a gift several months earlier. At any other time such a scene might be appreciated for its value and sentiment. With the president having this in his possession after this discussion was mind-boggling. The senator wanted to protest the raising of his window by the president.

"I mentioned once to you, Robert, that death is the penalty for clumsiness or cowardice. Perhaps you used a yellow cape this time, instead of a red."

"Mr. President, we're not talking about bullfighting here. We will simply dodge on the oncoming bull."

The president posed a grim chuckle, knowing that politics doesn't

545

come with forgiveness and second chances. He lifted one of the Remington revolvers from its case and examined the silver-plated barrel and gold-plated cylinder.

"Ever wonder what it would be like to be an actual cowboy with these fine pieces strapped to your waist, Robert?"

"No, Mr. President. I'm not sure I understand why you brought them with you tonight."

The president smiled, but the grin was not in humor. "Senator, you have no idea what they have on me. Do you know I planned on dropping a fuel-air bomb on top of that academy?" The president chuckled to some thought of his arrogance. "Damn if those kids didn't leave that morning! Then I gave orders to shoot their plane down. What kind of military do we have when pilots don't obey a direct order from the President of the United States? They know it all, Robert. They know it all."

"They can't indict you, Mr. President. You're vindicated, you know that. If you have to, resign, and those tapes will be sealed forever," Robert tried to reason.

****

Over a thousand yards away a half-dozen Secret Service agents were listening to this conversation and realized the magnitude of the president's decision of bringing a gun with him. Another time the men might have excused this as a show-and-tell, but not when the careers of both men hung in the balance.

Three of the agents leaped in their vehicle and spun off toward the beach. Several FBI agents had no idea what the emergency was, but they followed in close pursuit.

****

"Robert, do you think it's fair to our families to put them through an embarrassing impeachment process or criminal indictment?"

"Mr. President! You can't be serious about this?! You're way over-reacting. Trust me, this is not the way to save face. We all make mistakes, especially in the political circles we run in."

"Ah, yes, our political circles. I was looking forward to a retirement of leisure, a few lectures and book deals; maybe Jimmy Carter and I could challenge my father and Clinton to a tennis game. Can you see that, four old presidents chasing a yellow ball which would get the best of us?"

"Maybe we should head back to the house, Mr. President," Robert suggested.

"Here take one of these guns, Senator."

"I will not! I don't understand what kind of game you're playing here, but it scares me."

"It's supposed to, Robert. You're going to kill me. See, it's important I go out as a hero, a remembered man whose life was ended short by a disturbed senator. It's already written and planned. No matter what happens in the next few minutes you will be viewed as the villain. So, Senator, you have no choice."

Robert turned to exit the limo, only to feel and hear the pop of the bullet entering his side. The pain made his hands jerk to his waist.

"You're fucking insane!" Robert yelled in excruciating agony. He had trouble resisting the placement of the other revolver in his bloody hand.

"Your turn," the president calmly said. "Shoot me, you coward! Anyone who can do their own family in has to have the guts to kill in cold blood!"

The pounding on the windows distracted the president from his mission. Men were attempting to get in the car, one of them fumbling with the driver to get the keys.

"We don't have a lot of time, Robert. Let's go out as heroes, what do you say?"

"Fuck you! You're fucking crazy!" Robert dropped the gun on the floor of the vehicle. His eyes pleaded with the men in suit coats on the other side of the window.

"Have it your way," the president said and brought the other gun up behind Robert's ear and fired.

With the crack of the bullet the senator jerked to his left, blood splattered on the window, before the senator slumped forward and fell to the floor of the limousine.

The president smiled at the men he'd known and who had protected him for many years. Their pleas were almost humorous, faces of desperation and frustration, and there was a mass hysteria trying to get the key in the door. When the president heard the lock click, he brought the gun up and put it inside his mouth.

A split second before a Secret Service agent lunged for the weapon, the explosive blew through the back of the president's skull. The splatter of guts and blood repulsed the agents to literally freeze in their movements. Their role to protect the President of the United States was undermined by the one person they couldn't protect from himself.

Chapter Fifty
Philadelphia, Pennsylvania

With the morning headlines came a nation in shock. The deaths of the president and Senator Franco were blamed on mechanical failure of one of the Presidential helicopters used in a leisurely flight over the Maine coastline. The government had quickly crashed a perfectly good helicopter into the shore to simulate such a disaster. Such a cover-up would protect the American people from a notorious story of deceit, conspiracy, and murder.

There were five Muslin terrorists who had also read the morning headlines, as they kept undercover in wait of the upcoming Republican Convention. These deaths had literally taken the wind from these men's sails in their desire to bomb the convention center and wreak havoc on the political system of the United States with the deaths of the president, vice-president and the nominees for the Republican Party.

Selected from an elite cell of Muslim extremists, these men defied democratic values of tolerance, pluralism, and equal rights. Even though none of them were what one would call practicing Muslims they were dedicated to their limited reliability as indicators of individuals' alienation from mainstream society. This setback was not well received by these radicals because of their "divine" mission; one that would detonate two dirty nuclear bombs and disrupt America forever as we know it.

The ultimate goal of this group was to bring sharia law wherever they went, even though such an accomplishment would never be seen by their immediate death through this sacrifice committing their duty to Islam. Rather than retreat with this moment of grandeur taken from them by an accidental helicopter crash, these radicals saw something far more tangible than cancellation.

Being of this Muslim sect meant dishonoring all things Islam opposed: Jews, blacks, gays, capitalism and democracy. What they saw on the front cover of their newspaper was a picture of boys who had been raised into tall, strapping, healthy, multilingual young adults— veritable masters of the world for whom life would be safe, pleasant, and abundant in its rewards. Worse yet, they were all gay with a Christian upbringing. They appeared to have brought Western civilization to its utmost pinnacle in terms of freedom and the pursuit of happiness. Their future road ahead was very much like the actual roads in Texas, which seemed to stretch to the horizon, straight, flat, smooth, and with nary a bump. These Muslims frowned on any capitalistic optimism.

As if to insult their beliefs, Prince William had planned to welcome back these heroes in New York City with a ticker tape parade. This powder keg of human decadence, brimming with an alienation born of the immigrants' deep antagonism toward an infidel society that rejects them and their religion, was ready to redirect their anger and objective on this demonstration of evil in America's city of cities.

****

The birthday party for all thirty of the Kings was always a fun occasion. This was the second time in their sixteen years that they had had their party on foreign soil. It was true that all thirty boys weren't born exactly on the same day; a day was selected as sort of the mean of the dates the boys were actually delivered in various countries. The true birth dates of each boy had never been revealed, but the students knew that this particular day was either their birthday or close to it.

Mr. Kamito planned a grand party with a professional troupe that played a dramatic story out on a constructed stage at the resort. Their play was one of bravery and honor, components that define Japanese culture. Kami had witnessed this play as a boy himself, a true story that happened in 1868, when imperial forces attacked the Tsuru-ga-ju Castle in the city of Aizuwakamatsu. Twenty defenders escaped and retreated to a nearby hill. To a samurai, surrender was a fate worse than death. They made a pact and, one by one, committed ritual suicide. None of them was older than 17.

In many ways the play could put life in perspective as a treasured gift to relish and never to take advantage of. Not to cause such a celebration to feel grief, a group of lion dancers, usually done at major Shinto festivals, came forward and enlivened the atmosphere with harmony and robust. Soon there were fireworks, live music, and lots of food.

The boys were delighted with numerous gifts, always spoiled by their honored sensei, Mr. Kamito, and again his generosity was never challenged by his two favorite pupils in Kami and Reece. They wouldn't dare complain that their sensei treated their sons with way too much love and affection. Oshi had a habit of adding his own gift; this year it was a birthstone earring for each boy. The boys weren't allowed tattoos or gaudy jewelry, but Toy and Kami would make an exception for their sixteenth birthday.

The ladies were adorned in glamorous kimonos, traditional and beautiful. This was a boys' day, so the girls waited on them hand and foot, which Jay and Niqui appreciated and took advantage of. All the boys dressed up in flamboyant silk costumes after the entertainment had left. Of all the high-ranking samurai's cultural pursuits, none infatuated them as much as the tea ceremony as a badge of refinement.

It was at the start of this ceremony when Toy was handed an urgent e-mail from Artie. He found Debbie having a great time in all her ceremony garb and learning this ritual of 13th Century Zen Buddhist monks. She read the message with a frown replacing her giddiness. A pensive thought, possibly a reflection on a marriage that never seemed comfortable, yet good times had their memory, as well.

She handed the paper back to Dr. Kerho, then as an afterthought: "Robert wouldn't have survived the scrutiny and embarrassment of his peers. This wasn't fair, but fairness has little to do with birth and death. If it did maybe we'd choose another way or time to live our responsibilities to procreate the next generation to the same stresses and challenges we face. It's ironic that the same politics that he relished and adored killed him at a time of his greatest ambition and ultimate sorrow."

Toy wasn't sure if he should respond. He saw the hint of a tear from Debbie's eyes, but the woman was strength defined and had four children with the same fearless spirit. Mrs. Franco thanked Toy for his grief, then turned to take her part in this meditative act of making and drinking tea. In a matter of minutes he watched a woman who had faced death as a result of her husband's actions begin to smile and laugh when the boys and girls took delight and sometimes confusion in their utensils, ritual and admiration of 500-year old cups.

The sprig of pear blossoms flowed from a vase nearby, the smell of plum wood incense gave more to the meaning of life than it did in regret for the dead.

****

Two days later the Franco family flew back to Washington D.C. to escort the coffin of their husband and father back to San Antonio, Texas. The children grieved in their own way with memories of a demanding, yet motivating father who had given them a life better than most children had, with never having to worry about their next meal or having a roof over their heads. It was a life children can easily take for granted, a childhood with a swimming pool and brand name sneakers.

Having a reputation as a senator's kid had both its highs and lows. Politics is for the visionaries, and seldom is the politician appreciated until after his service and his deeds have been forgotten, whether for the good or the bad. The Franco children were old enough to understand their father's motives, his way of invoking God's name, wave America's flag, and carry his cross of the religious right a proof that God loves one people more than another.

Travis had tried his best not to cry, but he couldn't forget the words of the Kings' minister, Dr. H. Ray Harris, who had talked to the family before they left Tokyo. "We must always teach each other what it means to be aware of other people's feelings and respect of others' paths. Be a gift to everyone who enters your life, and to everyone whose life you enter. We can't judge your father, your husband, nor can we condemn, for we know not why a thing occurs, nor to what end. Your thoughts must be to the parent which gave birth to you. Remember him in good times, not in resentment."

And with these words Travis did cry; he cried for his brothers and sister, and for his own anger for not feeling loved and accepted because of his sexual orientation.   Mrs. Franco held him within her arms, and then she was surrounded by the arms of Niqui, Jay and Maria, who all felt the same hurt and sadness of loss.

****

When Artie Stenson left the British Embassy, he wasn't sure whether to duck and run or leave as a martyr.   Artie hadn't been told much, but he did read the Washington Post, so it didn't take a genius to figure out that two boys spying on their father had caused the eventual deaths of a president and a senator.

Artie didn't really believe the final disposition of the two men on vacation in Maine; he believed what most of Washington believed that the two men got together to commit an ultimate act of sacrifice to save disgrace on their families and on America, as a whole.   The blame went to China; their infiltration of our security system and undermining the decency of every American politician.   It all appeared too much as a dog and pony show to put the land of the big panda to shame.

With his dented crown on his vest he struck out to find a good meal before he ventured to an invitation from Mr. Redford at the Washington Post office.   Artie expected to be fired.   Why not?   He'd been gone over a month, had totally abandoned his assignment to critique films and theatre for the gay community, and any dimwit who had the patience and the time could do his job.

Since he wasn't shot upon leaving the embassy Artie felt somewhat assured there might not be a hit out on him with the vice-president taking over the helm.   His head was only a bit sore and the bruise on his chest made a constant reminder of how much he loved and owed Kings Academy.

After a meal, he wouldn't normally order with the money he was making as a peon for the Post, Artie left a fifty dollar tip because he could.   Before he considered how he would live the next few months of his life in freedom and with few worries, he took a taxi to the Washington Post to pick up his pink slip.   If Redford wanted to see him, Artie had a few words of departure on the tip of his tongue.   It's not that he needed the man's reference for a future job.

The Editor-in-Chief stood up from his chair when he saw Artie

553

approach his office. Other journalists, sitting in their various colored offices, eyed their returning peer with a glare of respect and reverence, though Artie thought they were gawking.

"Welcome back, Artie," Redford greeted, which returned a suspicious stare from a man with a golden crown on his chest—a dented one, at that.

"I feel like a man back from the dead," Artie joked and was ready for the hammer to drop.

"Damn good job, Stenson! I mean on the Kings' articles. I'm sorry we didn't do the whole gamble instead of hesitating. You were right, Artie, it was big! I also want to apologize for Wallace. What can I say? We've known for a long time that there's been a leak in the office, but even newspapers have their double-agents."

"Right, Wallace," Artie replied with hesitation and curiosity. "Pardon my ignorance, boss, but I've been out of commission for several days. What's this about Wallace?"

"Apparently Dr. Kerho called here to have a message relayed to you upon your arrival. Wallace took the call, told the White House, and you know the rest. I'm grateful you wore the crown there."

Grateful? Artie was running this through his mind. Since when did the editor feel grateful for anything but the praise he often received for other people's work?

"Oh," Artie commented. "Well, you know I have a nice nest egg now. I was thinking of relaxing for a while, maybe take a trip to Europe."

"That's what I wanted to talk to you about, Artie. I never appreciated your work ethic or your tenacity until this Kings' thing blew up to be headlines. Your reporting was damn impressive. I'd like to offer you a better position, say, seventy thousand, which is quite a raise for you."

Artie slumped down into the comfy chair in front of Redford's desk. His confidence just increased by a hundred fold. "I appreciate the offer, Redford, I really do. It's not like I need the money and I'm a bit burned out on the gay scene."

"Exactly, Artie, and you deserve better. I'm talking about a job where you can travel, keep an eye on this Kings Academy and bigger stories than the opening of a coming-of-age flick. Look, Artie, you're an ambitious man. You're going to get bored real quick with nothing to do, and, sure, you have a lot of bucks, but money depletes rapidly if all you do is play. I want you to explore your potential and your skills in journalism. What do you say?"

Artie had never felt wanted, never felt appreciated or given kudos for a job well done. The gay scene awaited him, but being a reporter for the Washington Post had been an important straw in his cap, even if it was for the gay community.

"Well, I do have evidence on China using this gene doping, and there are things I'd love to do to promote the Kings Academy upon their return to America. I think the Washington Post could get some great press by sponsoring some of these ideas."

"I'm all ears," Redford said and offered his hand for a deal. It was true he was getting incentive from the powers-to-be to keep Stenson, and Redford didn't wish to disappoint men who pushed buttons. He smiled like he'd never smiled when Artie accepted the offer.

"Excellent!" Redford complimented. "Get me that information on the Chinese cheating and your ideas on how we can use the academy to promote the paper."

"Now that sounds like the editor I remember," Artie spoke. "You know, I'm going to ask for the Lincoln Center for the Fine Arts."

Redford didn't react with negativity. "As a way to introduce these boys to the nation? I like it."

"Their talent might not be Carnegie Hall, but who wouldn't come out to watch Lincoln's son or Peter the Great in tights?"

****

Travis walked into his brothers' bedroom, just after the boys went to bed. He slid in next to Niqui, and Jay slid his bed over and moved in next to Travis. They snuggled in close, as if this bonding was a comfort they all needed with the death of their father.

"I discussed a few things with Mom, guys. I have a better understanding of our father now, so I want you guys to hear me out.

Mom's right, parents rarely let go of their children, but we seem to forget them real easy. I can see where all of us have gone our own ways, done our own thing, and kinda took for granted that our parents would always be here for us.

"It's quite possible that Dad felt really left out of our lives over the past few years. Sure he'd been busy with politics, but we know how interested he stayed in St. Bonaventure, the praises our accomplishments did for his ego. When we wouldn't let him in our lives, he found a way to do it on his level. Was it right? Not really but it wasn't the end of the world. For us to forgive him, we're forgiving ourselves. It's exactly what they're trying to teach us at the academy, giving and receiving. When we love others, we're loving ourselves; not for others, but with others. It's all experiencing our grandest idea about ourselves and who we really are.

"If I'd been more patient with dad on my sexual orientation, allowed him to experience my joy, the love I have for Alex, my feelings and thoughts, I think dad would eventually have seen me as the same son he always loved. Dad had beliefs that we are questioning with our new knowledge. Dr. Marion said that religions would not exist if the whole human race understood that God doesn't have preferences, because a religion purports to be a statement of God's preferences."

"I think I understand," Niqui spoke. "Do you think God might want us to love one person more than another, like in a marriage?"

"Alex and I discussed this is Japan. Marriage is like a social convention. A bargain is struck; you give me this and I'll give you that. It's a way for a female to guarantee her support and survival, and a way a male can guarantee the constant availability of sex and companionship. Yet, if all we are is love, wouldn't it be better to be unlimited, eternal, and free, by nature? If we sign an agreement or make promises to another, what happens if we want to love another? We've given up the very nature of our being in trade for some form of security, promised sex, or someone we can count on to do our meals, our clothes.

"See, marriage is an artificial social construction designed to govern each other's behavior. We shouldn't agree to give something of value which you have to another only if they had something of value to give you in exchange. What we must learn is that what we give to another, we give to ourselves. What goes around, comes around."

"Cool," Jay said and placed his head on Travis' chest. "So it's okay to love a lot of people."

"Yup! True love is always free, and obligation cannot exist in the space of love. If you see love as an obligation, you will eventually resent it. But if we see this as a free choice, made over and over, then we tell and live our truth. We can remember our love for our father without expecting anything in return. We know not why he did some of the things he did, but we can forgive him."

The boys felt better with some of their struggles and regret. They talked about Jay being a future father and Travis said that he is going to contribute a sperm sample so he and Alex can have children to raise at the academy.

Mrs. Franco and Maria knocked on the bedroom door, only to come in to find her three boys huddled together on the same bed. A year earlier the boys were often at each other's throats over some minor possession or entitlement. They didn't share and their favorite entertainment was to argue with each other or attempt to out dress their peers to appear cooler. Debbie stood for a few seconds and admired their casual acceptance and innocence of each other. They had grown to be more than just brothers, but friends, friends who could share each other's secrets and foibles. In many ways they were in love with each other out of respect and appreciation in sharing both loss and love. Mrs. Franco sat down at the foot of the bed with her daughter, now showing the beginnings of a rounded belly. Maria squeezed in between Niqui and Travis. She took her hand and swept it through Jay's sleepy locks of hair.

"I suppose this is as good a time to discuss our future as any," Debbie began with her family in one place. "We have a few options, so I'll put this up for a vote. Our home here in San Antonio is free and clear, so we aren't threatened to move or have an abundance of bills to pay. Your father has left us an inheritance that will keep you kids in sneakers for a few more years. I can call St. Bonaventure tomorrow and have you resume classes there. I'm sure they'll have open arms for the two of you and Maria, after the birth of her child. Stanford has let it be known that you're welcomed there in the fall semester, if you so choose, Travis."

"What are our other options, Mom?" Maria asked.

"Dr. Kerho has offered to accept you kids at the academy, and he has been kind of enough to offer me a job if I wish to run the nursery in the new dormitory for the expected additions. I also have my teaching credential and I can help out in the youth dorm during the week."

"Mom, the staff has to be naked to work there," Niqui reminded his mother.

"And you think your mother doesn't have a nice body? I've seen more naked boys in the last few months than in all my life. I'm beginning to like it."

"Gee, Mom, I think you're beautiful, and not because you think we're cute," Niqui said and got back in good graces.

"Good cover up, kid," Debbie joked. "Dr. Kerho has also suggested that I can build a home on the property set aside for residential. He's figuring many of the current students will be building homes for their future. That leaves us with your decisions. If you'd like to become students you have the option of staying with me at this new home or living in the dorm, or we can all stay here in this community."

"Let's vote," Travis suggested. "Those who want to stay here raise their hands. Okay, those who want to be a King, or Queen in this instance, raise their hands. That's four-zero, Mom. What's your vote?"

"Since your father is no longer with us I assume my vote counts five," Debbie announced.

"Booooo!" her children responded.

"Don't panic. We have a five-O count. I'll call the doctor and tell him we'll be moving. Better start looking through magazines and designing our new home, kids."

"How about New York City, Mom?" Jay said from his position of rest on his brother's chest and getting his hair soothed.

"We can't move there, silly," Debbie teased her youngest.

"Ah, Mom, that's not what I meant. Can we be part of the parade?"

"I never thought you'd ask. I want all you packed and ready to head for the airport by noon."

Her children didn't need words to express their happiness; they simply bounced on their mother and had her giggling in surrender.

****

On the other side of the globe the Kings Academy was just taking off from Tokyo to JFK in New York City. Their pictures were now on magazine covers from TIME to gay friendly periodicals like OUT, ADVOCATE, GENRE, and INSTINCT. The gay community had adopted these boys with fanfare and total acceptance.

Though the academy had traveled around the world to so many varied locations, they'd never visited the Big Apple to explore its wonderful sites and culture. Artie Stenson had done his job and prepared New York for this group of young men who had received more than their share of publicity in the past year.

In short order numerous agencies and venues offered to host this group with special events or shows. These became partly self-serving and respect for these "world citizens" the United Nations had so proclaimed them to be. The American Ballet Company didn't put too much faith in Artie's assertion that the Kings Academy had a dance troupe that could match any young people's repertoire in the country, if not the world. They gave in to a performance only after the Mayor of New York put his weight behind such a performance at the Lincoln Center. The New York Philharmonic was an added addition to this, just in case Artie was more bull than show.

The New York Times had urgently advertised the coming attractions, which stirred massive interest and ticket sales for the performance at the Lincoln Center. The tickets were sold out in less than two hours. All Artie had to do was interview for the New Yorker and give a review what he saw the Kings' dancers do in Shanghai. The gay community alone could have bought out the Lincoln Center ten times over.

****

Five suspected terrorists had managed to rent a U-Haul for their rapid mobility to New York City. They had also developed their own itinerary and decided that, given the tentative schedule of the ticker tape parade, they would devise several plans to assure their success. With unpredictable weather this time of the year most people didn't give much hope for a parade. Given this variable the Muslims decided to target the Lincoln Center, just in case the parade was cancelled. Both targets were accessible, and the Lincoln Center was very much like the confined area of the Republican Convention.

Islamic allies were never one to turn the other cheek when someone criticized their religious beliefs. To criticize someone else's beliefs was part of their agenda. Jews and gays have always been easy targets for extremists, even though their religious beginnings hardly denounced such cultures and orientation. The acts of pedophilia and same-sex relationships were more common in the Muslim world than in any other culture, save the Greeks of Ancient times. This was something easily ignored.

These terrorists had come from a European environment where they had instilled fear amidst various countries. Only a few weeks earlier England had emerged with a bill to protect Muslims and gays from discrimination. Sadly, "Downing Street," explained that they feared the Muslims might feel offended if they were lumped together with homosexuals. So gays had been dropped from the bill. Given that Britain's Muslim voters were thought to number about 1.3 million and gay voters around 3 million, one could only imagine what would happen when the rapidly increasing number of the former came to outnumber the latter. It was just another case of a world in fear of trampling on gay rights to curry favor with fundamentalist Muslims.

To annihilate this most recent threat of blatant acceptance of homosexuality was now the premise of this group of five. As they left Philadelphia in a snow squall they began counting their own hours left on this planet to one of pleasure with all the virgins of both sexes in paradise for their enjoyment.

Chapter Fifty-One

The 747 landed at John F. Kennedy International in the early evening. The reception was one the airport had never witnessed before or since. The Mayor of New York was the first to greet Dr. Kerho and Dr. Marion when they stepped into the terminal. Reporters flashed their cameras as if they were at Yankee Stadium.

Near the front of the crowd Dr. Kerho spotted two faces, two Russian citizens he had invited to America. Toy asked one of New York's finest to escort his guests to where the Kings Academy boys would be taken. Ms. Stravotsky and Sergei Feldekov had landed only hours earlier to be with the Kings during this festive celebration and return to the United States.

Sergei ran into the arms of Trevor Alekseyevich and the two kissed to the delight of a few photographers who had their faces pressed against the glass separating this holding area from the rest of the terminal. Within minutes another plane landed from Iceland and Mrs. Bjornsson stepped off her flight with her husband-to-be. They were quickly united with Leif, and a mother and son reunion brought more than a few tears to his brothers.

By motorcycle escort the group was taken downtown to the Waldorf Astoria where the top floor was prepared for just the Kings Academy.

Given the time difference and restless disposition of so many of the Kings' students, they desired a tour of the town at night and received precisely that. For security reasons this excursion was a godsend to the police who found it a lot easier to handle crowd control and the paparazzi with this late night tour. The boys went to the top of the Empire State Building and the Chrysler building, before being allowed to wander through the Guggenheim Museum and the Metropolitan Museum of Art.

As an added feature their bus stopped at the Shubert Theatre on

Broadway to take in the last show as special guests, receiving a standing ovation as they entered.   As the hours tolled into the late night the group was finally returned to the hotel in preparation for the next day's busy schedule.

As part of the city's welcome, a luncheon was organized the following day by the mayor.   This was great press for him and the exposure was timely, given adverse publicity in recent months.   The mayor apologized to the Kings Academy for the actions of the government, as if this was his role.   Since Senator Franco's demise had opened up the Republican ticket, the mayor had decided only the week before to throw his hat in the ring.

The late afternoon gave the boys time to digest their arrival and enjoy their own free time, though Dr. Kerho was reluctant to allow his teens to walk the streets without an escort.   One exception was Leif, who broke free from this restriction and all but ran to the New York Public Library.   He asked a librarian, who recognized him from his many visits a year earlier, if the lady had seen the boy he had often showed up with.   She said the boy had come quite often after Leif left New York, but then stopped altogether.   Leif appreciated the information and immediately departed to head for this safe house for runaways and abused adolescents, which had all but saved his life from the streets.

Father Haley smiled ear to ear when he saw Leif walk in.   It was no secret that Leif had been recognized by more than a few when his face appeared with the other Kings' boys on the cover of so many magazines.

"Leif!   God bless you!" the director yelled and kissed the boy on the forehead.   "You've grown three inches since I've seen you last and are as handsome as ever.   To realize I had the King of Sweden living here, it's crazy.   I remember when you were Erik Eldjarn, if I remember right," the director joked.

"Actually, sir, I'm not Erik Eldjarn, I'm Leif Marion.   I was Leif Bjornsson, but I've had my name changed, and I'm certainly not worthy of a king.   If you don't mind me asking, sir, have you seen Rodney Miller?   You know, the boy from Ohio."

Father Haley lost his smile and exuberance and gripped Leif by the shoulders. "I'm sorry to tell you this, Leif, but the boy died of a drug overdose a few weeks ago. If he'd just hung in there for a few more days he would have seen your picture in the paper. I just know it would have made a difference. He often talked about you as his idol, a boy who made him come to grips with his sexuality and being true to himself. I see it all the time, a boy who gets in with the wrong group and the reward is drugs for sex. I know he loved you very much."

Leif couldn't hold back his tears and lost control as he was taken into the arms of the Father, a man who had the courage to rescue this boy at the precise minute from also being encased in a world of sex for sale and street hustling.

When Leif returned to where the Kings were staying he was a different youth, a boy who had witnessed and survived far more than a boy his age should. His solemn demeanor was picked up by Alan, who relayed it to Dr. Marion. Another cry and release of emotions helped this young man deal with emotions too fragile to be kept within.

The evening was climaxed when Artie informed Dr. Kerho that a church in Greenwich Village wished for the boys' choir to sing that evening. Toy had given Artie a recording of their church choir, and Artie had recorded this CD to have it advertised, as well. When it hit the airways this CD had become such an instant success that stores couldn't keep up with the demand. With a quick dinner Toy had the boys dress up in their choir uniforms that they wore during church service at the academy: white yukatas over a fundoshi. Dr. Kerho wasn't out to acquiesce to the modest code of any church.

When word got out that the Kings Academy choir was a special guest for service that evening, the gay community packed the church to such a proportion that the police, literally, had to stand guard and turn people away. Leif and Tomas amazed this congregation with their tenor voices, and Leif also gave a testimony on his memories of Rodney and Sveinn. His tears were comforted when Oshi leaped on stage and helped his protégé through a tough recollection from the heart.

Rev. Harris had from the beginning made sure church service was one of spirit and celebration. The boys had the church rockin' and swayin' to this gospel singing, their beauty in these shortened robes were perfect as 30 gay young men had no restraint in expressing their love for each other and for the Lord. **LEANIN' ON YOU** had the men and women clapping to the rhythm and their final of **HOW GREAT THOU ART** had the mass majority in tears. Over two hundred committed their souls to Christ, with Rev. Harris' call to the altar. Given such an inspirational night, Leif had whispered to Oshi afterwards that he was thinking of entering the ministry after he graduated from college.

"I think you'd be perfect for such a calling, my boy," Oshi said with pride.

****

The Franco family arrived in the early morning and was welcomed with open arms by Dr. Kerho. An available room at the hotel was quickly booked. Special passes were arranged for the family to be a part of this spectacular at the Lincoln Center that evening. A cool rain had settled over the city and possibly put a damper on the next day's parade. Before the evening's agenda the entire group visited the United Nations, where they were received with dignity and class. Prince William was their escort through the many buildings and its history.

As a unit they had dinner at one of New York's finest restaurants, then departed for the Lincoln Center for the Performing Arts. Unlike the Concert Hall at the Kennedy Center in Washington, the Metropolitan Opera House could seat 3800, and every seat would be filled for the night's special performance. Several dignitaries and New York's finest citizens were to be on hand in this overflow crowd. A special section for the members of the Juilliard School had been arranged, though this talented school for elite students of music and dance weren't quite sure what to expect.

The Lincoln Center consisted of six buildings set around a water scape where one may enjoy the Metropolitan Opera, the New York City Opera, the New York Philharmonic, the New York City Ballet, a theatre company, an arts library and the Juilliard School. Sitting between West 62nd and 66th, the Lincoln Center had extreme security measures to screen visitors and gawking tourists. The mass of TV trucks and vans made this appear more like an Academy Awards show.   With international uproar dwindling and the realization that these youth didn't present a threat to any country or its political system, there was no longer a demand for retribution or the return of a teenager who was now as American as any other boy in the United States.

From Broadway to Manhattan, for all these teenagers, the city was a wonder to their little world in Texas, and a whole lot different than Tokyo and Beijing.   The drive around Times Square to view this entertainment capital of the world had its educational side with the viewing of George M. Cohn's statue.   The Irishman was a motivating factor for the boys, especially, because of his great versatility—actor, playwright, composer, singer, dancer.   For the Kings' students it was a time of reflection of preparing their minds and bodies for an evening of entertainment; the skills that paralleled their hours of athletic training, but the importance of art and music were no less dynamic.

One quick stop at Stonewall felt like an honor for all the boys and their sexual orientation.   They'd all been taught the significance of the Stonewall riots in 1969 and how far the gay community had advanced since addressing their rights for freedom and acceptance.   For all the boys it was a pleasure to touch the building and honor its heritage and history.

Their arrival at the Lincoln Center was a crowd pleaser in the early evening hours, even in the pouring rain they were received by hundreds of curious spectators.   They were escorted through the Juilliard School, where students of prodigious talents expressed joy in meeting these teens from Kings.   These weren't the boys and girls they'd met through other Texas schools, but polished, well-mannered kids like themselves who aspired to greatness.   The New York art students were polite and in awe when they met Dane Nureyev, with the realization that they were looking at the smitten image of one of the greatest dancers of all time. To think that this boy wished to dance on stage as a comparison to his forebear took enormous courage.

The beauty of the stage and orchestra pit were as prestigious as any performance arts center in the world.   Lance checked the floor and found it suitable for ballet, possibly the best he'd ever experienced next to the professional dance floor at Kings. Their blue suit jackets with the Kings' insignia in gold were an eye-opener in itself. Brittany had put a peach carnation in each of the boys' button loops, which she thought gave the boys a special touch and a classy look. She had always thought of herself as the mother to these boys and had no less love for anyone of them than what she felt for her biological children.

****

In a U-Haul a few blocks away five men sat in the pouring rain and began to deliberate their chances of success. Each terrorist had attached to their right thigh a ballistic nylon holster that held a Heckler and Kock USP-0 and five spare fifteen-round magazines. The left thigh supported six 30-round syntetic submachine gun magazines loaded with 155-grain Winchester Silvertip. Strapped to their back was a scabbard holding HK's ubiquitous MP5 submachine gun in 9-mm caliber, with a Knight wet-technology suppressor screwed onto the barrel, and a seventh full mag of Silvertips within easy reach. They also had six DefTec Model 25 flashbangs in modular pouches Velcro'd to their CqC vest, along with secure radio, lip mike, and ear piece, twenty feet of shaped linear ribbon charge on a wooden spool, primers, wire, and an electric detonator, a pair of eighteen-inch bolt cutters, an electrician's screwdriver, lineman's pliers, a short steel pry bar, and a first-aid kit, which seemed silly for men who figured they'd never make it out of New York alive. Actually every bit of their equipment was like boys playing soldiers when, in the cargo of this U-Haul, lay a gondola and fabric for a hot-air balloon, and two nuclear dirty-bombs, just in case one of them failed, the other was set on a timer.

The Kings Academy was frosting on the cake, what with the addition of so many high-powered political figures, celebrities and all at such a prestigious place as the Lincoln Center. It was ironic that it took the cloning of the dead to offer this opportunity to strike fear and horror to the very depth of the American heart.

To these Muslim fanatics there was a sense of frustration on whether they could inflate the balloon given the weather conditions and locality. As determined as their leader was, he had numerous obstacles at every turn. Deterred by this massive security detail and the weather,

567

these dilemmas had turned this demigod to an enraged beast. He'd even resorted to slugging one of his partners next to him in frustration. He turned down an alley and decided that this might give them enough cover to prepare their assignment.

****

The program for the night's performance was hastily printed with the help of Artie Stenson. He had become the brainstorm and chief director under everyone's noses. As this was progressing there was a common question among everyone at the Lincoln Center: Who's this Artie Stenson?

Artie's introduction to the Kings' dancers said it best: "This is a performance to human creativity, designed to bring people together through the shared discovery of a vast and varied existence of love and acceptance, in an environment of bold expression and natural beauty. No biases. No borders. Just a profound commitment to conquer convention, cure conformity, and embrace a relationship for common man and God."

****

The performance was opened up when Zach Porter, the offspring of Cole Porter, took the podium and conducted this renowned orchestra with several of his forebear's countless musical collaborations: You Don't Know Paree, I love Paris, I Happen to Like New York, Take Me Back to Manhattan, and Martinique. It inspired and invoked memories of this creative lyricist and composer.

No more had Zach stepped off than Alexie Tchaikovsky received a standing ovation when he was introduced. He spoke with tears when he admitted that he accepted the applause, not for himself, but for the great accomplishments of Peter Tchaikovsky, a man who had struggled with his orientation all of his life, only to succumb to the shame that was directed at him from a homophobic society. The composer had died of Cholera at such a young age in St. Petersburg. The 1812 Overture hastened the crowd to a frenzy, followed by Serenade for Strings and Mazeppa. The numbers were breathtaking and never had so many musicians felt such wonder and delight than performing for what they knew was an identical image of this wonderful composer.

Lance and Leif took the stage with a number they had practiced for three weeks on a Broadway play popular in 1953, a mature drama entitled Tea and Sympathy. In its day the play was hopelessly dated and silly, bringing up and tantalizing the audience with the subject of male homosexuality. Set in a boys' boarding school the play tells the story of Tom Lee, a sensitive young student, played by Leif, who likes to sing, read poetry, and play the female lead in school plays, and hangout with the headmaster's wife. The boys had made humor of this to Leif, as the boy who liked to hang out with the headmaster's psychologist. Leif took it in good stride and admitted he'd long needed a daddy figure, and he had several to choose from.

The truth being, Leif had matured enormously over the past six months. He'd learned to be true to himself and find the joy he had been looking for. The boy didn't receive what he wanted from Dr. Makawa—he'd received more. Oshi had reached inside this young soul where no man had dared to venture. Leif understood this conception of a "Penis Mind." In this state of mind, you can only think about one thing—getting someone into bed. In that moment, virtually nothing else matters. Dr. Makawa taught him that there was something far more important than just sex. Getting to know someone first was discovered to bring far more joy into Leif's life than acting on sexual urges as a quick way to undermine authenticity and a certain way to create some uncomfortable, if not downright painful relationships.

Leif grew emotionally and physically with his stay at Dr. Makawa's home. He had always given 110 percent in his dedication to whatever Oshi taught him. Their same interests and commitments made for an honest emotional connection. When the sex happened it was totally unexpected and the best night of Leif's life, according to him. He cherished and valued this man so much, Leif no longer felt the need to brag about any conquest. His loyalty and respect for Dr. Makawa was exemplified in everything Leif did, but these moments of privacy were kept between them. Being given United Nations' status and diplomatic immunity the rest of their lives also relieved the moral division.

On this magnificent stage the boys combined their ballet skills and dance numbers to musical numbers, but a whole different scenario.

Often using songs from the musical, Grease, the humor had the teens comfortably in their daily life with what they knew to be true in a life of a gay teen, a role of playing for friends more than being real to themselves.   The audience received both the titillation of thinking that Tom may be gay, while at the same time being allowed to feel morally superior to the bigotry of all the other characters.   In the original play, there was sympathy elicited on the basis of Tom not being gay, but, to these teens, there was no doubt of Tom's identity, his fight for respect and recognition.

The ballet and dances were exquisite with colors and a backdrop of nature and mosaics.   It was the combination of modern dance that set the tone for these teenagers who loved the true beauty of the arts.   There was a touch of nudity behind a blue sheet of plastic in the third act when Tom is caught swimming nude with one of the male teachers, played by Sergei, with a fake mustache and a tint of gray at the temples.   It was a beautiful scene and had the audience numb when Leif sang, The Wing Beneath My Wings (Did I ever tell you that you are my hero?)   Toy and Kami had no idea the boy could sing that well, yet his voice had many in tears.

The young man and teacher were caught in their encounter, then the rumors spread that Tom was a homosexual began to proliferate.   After a ghastly encounter with a girl who wanted to change him—Wendy--and a most humorous scene which every gay boy in that auditorium felt, Tom contemplated suicide.   It took a creative wardrobe to hide Wendy's stomach.   Through a collage and holography of photos of young teens who had killed themselves in years previous—supplied by Artie and all displayed on a gigantic screen around Tom—the boy gets the courage to be who he really wants to be, no matter what anyone else thinks.

It's when the headmaster's wife (Brittany) tells Tom that, yes, he is different, "But one day you'll meet a girl, and it will be right."   She begins to strip him and, in the original play the curtain dropped, as if the boy will change because he will be convinced sex with a female is what is acceptable.   But in the Kings' version Leif stops this undressing, examines the woman in front of him, turns and runs into the arms of his teacher.   The final dance is a celebration of life in the true energy and excitement of being gay and proud.

Out of all the boys at Kings Academy, only Leif could understand the total impact of a play like this. In order to discover the self, he had to first face his core of shame. He had to acknowledge that he had long held a belief in his own reprehension, and this belief had directed Leif's life, and not for the better. Perhaps this seemed obvious and logical in a gay boy's life as they objectively considered shame, but the subjective experience of facing toxic shame is utterly wrenching in its magnitude. It quakes even the most stable part of a boy's soul, and leaves them terrified by the knowledge that they know nothing of who they truly are. What a gift the boys at Kings have always had by their own acceptance of their parents, peers, and staff. Only by the acceptance of openness of a gay lifestyle without judgment had Leif felt truly blessed to be in his skin.

The audience applauded and rose to a standing ovation. Dr. Makawa was probably the happiest of them all. He knew Leif like the back of his hand and had deep feelings for this boy in his life. Oshi had taught the boy that there are other ways in life to control ones emotions than with just sex. It wasn't easy to tell a boy, who had never had a father that the first man a boy loves is his father, and this boy craves from him love, affection, and tenderness. Leif never received validation from his mother and, consequently, sought survival by learning to conform to the expectations of others at a time in his development when he should have been learning to follow his own internal promptings. Rejected by peers, Leif found happiness through sex and men.

Living in the environment of Kings Academy, Leif no longer felt that sense of shame and guilt. His secrets were revealed which helped him heal a wounded and troubled boy. Dr. Makawa validated this sense of worth as an adult in the boy's life, an adult who recognized that the boy accepted himself as a gay adolescent and one who was prepared to give as well as receive love. There were no promises of marriage, or secretive encounters, or secrets to be kept from everyone else. This love was built on honesty and joy for each other.

Oshi knew himself better by this encounter, as well. His passion was for his own sex and this he could never deny again.   As he taught Leif, he, himself, learned that to authentically connect with his world and achieve the contentment that he craved, he had to relinquish those old behaviors and break free from their suffocation.   The two of them had discussed taking the Makawa name for Leif.   Leif was so filled with happiness that he decided to accept Dr. Marion's, as not to violate something so special in his life that he didn't want to appear possessive and intrusive to a man who meant more than life itself.   His love for Dr. Makawa gave him a depth of protection that other boys couldn't understand or supply.   Nevertheless, Oshi had taught him patience and a future with a boy his age that would result in a lifetime bond.   It was a future Leif was willing to wait for as long as he was the protégé of this great Japanese gentleman.

In robes the boys and girls of the academy came out on the stage and sat over the edge to take questions from the multitude of teenagers. They even answered who their forefathers were and how they felt about being cloned from such icons.   From the audience this mystery had vanished with the realization that these children were no different than themselves, so the questions centered on what their academy life was like, their studies, expectations, dorm rules and limitations.   What was it like to be nude on a stage in front of so many people?   What were the boys' views on their sexuality?

Toy and Kami sat back in total absence of adults getting in the boys' way.   They too had cried with the singing by Leif, the expressions of their boys who had designed and choreographed this play with the assistance of Zach's conducting.   It was fun to hear the boys' answers to questions they took for granted, the rules they figured other children had, and the freedom of expression which defined their happiness and relationships.

There was one question and answer which caused another amazing round of applause, the one aspect that gained even more respect from the thousands of teenagers who were now in love with this academy.   It was from a boy about their age who wasn't so sure he heard right that there were no barriers to sex in the dorm.

"Are you saying that you guys can love each other and there aren't any adults shaking their fingers at you?"

Tad Barrie ran with his answer. "At our school love is absolutely, totally free. Like sunlight, like the air, love is within and around each of us at all times without any exception or conditions whatever. We have only to open our awareness to it to know that love is already there. There is no distance between us and love, ever. As young children we expected to share and experience what each other felt; we knew no other way because no one ever told us that it was dirty or secret to share our bodies and feelings. I can't tell you how many times I went to bed with one of my brothers in my arms because we wanted to share the warmth and skin of another. Actually, it's almost every night."

A Juilliard dancer was in awe of this freedom. "Is it true you do nude ballet numbers? Don't you feel......?"

"Naked?" Will Hawthorne asked and received laughs. "Yes, we definitely feel that, but that's how we live and sometimes it feels more uncomfortable and repressive wearing clothes. I like to share how my body expresses itself to others. It feels really cool. We're taught modern and classical dance, but we're also to be aware of our bodies' capabilities and its own intelligence. The impulses that are feeding and informing our movement on stage are electrical impulses. We work with the impulses, allowing them to come forward in the truest way. It's very improvisation, like an organic phenomenon. Often times we have physical arousal to our dance and our instant impulse to allow our bodies to express from the inside out, rather than remember some type of form or rehearsed technique."

"Can you do one for us tonight?" The same lady asked and received both laughter and applause for the boys to proceed.

"That's up to our fathers," Tomas stated.

Toy and Kami glanced at each other and gave a thumbs up, but one of the directors of the Lincoln Center cleared his voice to gain the mayor's attention.

"Sir, I'm not sure......I mean they're minors, which, given the current hysteria of....."

New York's mayor raised his hand. "Arthur, they're not having sex on stage. Don't panic. Who are we to rain on these boys' parade?

Dr. Kerho, the boys are most welcome to be themselves. This audience would love nothing better than to be entertained by such beauty."

****

An orange and white U-Haul circled Central Park, looking for a place they could turn off and attempt to inflate a balloon. All they saw were police barricades and an abundance of vehicles. Driving further down 5th Avenue they pulled into an alley that appeared vacant. No more had their leader stopped and began to open the back of their vehicle when a trash truck appeared from the other end of the alley.

Their leader, Hasad, had already made up his mind to kill whoever was in this truck approaching their position. He didn't need any more disturbances or delays. Reaching for his weapon in preparation, there was another stream of headlights that came in this alley to their rear. Confronting a patrol car was not in Hasad's plans. He quickly got back into the cab of this truck and began to back up.

Officer Hawkins and his partner noticed a parked truck in the dark alley as they routinely watched this sector for trouble or other problems. Hawkins noticed that the driver had a suspicious demeanor and the vehicle had out-of-state plates. He also noticed that the man's jacket bulged in different areas; yet, a sanitation truck was ahead of him and his radio dispatch had just called in to monitor 58th Street near the Lincoln Center for a car collision involving three cars. Having second thoughts and pressed for time, Officer Hawkins decided not to pursue this U-Haul, figuring the driver had simply turned down the wrong way and it wasn't like he wanted to get out of his patrol car on a rainy and cool night.

For Hasad and his men their night had turned into a disaster in preparation and planning. They considered other alternatives, including returning to the alley in a few minutes. Wet, desperate and hungry, the terrorists agreed they would all pray to Allah that the weather would change for tomorrow's scheduled parade. They drove to a shopping plaza where the driver parked the vehicle and the men slept as they sat-- miserable and ready for death.

****

574

The Kings' boys always came prepared to do what they loved, even during their leisure hours in the dorm. Music sheets were handed to the musicians and Zach was at his best when he had talented people to conduct. The dance started with a group dance to the throb of the rock organ, exuberant, demanding in its rhythm and pulse, persistent, rejoicing in companionship and celebratory of youth. It became the worship of Sunday with a number of solos and small ensembles as the mixture of rock and Baroque increased in intensity.

In the part called Summerland, the pulse of the score slowed down. The dancers rose from their circle and in various ensembles suggested the sweetness of young love between two boys. Couple after couple took the stage, one of them dancing a protracted adagio of affection. The accent was on high spirits, their glistening bodies with remarkable lifts in which the stronger boys reached the lighter boys to the skies. Trent and Dane bounded through the stage completely naked; their spirits brought ooohs and ahhhs from the audience, and even had Prince William with wide eyes and a glistening smile. In the huge blue ball to the right of the stage the beauty of the male body was exposed in huge dimensions. Trent was in the realm of comfort on stage, a boy who had danced long before he knew his forebear was Vaslav Nijinsky, one the greatest dancers in the history of ballet. Trent's genius was no less that of his progenitor, he embodied the sensuality and sexual ambiguity associated with full expression of movement. The boys at the academy knew that, when Trent danced, no boy could compare to his imagination, his unrivalled font of creativity. At seven the boy had mastered grand jetes, pirouettes, entrechats, and epaulement. To have both Dane and Trent on the same stage was the equivalent of putting Nijinsky and Nureyev together to share their mastery. Unfortunately Rudolph was only twelve when Vaslav died in 1950.

The Kings' directors never forbade the boys from being sexual beings. At thirteen-years of age, Trent had duplicated Vaslav's amazing autoerotic performance in the ballet L'apres-midi d'un faune. Between him and Dane their goals as young teens were to perform the CLOWN OF GOD, OLYMPIA, SEBASTIAN, and GEMINI before their peers and staff. Their magnificence was not for the masses, but their performances were saved for eternity on video.   Turning sixteen they were ready to team together in a tour of the world; their first revealing was to be with the Dallas Ballet Company, where the two would perform MONUMENT FOR A DEAD BOY, by Rudi Van Dantzig, the creator of the Dutch Ballet Company and author of FOR A LOST SOLDIER; a story of his own life when he was a twelve-year old during World War Two, and one in which the boy had fallen in love with one of the Canadian soldiers who liberated his village from the Germans.   The Kings' boys had loved the film during their youth, but couldn't understand why the Canadian left the boy stranded and confused.   The movie became one of Dr. Makawa's most challenging lectures to explain human nature and self-serving interest of adults.

The Lincoln Center was spellbound with this reality and beauty. Ballet enthusiasts knew this movement as representing Saturday, standing for the ancient Sabbath.   The dance began with two naked boys entering the darkened stage carrying candles.   The entire group followed, each bearing a lighted candle.   Ms. Stravotski was at the edge of her seat in the front row with the Franco family on her right.   She kept whispering Trevor's name to reassure her mind that this was a dream of someone dancing with the likes of Ninjinsky.   Her eyes were glued to the stage with a memory she possessed of this dynamo twelve-year old who had been, for a few minutes, the awe-inspiring student she had often envisioned teaching.   It was when this group departed that Trevor came forward to perform a dance, a version of the Ite, missa est that concludes the Latin Mass.   Dane joined Trevor as they danced to the rhythm and rock beat.   Their glistening bodies had the audience constantly looking toward the large screens that magnified this sexual essence and maleness of their genitals.   The elongated penises appeared to have their own ballet movements as they flipped upward and sideways, swaying to the bodies' leaps, twists and turns.   The boys were joined by other male soloists and the audience was given an impression that there was nothing on earth like lifting their arms and responding to the adrenaline of being naked and alive.   The score intensified in cracking crescendos of sound and licentious expressions of maleness in their desire to engulf each other in sexual positions and romance. The best thing, Trevor managed to do all this without it being obscene.   For Dane, he found no passion in restraint and his sex became a tool to communicate to an audience that he knew had their eyes locked on his sexual organ.   His sense of exhibitionism teased this audience with a near hardened member, as if this coordination with a projection of male virility directed the rest of his body into the next bound into space with his legs split to align this erection with his leading leg.

Ms. Stravotski was in her glory with the performances of Dane and Trevor.   Trent leaped forward to join his two peers. Their animal instincts defied gravity, their posture and jumps were above anything she had witnessed in St. Petersburg.   To realize that the likeness of Vaslav Nijinsky and Rudolph Nureyev were on stage, as well, was like a dream come true.   The boy's erect penis was something only Rudolph would have done during his day, but she wasn't insulted, only heated by its visual representation of male beauty.   Ms. Stravotski had just accepted a teaching position at the Carlisle Ballet School in Pennsylvania, but she would always remember this night and inspire other boys to be as daring and free with their bodies.

As all the dancers re-entered, each putting a candle on stage, the music gradually diminished and the lights slowly vanished so all the audience saw on stage was the presentation of candles. For those who understood ballet and the Trinity, this was a dance about man seeking his inner self. When the kids leave the candles on the floor, it's not meant to suggest just the literal peace marches.   In the end, each individual must express himself.   All he can leave is the light of himself.   The whole dance was heady and joyful, a celebration of a youth that can effortlessly encompass and transform the past, an experience that each boy had to go through to accept from his predecessor.

The splendid dancing was a delight to watch by the almost four-thousand men, women and children.   If the Juilliard director expected a mediocre group of boys he was quite surprised to see the absolute professionalism of young and phenomenal dancers.   He was whispering to his assistant that he wanted the two main dancers as part of the New

York Joffrey; his knowledge of their progenitors had of yet to be explained to him.

"Tell them no auditions are necessary. Just get them to come to New York," he begged.

When the boys dressed, Dane came on stage and introduced his brothers, the "infamous thirty" he called them in humor. They all sang YOU RAISE ME UP, as a tribute to the audience, but every boy was thinking about someone special in their life. As they gathered on stage Artie was invited on stage and introduced. He in turn thanked Prince William and had him come to the stage to say a few words. There was a standing ovation as the prince walked to center stage and hugged a few of the Kings' boys. He asked the eight English boys to come to his side, and then beamed with joy at his surroundings with the likeness of Britain's finest standing next to him.

"I certainly want to thank Dr. Kerho, my friend since we were rowdy teenagers ourselves. You are as handsome as ever, my dear friend, and it has been a pleasure being part of this discovery. To the dancers here, I am speechless. I would think you mimicked a nightly celebration from one of England's finest, Eaton." The audience cracked up at the joke about one of England's finest boys' schools.

"Ah, my English boys!" Prince William grinned and stared at the eight boys who were cloned from England's greats. "We've given the Americans so much of our heritage, now we adorn them with more beauty. I do hope everyone believes we English are as well endowed." Will waited for the laughter to subside. "We as a nation must not sneer at others who perhaps have not the same chances and opportunities as ourselves. Right, boys?"

The boys nodded and were happy to appease the future King of England.

"I believe we can all agree after viewing these kings tonight that hooliganism among teenage boys could be restrained if we, as parents, kept our sons in tights, or less, until the late teens. May I be the first to offer you boys an apology for so many things. Too many of your forefathers were shamed by a culture ignorant of sexuality and God's gifts to the human race. I'm sad to say such prejudice and bias exist even today as people fail to see the beauty and talent of boys and girls with a different orientation, by God's choice alone. I cannot take responsibility for a Victorian England, but I can honor the boys here tonight who are truly a blessing bestowed on the human race by visionary men. I leave you tonight my deepest appreciation to all the Kings Academy students and to my English boys. In your blood you will be English forever. This England never did, nor never shall, lie at the proud foot of a conqueror, nor these our princes have come home again, come the three corners of the world in arms. And we shall unite and shock them! Naught shall make us rule, if England to itself do rest but true."

The boys hugged the prince and carried him off the stage to a standing ovation. All of England would consider this the utmost praise and respect. Dr. Kerho was invited to speak and he moved to the stage with the encouragement of Dr. Marion.

"There are probably other ways of meeting the boys of Kings Academy, but what better way than how God created them." Toy began and received a hardy acceptance for this.

"These wonderful teenagers have no fear and think they can do anything. My sons don't experience lack of freedom. They think they can love anyone. These teenagers before you do not lack of life, as they believe they can express their happiness through dance and music. These boys before are no different than the faces I see throughout the audience tonight, as they believe they will live forever—and people who act like children think nothing can hurt them. I'm saddened when my children learn of ungodly things, and these are taught by grownups who should know better. I have no regrets bringing up my sons in all their glory, to run around naked as the day they were born and hug everyone, thinking nothing of it. If adults would only do the same thing, what a world we would have.

"Alas, when people refer to these boys as clones they are surely mistaken. We are all creatures and creators of originality, everything we create is original. It is not possible for any thoughts, words, or actions to be duplicated. We cannot duplicate, we can only originate. As there are no two snowflakes alike, here there are no two people who are alike, no two thoughts are alike, no two relationships are alike, and no two of anything are alike.

"I ask all of you here tonight not to exclude these boys for who you think they are, or for their love for each other. The world has religions, clubs, associations and agendas that are based on exclusions, and support these with a cultural myth of a God who excludes. I tell you tonight that the culture of God is based on inclusion. In God's love, everyone is included. Into God's Kingdom everyone is invited."

Toy invited his own right hand to stand beside him, and as Kami came on stage he received a robust kiss from his lover for so many years. They held their sons in a circle and waved to the crowd. Dr. Marion thanked the audience with his own gesture.

"Everyone knows who they are in their hearts. If you open your heart, if you share with others your heart's desire, if you live the heartfelt truth, you fill the world with magnificence."

With that closure came a cowboy honky-donk rhythm the boys knew only so well; Brittany had sat at the piano and went right into Wild West Hero. There were a lot of people whistling and singing: 'I wish I was a Wild West Hero,' when they left the Lincoln Center that night.

Chapter Fifty-Two

Michael Rush was now a divorced man with a boyfriend of six months that he loved dearly. They moved in with each other in an apartment overlooking Central Park. His lover was also a broker with the New York Stock Exchange, where the trading floor was also their office. Prices of commodities were largely set by supply and demand on this central area of Wall Street.

Michael never thought he'd ever see Leif again until his eyes glued onto a picture of this Kings Academy from Texas. He wasn't sure at first, figuring his eyes were playing a trick on him since he was so enamored with this boy to begin with. His own brother had basically ousted him after Leif's visit to Tampa, a gesture by his brother to send Michael a picture of Leif by the backyard pool, but a photo his wife saw with a dozen questions on why it was sent. Michael couldn't continue to lie.

It was a most difficult time to tell his wife that he was gay and had never wished to victimize her through a marriage that appeared happy but, underneath, was stressed with a need to meet and have sex with men.

The possibility of seeing Leif again was too overpowering to not take the opportunity to catch a brief glimpse of this delightful teenager who had, for a few days, become so much of his life. He had tried desperately to acquire tickets to the performance at the Lincoln Center, only to fail miserably to acquire even the highest priced ticket on the street. Michael had also heard that the boys' choir had sung at a church in Greenwich Village, only to find the church packed with standing room only. His only hope now was that the weather would cooperate and the city could celebrate these boys through a ticker tape parade.

The Friday morning began with drifting dark clouds, then the skies began to clear up by mid-morning into a bright and sparkling spring day. The sun dried the streets and sidewalks and warmed the temperature to a mild 65 degrees. Every radio station around the Big Apple was giving this a gorgeous day to have a parade. Scheduled to begin at one in the afternoon Michael and his significant other left the stock exchange at

11:30 a.m. and began their daily jog up Bowery to Park Street and ran until they were near Central Park. Michael's idea was to wait until the parade stopped around Columbus Circle, where he hoped to catch Leif's eye and have a few words with the boy. Hopefully the young man would remember, if not acknowledge him. He still couldn't believe that Leif was actually Charles the Twelfth in the flesh, though he was smart enough to know of the connection and not the myth.

****

The five terrorists awoke hungry and agitated after blowing their opportunity from the evening before. Their only encouragement was that the sun was shining and the perfect weather conditions would enable them to finally resolve this mission and wreak havoc on the United States.

Sending a gofer for coffee and donuts helped relieve the morning's tension as the hours dragged in wait for this parade to start. Now that this shopping center parking lot was beginning to fill, the men began to regain the confidence that they wouldn't be detected.

At 11:30 they began to move their vehicle to a location that best fitted their needs. The idea was to blend in and make it look like their balloon was just another feature in preparation for the parade. Traveling on 5th Street they circled Central Park twice before deciding that The Pond, close to the children's zoo, would be the best location to work from.

****

The Kings' boys, staff, Franco family, guests, and occasional visitors met for an early morning breakfast. The morning papers had nothing but praise and interesting reviews of the previous night's performance. Criticism was often the norm when nudity encompassed the many plays throughout the city. The boys' performance had been so discreet with lighting and color that it turned from blatant nudity to beauty and art. In the gay community there was rampant jealousy of anyone who had witnessed Trevor, Trent and Dane and their display of maleness. Men who had captured this erection and dance on their cell phones were selling pictures and video for several hundred dollars on the street. It had made instant riches for several proud ticket holders.

582

New York City officials decided that the best form of transportation for the participants of this parade was in convertibles, twenty of them to be exact. Two boys would ride in the back seat of each car, with no names or countries attached to their identity. The parade would commence at one end of Broadway and end near the Lincoln Center, nearly five miles away.

Many groups had asked to march as part of this spring welcome, but few were accepted. The gay community was out in full force and was ready to give the boys from this academy a warm welcome to their city.

By 12:30 p.m. and after a hearty lunch the boys and staff were escorted to the starting point of this momentous occasion. The Mayor of New York was in the first vehicle along with the two doctors as his guests. Broadway was lined shoulder to shoulder and, when the cars began to move slowly down this illustrious street with a high school band leading the procession, the litter of paper began to fall in bucket fulls upon these smiling faces.

****

Michael Rush was glad he had avoided running through the masses that had gathered for this special occasion. To allow time for the parade to reach his designated area, where he'd planned on catching Leif's eye, he and his partner decided to enjoy Central Park in all its beauty and grass fields while waiting. They would rather run on soft grass, compared to the hard pounding their feet took on the pavement.

As they rounded the grounds of the children's zoo they noticed a parked truck alongside the road. The truck's occupants were unloading a heavy crate which crashed to the pavement and broke open with such force that the balloon's fabric spilled out onto a vast area of earth. The abundance of color and silk were clear evidence of a hot air balloon, with a gondola on the ground close by. It wasn't the fabric or the accident that caught Michael's attention, though; it was the look he received by one of the men with a Middle Eastern face. The expression wasn't so much in disgust of the incompetence of these men removing this from the truck, but it was one of suspicion that Michael had maybe seen too much.

Michael was ready to dismiss this as nothing more than someone preparing a balloon for the parade. It did seem odd that music could be heard in the distance and spectators' yells and cheers began to fill the air with excitement, denoting that the parade had begun. If this was part of the parade it was definitely behind schedule.

In seconds their jog had taken them into a wooden area and out of sight of this setting. Michael could have easily put this incident to the back of his mind but, as the two of them reached Columbus Circle, the Wall Street broker thought precaution was more fitting than dismissal. As both of them slowed to a walk there were so many officers Michael simply picked one.

"Sir, I don't mean to bother you, and I can see you're busy, but there's this truck down below near The Pond by the Children's zoo. What bothers me is, they were unloading a crate that appeared to be a balloon; I mean, a hot air balloon. Now they might have some of these in this parade, so maybe it's part of the coverage, but these men were Middle Eastern in appearance and one of them didn't seem too happy that I saw this crate collapse."

The blue-suited officer eyed this jogger in front of him and didn't see a kook or someone trying to mislead him. "Where did you say this was?"

After a few more questions that appeared not to go anywhere, Michael and his friend strolled off, but only after they glanced back and saw the officer begin to speak to someone from his headset.

Homeland Security was not remiss in any demonstration or gathering in New York City. They also scrutinized all calls throughout the city from police to fire. The call in to headquarters for someone to check on a U-Haul in Central Park gave these agents immediate concern. They sent two cars directly to the area in question.

****

Hasad had yelled and cursed at his men for the last ten minutes in their haste to inflate this balloon and load its cargo in the gondola. The other nuclear device was planted behind a bush and was now being readied to go off in less than ten minutes. When two men approached with suit coats on and flashed badges, Hasad made sure their lives were dwelt a quick death by guns that came out of nowhere. What the terrorists did not see was the other vehicle in close proximity that watched this killing in horror. In a second they had radioed in and would have half the city's police force at this location within five minutes.

Hasad had just begun to give the orders to lift when the first round of bullets struck him in the back and head. Officers and undercover agents swarmed literally out of the woods to send a fusillade of bullets streaming into this balloon and the instant death of three terrorists. The partially deflated balloon swayed and changed directions into a tree with limbs bare of leaves but sharp enough to snag this balloon from reaching toward the skies. It took a sharp eye from an agent to spot one of the last living terrorists attempt to recover from his wound, his body hung halfway out of the gondola. One more accurate shot to the head eliminated any last chance to detonate the bomb inside. On the ground and a fraction of a second before the final terrorist armed the second nuclear bomb, he also was splattered with a barrage of bullets that made sure his fingers never finished the job.

****

With all the commotion and thousands upon thousands of spectators lining Broadway, few people heard the 30-second gun battle that was very one-sided after the first two agents were shot in cold blood. Others thought it was simply a fireworks display. Sirens and an enormous number of emergency vehicles breezed through the intersections surrounding this parade. No less than 200 men and women from every emergency service available were on the scene within fifteen minutes.

Michael Rush had no idea what he'd done until that evening when the news hit the networks. His recognition would never receive the praise it should have if it wasn't for a young man who decided there was one person in this whole crowd he wasn't going to ignore.

Leif had smiled more in the last thirty minutes than he had in his whole life. At least that's what his mouth told him. His body was covered with confetti of all colors, and he and his brothers laughed when they had to pick tiny pieces of paper from their hair, eyes, nose, and mouth.   They all had a back seat of gifts, from teddy bears to jewelry to invitations for marriage and an array of dates.   From the very beginning the boys had decided to sit up near the trunk, instead of actually in the seats.   This didn't fare well with the officers trying to protect these boys, but the parade was completely without demonstration, save a few signs that expressed an archaic biblical prejudice against gay people.   This hardly phased the excitement and adrenaline the boys felt.

As Leif's car approached Columbus Circle he thought sure he'd heard his name called by someone in the crowd.   His eyes scanned the many faces of cheering people of all ages and sexes.   Several men wore outfits that had Leif laughing, but then he spotted the face that brought back a memory of love and protection.   Leif wanted to tell the driver to stop; instead, Leif jumped from the back seat and ran into the arms of this man who had meant so much to him for one weekend.

His affection had an immediate stop put to it by two policemen who insisted Leif return to the car.   "Waldorf Astoria!   Tell them to notify Dr. Kerho!" Leif yelled as the car caught up to the others ahead.

Michael was beside himself that Leif not only remembered him but was the same lovable young man he had gotten to know for several precious days.   When the parade had concluded and the Kings' group had returned to the Waldorf, Leif stayed in the lobby and waited patiently for the face of this adorable man in his life.   And then he came.

Michael had brought his companion to the hotel, as well, and introduced him to Leif.

"Leif, I owe you my life for everything that happened that wonderful weekend.   Obviously you're not eighteen, but I knew that." Michael broke a smile which put Leif's lie at rest.

"I'm sorry about that.   It was a tough time to be a boy in New York City.   If I've caused you discomfort by doing that I'd love to do something for you.   Just name it."

"Leif, you've made me the happiest man in New York.   Jeff and I love each other and I am allowed to see my children.   I'm not sure any of this would have happened if we'd never met."

"I might not be here talking to you if some guy hadn't reported seeing terrorists in Central Park this afternoon. It was just on the news. I'd love to thank him."

"That was you, Michael," Jeff spoke right up excitedly.

Michael blushed and wasn't sure how to act, or even to take credit. Leif called over Dr. Marion and told him what had just happened. In moments several people had overheard this conversation and were talking on their cell phones to people of interest. An officer came running in and asked Michael to stay where he was until his face was verified by another officer who reported this information.

By the time the evening had come around Michael Rush was the toast of New York City and of the Kings Academy. Dr. Kerho made a special invitation for the man and his companion to visit Texas at their convenience, and even Prince William offered this hero a trip to London to be honored for his quick thinking and bravery.

"And here I thought life couldn't get better," Michael said with tears gushing down his face.

Epilogue

The flight back to Texas was long overdue for the Kings Academy. This was not the vacation that was planned, but the experiences and memories would last a lifetime. The two doctors were in the first class section along with Oshi, Brittany, Debbie, Reece and several of their coaches and staff. Occasionally one of them would wander through the aisle and glance at the faces of the boys who had seemed to have gone from teenagers to young adults in the span of six months. Whether they really realized the precariousness of this journey and the international significance weren't revealed on their faces.

Toy and Kami had many new challenges to decide upon as they resumed their duties at the academy. A leading sports company wanted to market Bobby's invention of the Kings' swimsuit. They wanted to call it the Newton Suit, but the doctors were against having their boys' names on any merchandise. Instead there was a compromise; the academy would build their own plant on the outskirts of the academy grounds for the suit's development. The components were still very much a secret and this silicon garment would be called the Nudeton Suit. This venture would be the first money-making enterprise for the academy and promised the institution millions of dollars each year in revenue.

Five new students would be added to the student roster, and Sergei had no intention of ever returning to St. Petersburg. Mrs. Bjornsson and her new husband were moving to Tucson and were expected to visit the academy quite often. She was now in full acceptance of her son's orientation and his decision to be a King forever. Leif was the first boy to tell Dr. Marion that he wanted his first born to be called Rodney Sveinn Marion. The doctor promised to honor this request.

The academy's new dormitory was to be named in honor of a most deserving man—Rush Dormitory. Already the academy had schools from throughout the country wanting to compete against this prestigious institution. Dr. Kerho was establishing new guidelines for visitors and the type of school that would be accepted for competition. Stanford was biting their nails in anticipation of spending a few days training with the Kings' boys.

There was one aspect that required review. The decision to begin a new generation of students for the academy brought an interesting question. Though a collection of sperm samples were ready, several students had other ideas. Jon Lincoln represented his brothers in their thoughts on this subject.

"Dad, we're concerned that, if we do this insemination with our sperm, the likelihood of having gay children will be barely ten percent, if that. This will change the entire nature of our academy. It won't be like Kings Academy anymore. On the other side of the coin if we were to clone ourselves we'd have fifty or more gay sons with your idea of twins or triplets; though, I'd be cautious on too many more Alex's."

The directors smiled but their faces accepted this rebuttal for its intellectual insight.

"Well, that would ruffle a few gay feathers, wouldn't it?" Dr. Kerho humored. "What do you think we should do about this?"

"The guys and I were thinking it should come to a vote."

"I think that's fair. What do you think about it, Kam?"

Kami contemplated this idea and nodded.

The three of them moved to the center cabin and interrupted the movie that was playing for a few minutes. Dr. Kerho first asked for any commentary, which had Alex rise from his seat to address his brothers.

"Guys, can you imagine a dorm with girls? Can you imagine having a son who is henpecked and feels he has to be prim and proper to impress a girl? What's the world coming to when you can't pat a boy on the butt with a wink, knowing you'll going to have sex in less than a minute? Can you imagine…."

"I think they get the message, Alex," Dr. Marion stepped forward and helped Alex back into his seat with a stern hand on the shoulder. His brothers laughed and applauded the aberrant halt to Alex's histrionics.

Dr. Kerho resumed. "Okay, I think we have one side of this debate. Do we have an opposing viewpoint?"

Jacob Tilden stood. "Having girls in the dorm would allow our sons to experience more female emotions, like tears, whining, shock, spasmodic outbreaks, period moods, jealousies, gossip, in-fighting for boys' attention, and squabbles about who did what with whom. In addition our sons will be blessed to have in the bathroom mascara and lipstick stains, curlers, tampons, and the female sex will love to complain about our sons' bathroom habits. I say not!"

The boys unanimously cheered and patted Jacob on the back. Dr. Kerho and Dr. Marion chuckled and shook their heads with this "supposedly" female affirmative speech.

"Are there any other advocates for women's rights? Real ones, I mean. Wendy? Does my daughter have a say in this?"

Wendy patted her stomach. "See what happens when girls get their say? I personally think that my best lovers have been a girl who understands how to love another female and boys who have had sex with boys. Niqui and Jay are much improved with their experiences with my brothers. I say gay boys rule!"

"Yeah!" The entire plane came alive with a girl's perception.

Dr. Kerho held up his hand. "Enough! Time for a vote. Those who want to roll a dice with nature, raise their hand. Okay, those for duplication of hormones."

Dr. Marion didn't really have to count but it gave him a comedy routine to the laughter of his sons. He turned to his significant other and whispered. "Déjà vu."

Dr. Kerho faked a grimace as if it had been a close vote, then held two fingers to the bridge of his nose in pain. He raised his head slowly to allow his eyes to bore into these excited sixteen-year olds.

"So, what you're telling me is that your fathers have to prepare for another twelve-year old boy who looks like Alex, with the same precocious nature for mischief and is prepared to get the spanking of his life by lying spread-eagle on top of the water slide, with a raging hard-on, mind you, yelling, "'Omnia vincit amor?!'""

THE END

590

Kings Academy: (3rd in a trilogy: The Hyacinthus Project and The Art of Loyalty)

Human cloning is a science that humanity would rather avoid, until a renowned geneticist decides to create his own genetic utopia. Stricken by cancer and unable to realize his dream to completion, he selected two unlikely teenagers to inherit his fortune with instructions to build an academy and be parents to 29 boys.   Fifteen years later Kings Academy's fascinating secret begins to unravel with the discovery of tampered tombs in Europe and an investigation by a wrathful Texas State Senator.   Faced with banishment or death Kings Academy forces the world to deal with its mysterious and wonderful gift to mankind.

The Author:

Alan Stroup is a graduate of Slippery Rock University, with a B.S. in Health Science.   His graduate work was at Stetson University in Florida.   The author has taught school and coached for 22 years. Coaching Track and Field at UCLA, the school won the NCAA Championships in 1987 and 1988.   As well as coaching numerous Olympians, the author also coached the current world record holder and Olympic champion in the 400-meter hurdles.   Mr. Stroup began a movement to start an American Youth Academy, much like the Eastern European sports schools, but the concept was never funded.   He has written twenty books, mostly dealing with gay culture and athletes.

The author welcomes input from readers: **www.Af71vet@yahoo.com**

Other books by Alan Stroup:

A Blue & Gray Perspective By A Boy Soldier   (*LGBT Award Winner)*

The '39 DiMaggio

The President's Boy

First Boy – The Crusade

The Hyacinthus Project

The Tennis Kouros

The Art of Loyalty

The Huckleberry Pirates

XY Minus Three

The Neurokid

On Second Thought

The Last Castrati

Whispers In An Italian Restaurant

Age of Dissension

Director's Cut

Kings Academy II Saving the Planet

Time Bomb

www.ingramcontent.com/pod-product-compliance
Lightning Source LLC
Chambersburg PA
CBHW051847170526
45168CB00001B/11